A Sustainable Bioeconomy

Mika Sillanpää • Chaker Ncibi

A Sustainable Bioeconomy

The Green Industrial Revolution

Springer

Mika Sillanpää
Laboratory of Green Chemistry
Lappeenranta University of Technology
Mikkeli, Finland

Chaker Ncibi
Laboratory of Green Chemistry
Lappeenranta University of Technology
Mikkeli, Finland

ISBN 978-3-319-85714-5 ISBN 978-3-319-55637-6 (eBook)
DOI 10.1007/978-3-319-55637-6

Softcover reprint of the hardcover 1st edition 2017

Printed on acid-free paper

This Springer imprint is published by Springer Nature
The registered company is Springer International Publishing AG
The registered company address is: Gewerbestrasse 11, 6330 Cham, Switzerland

Preface

Nowadays, the sustainable production of food, energy, chemicals, and materials is the major challenge facing modern societies and future generations, after decades of reliance on fossil resources which, on the one hand, did generate economic growth and prosperity but, on the other hand, has left heavy environmental, geopolitical, and social legacies. In this alarming context, the concept of bioeconomy has been developed and promoted as a new sustainable and knowledge-based economic model centered on the use of renewable biomass and derived agro-industrial and municipal wastes using various supply chains and pretreatment, conversion, separation, and purification procedures and technologies. Thus, as a multidimensional concept, bioeconomy has the delicate task to replace the declining fossil-based economic model and manage its global and complicated legacy, while facing its own set of challenges, especially during the delicate transition phase toward the full-scale implementation of a biomass-based economy.

Throughout this book, the authors presented, analyzed, and discussed the concept of bioeconomy from various angles in order to provide basic and advanced knowledge about bioeconomy for students, researchers, industrialists, decision makers, and the general public by showing opportunities, discussing R&D findings, analyzing strategies, assessing the impacts and challenges, showcasing industrial achievements, criticizing policies, and proposing solutions. The task was indeed challenging for one book, and we sincerely hope that we were able to accomplish it.

Hence, this book, which is divided into nine chapters, started in Chap. 1 by analyzing the current situation resulting from the petroleum-based economy, showing its deficiencies and disastrous legacy, which is one of the major driving forces toward the shift to a new model: biomass-based economy. Chapter 2 analyzed the concept of bioeconomy and its sustainable dimension by discussing the proposed definitions and key issues related to the current transition phase such as raw material change and sustainable profitability. The expected role and impact of sustainable bioeconomy on the two main economic pillars, agriculture and industry, are also presented.

In Chap. 3, renewable biomass was discussed, as the core element in the bioeconomy concept, in order to provide the readers with information about its definition, classification (woody, herbaceous, and aquatic biomass, along with derived wastes), composition (cellulose, hemicelluose, lignin, proteins, lipids, etc.), as well as the various opportunities for their industrial valorization into strategic and added-value products.

Then, the opportunities to produce a multitude of bioproducts from biomass were showcased and thoroughly discussed in three consecutive chapters: Chap. 4 for biofuels and bioenergy, Chap. 5 for biochemicals, and Chap. 6 for the production of biomaterials. In each one of those chapters, a theoretical background was presented, followed by a detailed analysis of the various mechanical, thermochemical, and biological conversion procedures applied to transform raw biomass into value-added end products including bioethanol, biodiesel, biogas, organic acids, food and fuel additives, biocosmetics, biopesticides, as well as pulp and paper, bioplastics, biochars, and activated carbons.

One of the main challenges facing bioeconomy is to develop viable and efficient industrial-scale production schemes. Thus, Chap. 7 was devoted to analyze the industrial dimension of the bio-based economic model and its sustainable and integrated biorefining activities. In this chapter, the implementation of bioeconomy on the ground was examined by illustrating the various designs of biorefineries, the obstacles facing the implementation scenarios, as well as some study cases of green biorefining technologies. The knowledge and experiences of key countries in the field of bioeconomy were detailed and discussed in Chap. 8. The objective was to provide readers from different backgrounds with the strategic visions of the USA, many Eastern European countries, and China toward adopting bioeconomy and its various sustainable industrial-scale production processes and technologies. As well, the available bioresources, opportunities, and challenges in the studied countries were also investigated, along with some interesting industrial study cases. A special focus was made on the industrial achievements and prospects in Finland.

In Chap. 9, the various impacts of bioeconomy and the prospects of its worldwide implementation were thoroughly discussed from a multidimensional outlook including industrial, environmental, social, and geopolitical perspectives. This includes reflections on the need for a continuous monitoring of the sustainability of bioproducts and biorefineries via various indicators, as well as the assessment of key environmental and social factors such as greenhouse gas emissions, land-use change, biodiversity, employment, and food security.

Finally, we sincerely hope that our contribution to promote sustainable bioeconomy in this book will benefit researchers, industrialists, decision makers, professionals, and students around the world and thus create a momentum behind biomass-based economy and sustainable development. The authors thank Springer International Publishing for supporting our book from the preparation phase until its final publication.

Mikkeli, Finland

Mika Sillanpää
Chaker Ncibi

Contents

Chapter 1
Legacy of Petroleum-Based Economy

Abstract During the industrial revolution of the nineteenth century, the use of coal as fuel set the "train" of progress in motion, which definitely induced a significant improvement in the living standards. After several discoveries, inventions, and innovations, the use of crude oil, the so-called black gold, enabled humanity to reach a higher level of prosperity, especially so between the end of the second World War and the oil embargo crisis. Currently, crude oil is the most traded commodity in the world market and is the main feedstock to produce a wide range of fuels and products such as plastics, textile fibers, dyes, etc.

The heavy reliance on petroleum and other fossil fuels for decades caused many environmental disasters around the world and major geopolitical tensions especially in oil-producing countries. In this chapter, the environmental (water, soil, and air) and geopolitical legacy of the petroleum era as well as its impact of human society are thoroughly discussed in order to highlight seriousness of those issues and the necessity for an alternative sustainable economic model for the future.

1.1 Introduction

Different economic systems were and are being implemented worldwide depending on the degree of governmental involvement, on the one hand, and the manufacturers and consumers freedom to decide what, when, and how much to produce, on the other. Nonetheless, although Humanity developed different economic systems from ancient history until our current era, the straightforward objective was always the same: GENERATE and GROW WEALTH.

Basically, economic systems are founded on four major activities: (1) resources exploitation, (2) commodities production, (3) trading (resources or commodities), and (4) consumption. It is indeed amazing how the Human history could be summarized into those four activities. First, men exploited the natural resources for consumption. Then, they used those resources as feedstock to produce commodities for themselves. Later, they were able to exploit more resources thus increasing and diversifying their products. At this point, they started selling those products gradually for other tribes, other provinces, and other countries. After centuries of inventions and industrial revolutions, men are now able to exploit, produce, and sell products in every corner of this earth.

M. Sillanpää, C. Ncibi, *A Sustainable Bioeconomy*,
DOI 10.1007/978-3-319-55637-6_1

This tremendous industrial progress was based on two main factors, invention and energy supply. It all started with wood for which the discovery of fire enabled Humans to exploit the energy stored in biomass. The invention of the wheel helped exploiting another form of energy, animal traction, as well as the sail which exploited the wind and enabled travels and commerce through vast seas. Then came the industrial revolution and the exploitation of coal along with invention of the steam engine by James Watt (1781) and steam turbine which opened the door for a new era of industrial progress and economic growth in the nineteenth century. The twentieth century was the age of petroleum. Innovations were abundant and mainly related to its extraction (various drilling techniques), refining (atmospheric and vacuum towers, cracking techniques...), and utilization (petrol engine).

All those historical developments led to the increased dependency of countries on energy resources. The production systems were almost entirely run on nonrenewable supplies of energy (petroleum as well as coal and natural gas). Overall, the generated economic growth seemed to have blinded Humanity for "a while" (a century and a half or so) about the obvious fact that we were feeding our infinite hunger for wealth from a limited provision.

1.2 Fast Facts About Fossil Fuels

Fossil fuels are the hydrocarbon-based matter formed through the anaerobic decomposition of living organisms (plants and animals) buried under thick layers of sediments. Over millions of years, the combined effect of pressure and heat is believed to induce the transformation of the formed organic matter in sedimentary rocks into liquid (petroleum), solid (coal), and gaseous (natural gas) hydrocarbons via catagenesis [1].

Fossils fuels have been the world's primary energy supply for deceases. According to the International Energy Agency (IEA) 2014 statistics [2], petroleum, coal, and natural gas accounted for more than 80% of the energy supply for the last 30 years (1972–2012). Overall, a slight decrease occurred during the last three decades (86.7% in 1972) and (81.7% in 2012), mainly related to the decrease in the petroleum share from 46.1 to 31.4%. However, the shares of the two other fossil fuels were increased (from 24.6 to 29% for coal and from 16 to 21.3% for natural gas).

Another important remark has to be made from those historical statistics. During the last three decades, the share of biofuels in the total energy supply decreased (10.5–10%). It is a slight decrease one might say, but considering the numerous breakthroughs made in the R&D field of biofuels, the increased awareness about the environmental risks of fossil fuels, the momentum behind global warming, and more importantly the involvement of countries and international bodies, we should have expected an increase in the share of biofuels in the world's primary energy source over the last 30 years. But no, it decreased. What happened then? Where did all this effort go? More importantly, if governments are involved in developing the biofuels sector, so the real question is who (or what) is more powerful than governments so that he (or it) could oppose the increase in biofuels share in the total energy supply or at least stagnate it for the last three decades? We shall address those important questions and many others later in this chapter and in Chap. 2.

1.3 Petroleum: The Fossil Fuel that Changed the World

1.3.1 Petroleum Composition and Classification

Petroleum, or crude oil, is a viscous, dark-colored liquid trapped in deep reservoirs in the crust of the earth formed by porous or fractured rock formations [3]. Petroleum is composed of a mixture of various types of hydrocarbons, organic compounds, and trace metals. Nonetheless, hydrocarbons remain the primary component of petroleum (largely alkanes and aromatics). They can be classified into four groups:

1. *Paraffins*: entirely made of straight or branched alkanes chains with a carbon-to-hydrogen ratio of 1:2. They can make up 15–60% of crude oil [4] and the shorter the paraffins are, the lighter the petroleum is.
2. *Naphthenes*: cyclic hydrocarbons with a carbon-to-hydrogen ratio of 1:2. They could make up 30–60% of the petroleum composition. These cycloparaffins (if $C > 20$) are more dense and more viscous than equivalent paraffins.
3. *Aromatics*: They can make up 3–30% of crude oil. Aromatic rings are formed by alternating double and single bonds between carbon atoms. Aromatic hydrocarbons can be monocyclic (MAH) or polycyclic (PAH). Compared with paraffins, aromatics possess much less hydrogen to carbon. Their incomplete combustion generates soot, impure carbon particle believed to be one of the causes for global warming.

Petroleum is commonly classified based on its density (light to heavy) and sulfur content (sweet to sour).

Density is classified by the American Petroleum Institute (API) [5]. API gravity is defined based on density at a temperature of 15.6 °C. The higher the API gravity is, the lighter the crude is. Light crude generally has an API gravity $\geq 31.1°$ and heavy crude an API gravity of 22.3° or less. Crude with an API gravity between 22.3° and 31.1° is generally referred as medium crude.

Sweet crude is commonly defined as oil with a sulfur content of less than 0.5%, while sour crude has a sulfur content of greater than 0.5%. Since sulfur is corrosive, Sweet crude is easier to refine and safer to extract and transport than sour crude. Like light crude, sweet crude causes less damage during the refining process, thus resulting in lower maintenance costs. Regarding the sour crude, in addition to the higher sulfur content, the possible formation of high levels of hydrogen sulfide can pose serious health problems, hence the need to remove it before the transportation of sour crude oil.

Benchmarks are crude oils from various regions used as pricing references for petroleum trading. As the most actively traded commodity, petroleum is bought and sold in contracts usually in units of 1000 barrels of oil. Thus, benchmarks help to determine the price of an oil barrel in a contract. There are three major benchmarks upon which is based the pricing of most crudes, namely: Brent, West Texas Intermediate (WTI), and Dubai–Oman.

1. *Brent Blend*: This waterborne crude is used in Europe and in OPEC market basket, making it the most widely used marker (almost two-thirds of all crude contracts around the world). This benchmark is a mix of crude oil from 15 different fields in the North Sea including Brent, Forties, and Oseberg. Crudes from those fields and others are light (API Gravity of 38.3°) and sweet (about 0.45% sulfur), therefore ideal for refining gasoline and diesel fuel, along with other high-added value products.
2. *WTI or US crude:* It refers to oil extracted from fields in the United States and sent via pipeline to Cushing, Oklahoma, the price settlement point for this crude. This oil is also light (API Gravity of 38.7°) and sweet (around 0.45% sulfur). Those properties make this crude ideal for gasoline refining.
3. *Dubai–Oman:* This Middle Eastern crude is a useful reference for oil of a lower grade than WTI or Brent (i.e., slightly heavier and sourer). Originally, this basket consisted of crude from Dubai (around 31 API and 2.13% Sulfur). Then, when its production plummeted to less than 100,000 barrels per day, crude from Oman was added. Starting from June 2007, the Dubai–Oman crude oil became the pricing benchmark for the Middle Eastern oil in the Asian market.

1.3.2 Worldwide Production and Consumption

The worldwide production and consumption statistics of crude oil, reported in Fig. 1.1, reveals how much the present economic systems are dependent on so called black gold.

Indeed, the reported data is showing a steady increase (almost linear) in both production and consumption between 1983 and 2013. Thus, for the last three decades, the petroleum production was increased by 13.7 (1983–1993), 15.7% (1993–2003), and 12.4% (2003–2013). The consumption also increased during the same period by 13.1, 15.6, and 12.2%, respectively.

Let us now analyze the production and consumption of petroleum by country. For this, we will compare the 2013 statistics [6] of two main groups: the 12 OPEC countries (Organization of the Petroleum Exporting Countries), on the one hand, and the top 12 countries having the largest economies (GDP-based ranking).

The related results are depicted in Fig. 1.2.

As shown, OPEC is responsible for almost 40% of the world production. Saudi Arabia is by far the highest producing country with 11.73 million barrels per day.

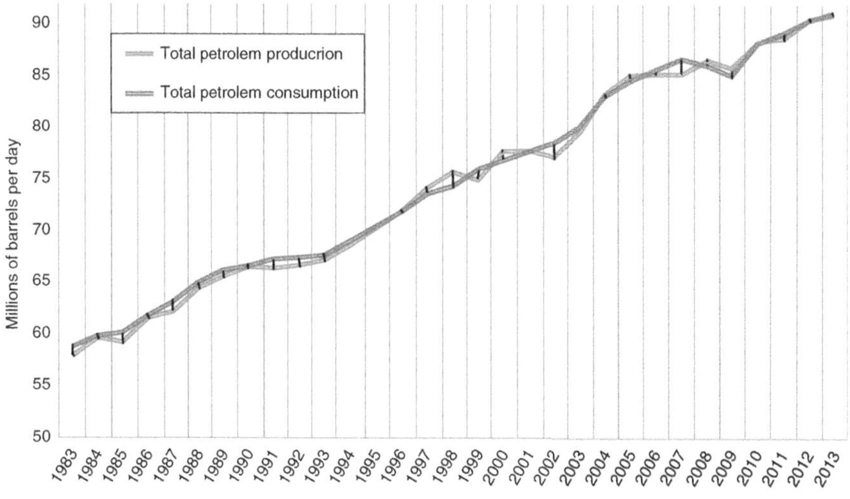

Fig. 1.1 Worldwide petroleum production and consumption between 1983 and 2013 (Data source [6])

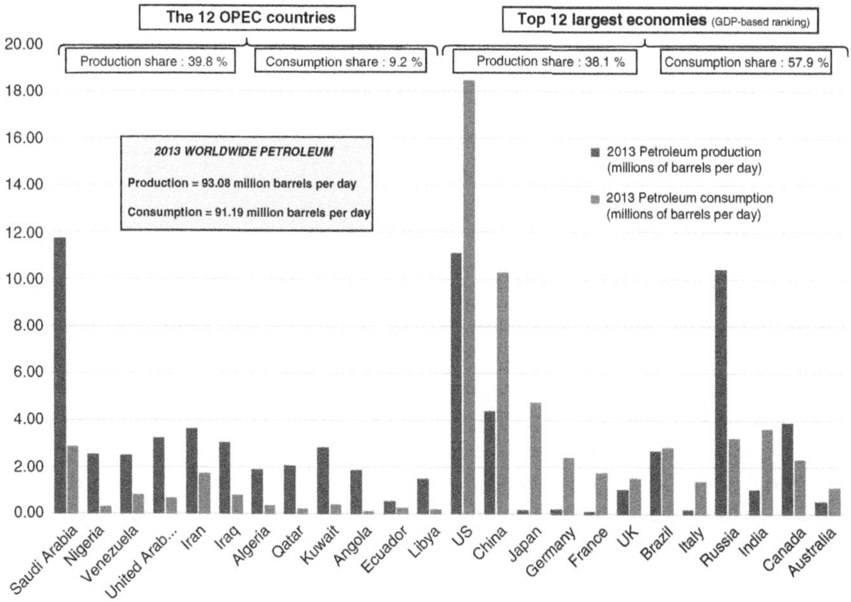

Fig. 1.2 Petroleum production and consumption: OPEC versus largest economies (Data source [6])

Among the largest economies in the world, the United States and Russia are the highest producing countries with 11.11 and 10.40 million barrel of petroleum per day, respectively. Thus, the crude oil world production share of the 12 members of

OPEC is slightly higher than the one for the 12 countries having the largest economies (39.8 and 38.1%, respectively). However, when it comes to consumption, the share of the 12 largest economies is 57.9%, six times more than the consumption share of 12 OPEC members (only 9.2%). Thus, we have two distinct models: countries producing petroleum to generate wealth and countries consuming petroleum to generate wealth. Is one of those models better? What are their respective repercussions on societies and the environment? We shall address this important issue later in this chapter.

1.3.3 Petroleum Refining Processes

As we have seen in the previous section, petroleum is composed of a mixture of hydrocarbons (paraffins, naphthenes, and aromatics) in varying proportions depending on the location of the extraction field. Other elements are also present in the crude oil including sulfur, nitrogen, oxygen, trace metals, and salts. The straightforward objective of a refinery is to purify petroleum, remove the impurities, and fractionate its hydrocarbons content to marketable products (gasoline, diesel, jet fuel, etc.). Further processing could be included to produce specialty end products such as lubricants, asphalt, wax, and other petrochemicals feedstock.

The refining process is based on the three major stages: distillation, cracking, and reforming/isomerization:

1.3.3.1 Distillation

After purification, the petroleum is preheated at 343–399 °C in pipe furnaces and transformed from liquid to gaseous phase (evaporation rate approximating 80%) [7]. The resulting hot gas is fed into the bottom of a distillation tower. As the heated gases move up the column, the temperature decreases and the various hydrocarbons gradually start to condensate based on their respective boiling points and molecular weights, hence the designation fractional distillation. As a result, three different categories of distillates are produced. The light distillates include liquefied petroleum gas (LPG), gasoline and naphtha, the middle distillates (kerosene, diesel), and heavy distillates and residuals (heavy fuel oil, lubricating oils, wax, and asphalt). The characteristics of each distillate (carbon content and boiling points) are presented in Fig. 1.3. In practice, atmospheric distillation columns are configured to stop at this level. More advanced vacuum distillation columns further refine the heavier fractions into lighter products in order to increase the production of high-value petroleum products.

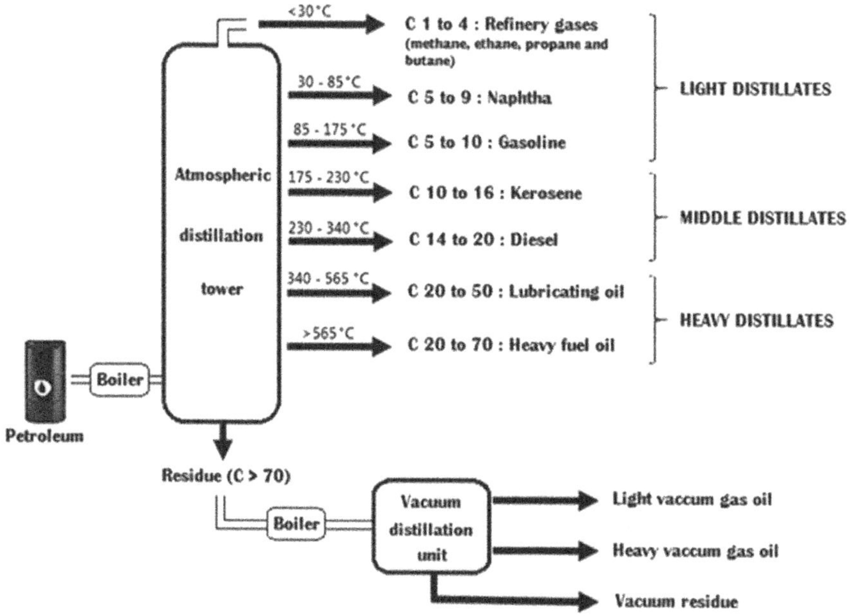

Fig. 1.3 Fractional distillation unit and related petroleum-derived distillates

1.3.3.2 Cracking

Among all petroleum-distilled products, the greatest demand is for gasoline. One barrel of crude petroleum contains only 30–47% gasoline. Transportation demands require that over 50% of the crude oil be transformed into gasoline [8]. To meet this demand, some petroleum fractions must be converted to gasoline. This may be done by cracking, i.e., breaking down large molecules of heavy hydrocarbons into smaller thus lighter hydrocarbons. Cracking is accomplished using high pressures and temperatures without a catalyst (thermal and stream cracking) or lower temperatures and pressures in the presence of a catalyst (fluid catalytic and hydrocracking). In practice, fluid catalytic cracking produces a high yield of gasoline and liquid petroleum gases, while hydrocracking is a major source of jet fuel, diesel fuel, and naphtha.

1.3.3.3 Reforming/Isomerization

Those procedures are basically applied to generate more useful hydrocarbons. Reforming is set to rearrange hydrocarbon molecules into other molecules, usually with the loss of hydrogen. An example is the conversion of an alkane molecule into a cycloalkane or an aromatic hydrocarbon, for instance hexane to cyclohexane.

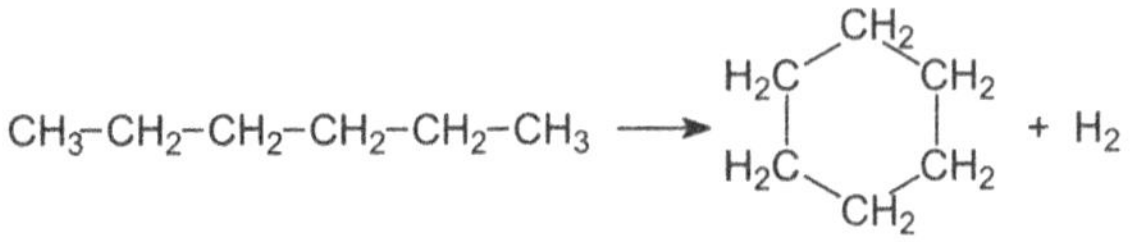

hexane cyclohexane hydrogen

As for isomerization, it is the mechanism with which the same hydrocarbon molecule is rearranged into a more useful isomer (i.e., same chemical formula but different structure). For instance, this process is particularly useful in enhancing the octane rating of gasoline, as branched alkanes burn more efficiently in a car engine than straight-chain alkanes. A related example is the isomerization of butane to 2-methylpropane (isobutane).

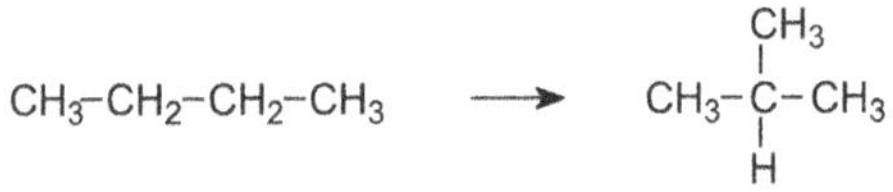

butane 2-methylpropane

1.3.4 Petroleum-Based Products

1.3.4.1 Products from the Refining Industry

As illustrated in Fig. 1.3, the refining process produces various commodities such as refinery gases (*aka.* liquefied petroleum gas, mainly propane and butane), gasoline, diesel, kerosene, jet fuel and fuel oils. The amount and quality of refined petroleum products is mainly related to the type of crude oil used as feedstock as well as the configuration of the refinery. In general, lighter and sweeter crude oils are more expensive but generate greater yields of higher value refined petroleum products including gasoline, kerosene, and other jet fuels. Heavier and sourer crude oils are less expensive and generate greater yields of lower value petroleum products, such as diesel and fuel oils.

In average, a single barrel of petroleum could produce 25–50% gasoline, 10–25% diesel, 10–40% fuel oil, 7–12% jet fuel, and 6–8% gases. Figure 1.4 represents the breakdown of a barrel of US oil (42 gallons ≈ 159 liters) into various refining products.

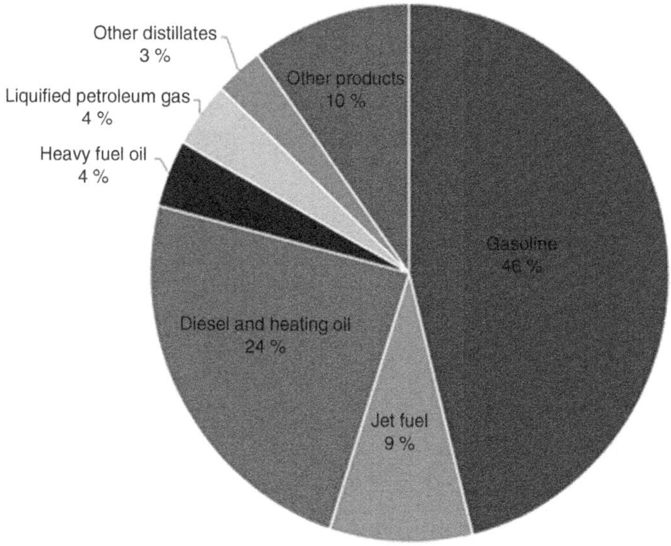

Fig. 1.4 A breakdown of barrel of crude oil into various products (Data source [9, 10])

1.3.4.2 Products from the Petrochemical Industry

The petrochemical industry uses a fast array of hydrocarbons as feedstock, belonging to two major groups: olefins and aromatics.

(i) *Olefins*: are unsaturated aliphatic hydrocarbons containing one or more carbon–carbon double bonds (alkenes), mainly produced from steam cracking and catalytic reforming. It includes ethylene (C_2H_4, the smallest olefin), propylene (C_3H_6), and butadinene (C_4H_6)
(ii) *Aromatics*: are unsaturated cyclic hydrocarbons containing one or more rings mainly produced by catalytic reforming processes. This group includes benzene, toluene, and xylene isomers.

Both olefins and aromatics are feedstocks for a multitude of chemical products and commodities. The flow diagram (Fig. 1.5) gives an overview on those petrochemicals.

Regarding the industrial applications, petrochemicals are the building blocks for the production of diverse products, thus providing end markets and consumers with various commodities throughout the world. The extent of utilization of petroleum-derived products is just staggering as it affects every aspect in our today's life.

The illustration in Fig. 1.6 gives a clear assessment on the degree of dependency individuals and societies alike have on petroleum, a nonrenewable depleting supply.

This list of end products is far from extensive. In the public mind, when the word petroleum is heard, most people will think of gasoline, diesel, plastics, and some textile fibers and dyes. The current situation is far from being restricted to fuel our

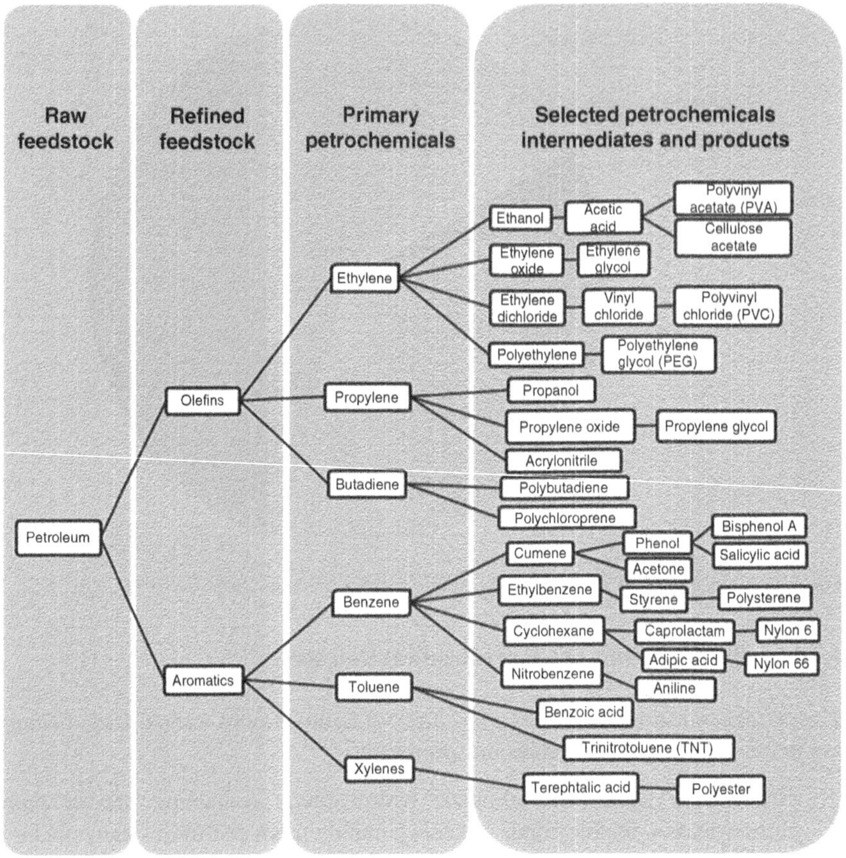

Fig. 1.5 From raw feedstock to petrochemicals: a flow-diagram illustration

cars, make our clothes, and produce some plastic bags out of petroleum derivatives. Indeed, as shown in the previous figure, we use petroleum to protect our crops, to take care of ourselves, and even to medicate ourselves with petroleum-derived pharmaceuticals.

But, the dependency does not end there. Ironically, we depend on nonrenewable petroleum to produce renewable energies. Indeed, in the petrochemical products related to the energy sector in Fig. 1.6, we have purposely mentioned two specific products: protective films and lubricants. The first, protective front and back sheet films are used in the solar panel industry as the outermost layer of the photovoltaic module to protect the inner components from weathering and also act as electric insulators [11]. Those films are mainly made from ethylene-tetrafluoroethylene (ETFE), polyvinyl fluoride (PVF), or polyethylene terephthalate (PET), petroleum-derived thermoplastic polymers. In addition, solar cells contain layers of encapsulants made from the copolymer ethylene-vinyl acetate (EVA), polyvinyl butyral (PVB), or thermoplastic polyurethanes, all petrochemical compounds [12].

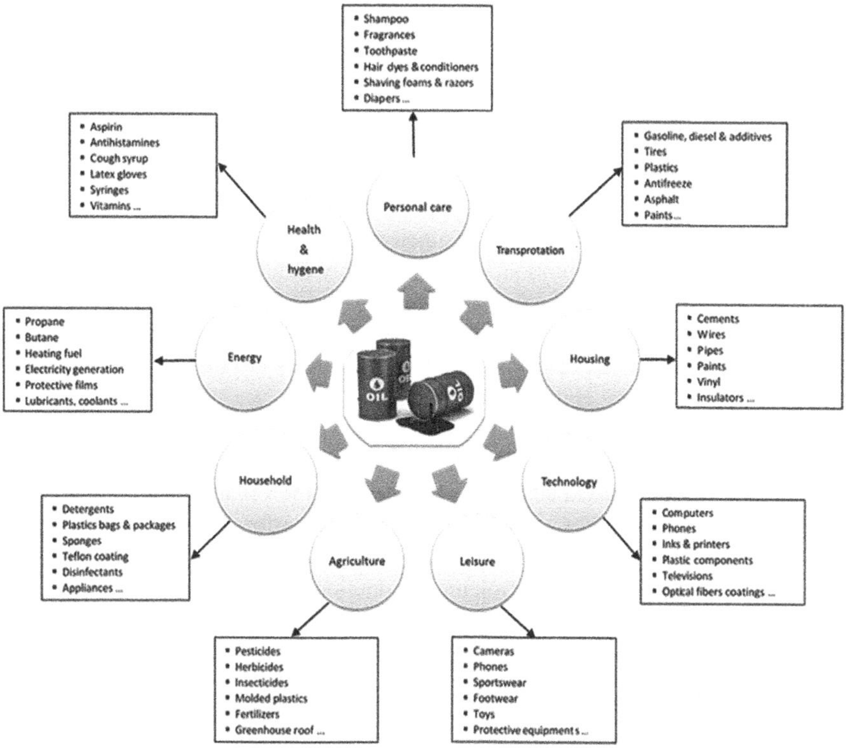

Fig. 1.6 Petroleum-derived end products and markets

On the other hand, petroleum-derived lubricants (oils and greases) and coolants have a very important impact on the wind energy sector. Indeed, wind turbines are very expensive machines generally installed in remote areas; thus, the use of lubricants and coolants (for gearboxes and blades) will help maintaining peak conversion performances and reducing the costly and time-consuming maintenance interventions.

Thus, petroleum is much more deep-rooted in our everyday life than most of us think. This situation has to be seriously taken into account when proposing bioeconomy as an alternative model. To "compete" with petroleum, the sustainability factor alone is not enough. The focus should be on establishing *equally* versatile and efficient production systems. At this stage, and this stage alone, that sustainability will intervene as a decisive factor.

1.4 Prosperity from Black Gold, to Whom and at What Price

1.4.1 Petroleum and Economic Prosperity: Producers Versus Consumers

During the industrial revolution of the nineteenth century, the use of coal as fuel set the "train" of progress in motion, which definitely induced a significant improvement in the living standards. After several discoveries, inventions, and innovations, the use of crude oil, the other so-called black gold, enabled humanity to reach a higher level of prosperity, especially so between the end of the second World War (1945) and the 1973 oil embargo crisis.

Crude oil is the most traded commodity in the world market. In Fig. 1.2, a comparative analysis was carried out based on the production/consumption data for two groups of countries: (1) the 12 members of the organization of petroleum exporting countries (OPEC) and (2) the 12 best performing economies in the world (based on the 2013 GDP statistics [13]). The main findings were that, for the 12 OPEC members, the petroleum production far exceeds the consumption, leading the straightforward strategy of petroleum exportation. As for the 12 strongest economies, the consumption exceeds the national production, except for Russia and to a lesser extent Canada.

Now, in order to better understand the situation, a simplified yet straightforward assessment of the petroleum impact on economies is presented via analyzing the relationship between one important economic indicator, gross domestic products (GDP), and the statistics data about crude oil (production and consumption). Basically, the comparison is between petroleum producing/exporting countries and consuming/importing ones.

First, the analysis of the GDP on the one hand and petroleum production on the other (Fig. 1.7) clearly shows that economic prosperity is not linked to the produced amount of crude oil. For instance, the combined GDP of all OPEC members almost equals that of Germany alone, although they are producing petroleum 218 times more.

It is therefore obvious that mono-product export-orientating economic model adopted by most OPEC members is not only vulnerable but also inefficient in generating national wealth. As for the countries with strong "mixed" economies, petroleum production seems to play a promoting role in the economy via boosting the diverse production activities in countries like the United States and China. Other countries with highly performing economies do not even produce significant amount of petroleum including Japan, Germany, and France. This means that petroleum production is not an affecting factor, so what about its consumption?

To answer this question, let us analyze the correlation between the GDP and the petroleum consumption data depicted in Fig. 1.8.

Contrary to its correlation with production, the GDP had a good fit with petroleum consumption. Therefore, for the selected countries, economic growth is

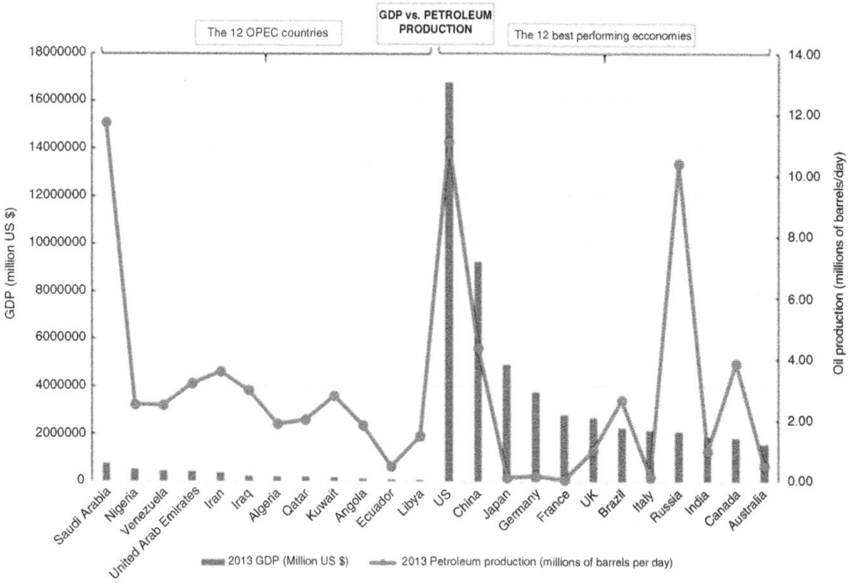

Fig. 1.7 Correlation between petroleum production (Data source [6]) and GDP (Data source [13])

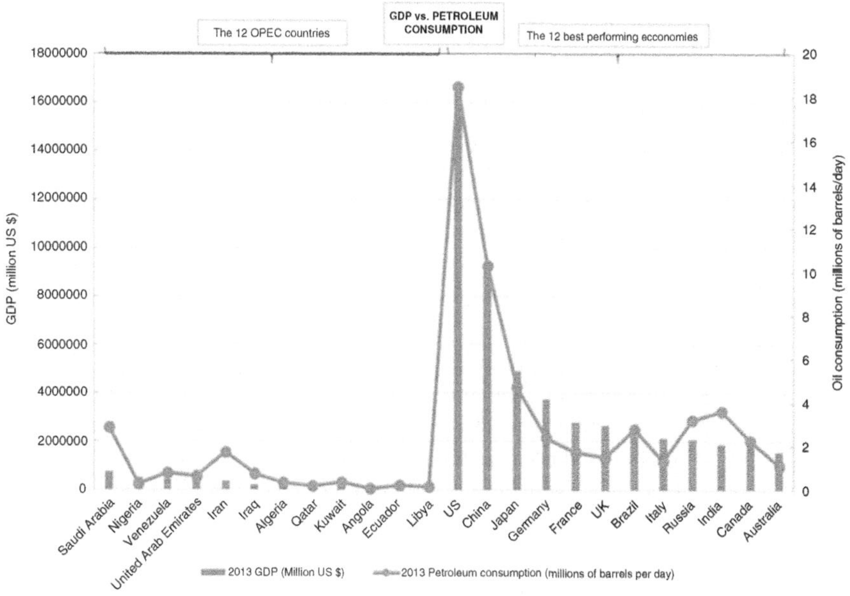

Fig. 1.8 Correlation between petroleum consumption (Data source [6]) and GDP (Data source [13])

linked to petroleum consumption and, as the most traded commodity in the world, induces economic prosperity for the consumers but not for the producers.

The question now is whether the petroleum consumption generates growth or the economic growth increases petroleum consumption. Several investigations were carried out on this matter to find out the causality relationship between oil consumption and GDP. Most economic analysts used the Granger causality model [14, 15] or the modified Toda and Yamamoto version [16]. It was found that this relationship is more prevalent in the developed OECD (Organization for Economic Co-operation and Development) countries compared to the developing non-OECD countries [17].

For the BRICS countries (Brazil, Russia, India, China, and South Africa), the analysis showed that oil consumption and economic growth are not sensitive to each other for the studied panel. However, for the distinct study case of China, a bidirectional causality was proposed [18], as energy supply is needed to "fuel" the industries, but rapid growing industrial activities will put pressure on energy demand thus increasing the oil consumption as well as coal and natural gas. Thus, although the industrial output of China only accounted for about 40% of GDP, the industrial energy consumption accounted for almost 70% of the energy consumption [19].

1.4.2 *Prosperity from Petroleum: The Other Side of the Story*

In the previous section, the simplified analysis showed that petroleum is more profitable for consumers than producers. Indeed, most exporting countries produce and sell crude oil to generate wealth, contrary to the industrialized countries, which import petroleum and use it to generate wealth. The common, but misleading, question asked regarding the most important raw materials on earth is: Do you have petroleum? The real question though is: what are you going to do with it?

In most nations, economic strategies and foreign relations policies are planed based on their reply to this question. Overall, countries could be divided into two main groups: the exporting producers and the importing consumers. "We are planning to sell petroleum and generate prosperity and economic growth from its revenues," replied the first group. The reply of the second group would be: "we are going to buy petroleum and use it as a feedstock to produce various kinds of value-added commodities. The commercialization of those products will generate wealth from various sources, thus generating economic growth and sustaining it."

Let us now analyze the situation for representatives of those two groups: the OPEC 12 members for the first group and the 12 best economies in the world for the second one. The relationship between those two entities has been suspicious most of the time and even nervous some of the time. The price fluctuations of crude oil throughout the last decades say it all. Both groups know very well the importance of the raw material being traded, the so-called black gold.

For OPEC, most of its members rely heavily on petroleum export. According to the organization statistics, the contribution of crude oil in the total export revenues could reach very high levels as it is the case for Saudi Arabia (85%); Nigeria (90%); and Venezuela, Kuwait, and Libya (95%). Thus, petroleum is the backbone of OPEC members' economies since its export is the primary income for member countries to generate revenues (average of 75% of total export revenues) [20]. To name a few, the revenues are being used for decades to build and maintain roads and bridges; construct and equip housing complexes, schools, and hospitals; and import many commodities, most of them derivatives from the refining and petrochemical industries. So, the diagnosis in this case is a "chronic" dependency on depleting fossil resources.

For the industrialized countries, petroleum plays another role yet equally important. It is the backbone of many industrial activities as a feedstock for the production of many value-added products or as the major energy supply for strategic sectors like transportation (cf. Fig. 1.6). For this group, the dependency on petroleum was gradual and discrete (but not for the big oil companies). With the development of the lucrative petrochemical industry, the dependency became chronic.

The actual relationship between petroleum exporting producers and importing consumers is a *mutual dependency*. The first group needs the second to buy the supply. The second needs the first to provide him with these resources to "feed" its large industrial complexes, always hungry for energy. Thus, instead of tense and nervous relationship between OPEC and the industrialized countries, the situation should be analyzed in a wise manner because the current situation is both precarious and escalating, especially considering the geographical distribution of petroleum reserves in the world (politically unstable regions) on the one hand, and with the orientation towards the exploitation of more fossil resources like oil shale and tar sands.

To give a very simple metaphor portraying relationship between OPEC and industrialized countries, let us tell you the story of two farmers, one with a cow (OPEC) and the other with a fruit orchard (industrialized countries). The first has a very productive cow giving him far more milk than his needs for his family. The second has a highly productive orchard with many kinds of fruits giving him decent revenues. In the beginning, the farmer with the cow did not know that it has milk but the second farmer knew that. So, he helped his neighbor to extract milk from the cow with the condition to sell him a daily amount at a "special" price in addition to some fruits from his orchard. The deal went on for decades.

The family of the farmer with the orchard liked milk very much, especially that they were able to produce delicious butter, cheese, and yoghurt out of it. Those products become a staple in their daily diet and they did not seem to get enough of it. The farmer with the cow, now aware of the importance of milk, started taking care of his cow to produce more and more. Overall, both farmers benefited from the precious milk. The first was satisfied with the revenues from his cow and took care of his entire family from the milk revenues only. The second farmer become too dependent on the delicious milk to a point that he is not taking care of his orchard anymore and started to think "why buying milk if I could get the cow or, at least, I can give its custody to a friend." From this point, the relationship between two farmers became very tense and all the neighborhood became anxious and fearful.

One day, a wise veterinary came to see the troublesome farmers. He told the first farmer (OPEC): "open your eyes, the cow is suffering and it has only couple of more years to life." Then, he turned to the second farmer (industrialized countries) and said: "if you like milk this much, I've heard of another farmer producing Soy milk, it's similar to the conventional milk and made from renewable resources." The two farmers looked to the vet, the meager cow and the deserted orchard. Then, they looked to each other for seconds and the first (OPEC) said to the second (industrialized countries): "Still want your usual dose of milk?," and the latter replied: "of course my friend," and the story goes on.

1.4.3 The Petroleum Paradox

The dangerous facet of the story is that both groups (exporting and importing countries) are establishing strategies for entire nations and given hopes for entire populations, based on a precious yet depleting resource. Every one of them is trying to ensure an ever increasing prosperity for their countries and well-being for their citizens (a quite selective as we shall see later). The goal in itself is worthwhile but the adopted strategy is short-sighted, hasty, and unwise.

Normally, strategists, policy makers, and decisions makers are wise and insightful men and women, the best a nation can give. So, why we ended up with unwise strategies leading to excessive dependency on a nonrenewable resource like petroleum?

For decades, political economists and experts analyzing the petroleum production, consumption, and trading proposed several concepts to explain the abnormality that oil-rich countries have weak economies, namely, the "petroleum curse" [21] and the "Dutch disease" [22].

The constant flow of petroleum to markets throughout the entire world, its "affordable" price range (most of the time), in addition to its huge revenues, gave the sense that it is unlimited resource. This illusion led to the gradual surge in dependency, and it is not only affecting the known oil exporting countries where the apparent abundance of petroleum strangely led to a weak industrial manufacturing and agricultural activities (the so-called Dutch disease) but also developed nations. Indeed, the largest concentration of refining capacity is not in the Middle East, the largest producing arrear and where most of the reserves are, but in North America and Europe. The Unites States alone accounts around one-quarter of the crude oil distillation capacity of the world [23].

A comparative example from the European Nordic region gives a clear illustration of this syndrome also affecting developed nations. The discovery of petroleum in the North Sea in late 1969 transformed the core of the Norwegian economy. Norway is among the top ten largest oil exporter, and petroleum accounts for 30%

of the country's revenues [24]. Since the so-called 1970's boom, those revenues were channeled to cover the high spending in wages and welfare benefits to unsustainable levels, making Norway one of the world's most generous welfare model, second only to Sweden [25]. Thus, the Norwegian economy gradually became highly dependent on its fossil resources and the economic growth highly susceptible to currency evaluation, pricing, and demand fluctuations. During the early stages of the oil boom, little governmental initiatives were instigated to develop the established industries and encourage new enterprises in the private sector. In the meantime, neighboring Finland, with no petroleum resources, managed to build a vibrant and highly industrialized economy involving various sectors such as electronics, machinery, mining, forest products, and chemicals [26].

So, the paradox is undeniable and valid for most oil-rich countries, mainly in developing nations (most OPEC members) but also, and to a lesser extent, developed countries (UK, Netherlands, Norway…). The general tendency is that, once a country discovers petroleum, it will become visible on the world map (even if too tiny), the economic growth will go up rapidly, the manufacturing and agricultural activities (if any) start to go down, and the whole country, if not the entire region, becomes vulnerable, unstable, and insecure. Some call it a disease, others a curse.

Either way, the fundamental question is: who brought the disease or spelled the curse on those countries? We shall address this important issue in an unbiased manner in the last section of this chapter and we shall point to the source of the problem directly. A straightforward and frank diagnosis is half the cure, and it is about time to begin the healing process.

1.5 End of an Area: Environmental Disasters and Geopolitical Instability

Petroleum is the cause of many environmental disasters and major geopolitical developments in the world. Many will agree and some will disagree, depending on the impact of this highly addictive commodity on one's life. Yet, petroleum is just the superficial cause. It's merely taking the place, for a determined lapse of time, of many other precious commodities before it such as gold, ivory, cotton, and minerals. The real and deeply rooted problem is in the mind of men, greedy and corrupt men.

This is not to diminish the role of petroleum exploitation in the current disastrous state of the planet. Rather, it is crucial to make a proper diagnosis and know the real source of the problem. If not, proposing bioeconomy as new sustainable model of exploitation and production, under the same circumstances, will be futile. We have to break up this cycle of bloody and costly struggle over resource. It has to end, too much is at stake. The global situation is not just precarious for next generations, it is already so for the current generation. The earth is suffering, soil, air, and water alike, and humanity too.

Thus, analyzing the legacy of the petroleum era would be a very important endeavor to draw lessons for the future and be a constant reminder for those of us with short memories.

1.5.1 *Serious Environmental Degradation*

Environmental pollution caused by petroleum extraction and refining, along with its derived products manufacturing, utilization and disposal, has led (and still do) to major ecological disasters. The impact is not restricted to industrialized countries where the bulk of the refining, transformation, and utilization occurs, but also in less developed countries where petroleum is extracted and in even less developed countries where most of the petroleumderivative wastes, among many other contaminants, are dumped. Thus, during the last decades, the only feature that equals the worldwide importance of petroleum in human societies is its worldwide mess in human environment.

1.5.1.1 Toxicity of Petroleum and Derived Products

Public Health Risk

Petroleum and its derived intermediates and products constitute one of the most widespread sources of environmental degradation in the world due to their wide and extensive extraction, storage, transportation, and use. Intrinsically, petroleum has various toxicity levels as it is a complex mixture of hydrocarbons, in addition to trace amounts of sulfur and nitrogen compounds, some of which are toxic, poisonous, and carcinogenic.

Petroleum products are persistent and highly mobile contaminants making them difficult to remove from soils and groundwater. Besides, several petrochemicals are known or suspected carcinogens or mutagens which can pose serious human health risks (e.g., cancer, birth defects, and other chronic conditions) at 10 ppb and below in groundwater resources [27]. Inhalation of high concentrations of crude oil may cause central nervous system effects, and a prolonged or repeated skin contact may cause dermatitis.

The major concern is about the potential carcinogenicity of some polycyclic aromatic hydrocarbon (PAHs) [28]. Benzo(a)pyrene is considered one of the most potent carcinogenic PAH. Along with benz(a)anthracene, they are classified as probable human carcinogens inducing skin tumor development [29]. At relatively high levels, the volatile organic compounds benzene, toluene, ethylbenzene, and xylene (BTEX), commonly found in petroleum derivatives, can be associated with serious neurotoxic [30, 31] and hepatotoxic complications [32].

Soil and Ground Water Contamination

Soil and ground water contamination by petroleum and its derivatives is related to two main pollution source: point and nonpoint sources [33]. The first could be caused by spills and leaks from petroleum wells, pipelines, and underground storage tanks. It could also be generated in landfills and industrial waste disposal sites. Either way, the leaking can seriously contaminate large areas of soil, deteriorating their fertility and even making them nonexploitable and source of serious health risk from organisms living in them or feeding from them. For instance, 25% of the content of underground petroleum storage tanks is estimated to be leaked into the ground [34]. If the treatment of soils contaminated by petroleum is not both quick and efficient, toxic hydrocarbons can leach into the groundwater and contaminate vital water resources, particularly if the aquifer is shallow and not sealed by an overlying layer of low permeability material like clay.

The second source is related to the commonly used petroleum products. For instance, the agricultural sector is highly affected by any deterioration of the quality of both soils and water resources. Thus, the various petrochemicals used in agriculture including fertilizers, pesticides, and herbicides constitute a nonpoint source of pollution especially considering their worldwide extensive applications.

Air Pollution

During the petroleum refining process, several toxic gaseous compounds are emitted into the atmosphere. As well, the utilization of petroleum distillates and derivatives in the transportation, industrial, and agricultural sectors is partly responsible for the worldwide degradation of the air quality.

The U.S. Environmental Protection Agency (EPA) designates six criteria pollutants for determining air quality including carbon monoxide (CO), nitrogen oxides (NO_x), sulfur dioxide (SO_2), ground-level ozone (O_3), and particulate matter (soot, dust, pesticides, and metals) [35]. In addition, petroleum-fueled vehicles and industrial machinery running on fossil fuel directly produce considerable amounts of toxic CO, NO_x, and volatile organic compounds (VOCs) in the atmosphere. The combination of VOCs and NO_x in sunlight create O_3, well known for blocking ultraviolet rays in the upper atmosphere, but could cause human health problems when present in the lower atmosphere (tropospheric ozone) [36].

Sulfur dioxide is a trace component of petroleum. When released into the air with the refineries emissions, it could cause acid rains [37]. Indeed, SO_2 and NO_x react with the water molecules in the atmosphere to form corrosive acid rain which could be harmful to exposed living organisms and even buildings and other infrastructure.

Incomplete combustion of petroleum occurring on a burning oil rig or when an oil spill catches fire generates smoke plumes containing a mixture of toxic gases (CO, CO_2, NO_x...) and particulates matter (soot and acidic aerosols). The particulate matter is also constantly being emitted by vehicle exhausts along with other gases [38]. The use of lead as additive to gasoline, to boost the octane ratings, contributed in increasing the lead levels in the atmosphere.

All those hazardous compounds in petroleum and its derivatives are suspected to induce serious health issues if exposed to high concentrations or for long period of time. According to EPA [39], the health complications include:

- Array of adverse respiratory effects, airway inflammation in healthy people, increased respiratory symptoms in people with asthma (inhalation of SO_2 and NO_x).
- Harmful health effects associated with the reduction of oxygen delivery to the body's organs (heart and brain) and tissues (inhalation of CO).
- Increased respiratory symptoms, irritation of the airways, difficulty breathing, decreased lung function, aggravated asthma, development of chronic bronchitis, irregular heartbeat, and nonfatal heart attacks (inhalation of particulate matter).

1.5.1.2 Global Warning and Climate Change

According to the intergovernmental panel on climate change (IPCC), global warming or the increase in global average temperatures is "very likely" due to the observed increase in anthropogenic (i.e., due to human activities) greenhouse gas concentrations. Among those greenhouse gases causing global warming are carbon dioxide (CO_2), methane (CH_4), and nitrous oxide (N_2O), along with perfluorocarbons (PFCs), hydrofluorocarbons (HFCs), and chlorofluorocarbons (CFCs). The investigations performed by the IPCC's experts revealed that the global atmospheric concentrations of CO_2, CH_4, and N_2O have increased markedly as a result of human activities since 1750 and now far exceed pre-industrial values [40].

Over the past half century, petroleum-derived fuels used in transportation and coal used in power plants are considered to be the primary causes of global warming, along with severe deforestation, intensive agriculture, and other human activities. Scientists predict that around one-third of the CO_2 emitted into the atmosphere every year comes from vehicle exhaust. Methane, although usually associated with natural gas, could also be emitted during petroleum extraction, transportation, refining, and storage.

Scientists from the IPCC examining global warming have predicted that, by the year 2100 average, the global temperatures could increase between 1.1 and 2.9 °C for the low scenario and between 2.4 and 6.4 °C, based on the high scenario [41].

Global warming was accompanied with long-term continuous changes in weather and climate. Worldwide climates changes are becoming more and more apparent. If researchers' investigations and scientists' predictions are enough for some, visible consequences of climate change and its impact on the environment, and the planet in general, will convince the rest. Storms are becoming more frequent, heat waves more intense, and droughts more severe. Ice caps are melting, oceans and sea levels are rising, threatening the ecological equilibrium of many marine ecosystems [42]. Sea-level rise can also contaminate groundwater supplies in coastal aquifers, thus endangering human populations and coastal ecosystems [43].

1.5.1.3 Major Oil Spills

In its journey from the well to the market, petroleum could leak into the environment during extraction, transportation, and consumption, thus causing serious disasters. The first two are called point sources of pollution (i.e., one identifiable source) considered as accidental spills from oil wells, pipelines, and tankers leading to tragic releases of large volumes of toxic hydrocarbons in the immediate environment. The other kind of petroleum-related pollution comes from nonpoint sources (i.e., multiple sources).

An explosion during the oil drilling process is one of the major causes. It mainly occurs when the gas trapped inside the deposit is at such a high pressure that oil suddenly erupts out of the drill shaft, ignites, and explodes the drilling platforms. Two infamous accidents of this type (1979 Ixtoc I and 2010 Deepwater Horizon) happened in the same region, the Gulf of Mexico (USA). The other point source spills occur during its transportation from oil from exporting countries to importing ones. The marine and/or land journey of petroleum is fulfilled by oil tankers and pipelines or by railways and roads. Some of the transportation are safer than others, but all are susceptible to unexpected accidents. The following accidents represent the major recorded oil spills from (1) tankers and (2) petroleum fields.

Disasters from Oil Tankers

- In the 1960s (March 1967), the *Torrey Canyon* tanker grounded on the Seven Stones reef between the Cornish mainland and the Isles of Scilly (UK). All its cargo of crude oil (119,000 t) leaked into the sea [44]. This supertanker, like many others, perfectly illustrates the international dimension and complicated liaisons (to say the least) in the petroleum sector. It was built in the United States in 1959 with an initial capacity of 60,000 t, later enlarged in Japan to 120,000 t. The crude oil came from Kuwait and was going to Wales. At the time of the disaster, the tanker was owned by the Union Oil Company of California (USA), registered in Liberia and chartered to British Petroleum (UK).
- In the 1970s (July 1979), the Greek crude oil carrier *Atlantic Empress* collided with the Aegean Captain, another Greek supertanker, 30 km east of the island of Tobago. A week after the collision, the tanker sank after spilling 287,000 t of crude oil into the Caribbean Sea [45].
- In the 1980s (August 1983), the Spanish supertanker *Castillo de Bellver* Spanish tanker caught fire about 80 km northwest Cape Town, South Africa. Explosions broke the vessel apart and the entire cargo of light crude oil estimated at 250,000 t spilled into the sea [46].

- In the 1990s (May 1991), the *ABT Summer*, owned by Saudi Arabia and registered in Liberia, exploded and sank with its cargo of 260,000 t of crude oil, 900 miles off the coast of Angola [47].
- In the 2000s (November 2002), The Greek tanker *Prestige* sunk in the territory of Galicia (Spain) and spilled some 716,000 t of crude oil [48]. The pollution affected coastal regions in Spain, Portugal, and France.

Disasters from Oil Fields and Pipelines

- In the 1970s (June 1979), the *Ixtoc I* oil well exploded in the Gulf of Mexico. The resulting fire caused the oil drilling platform to collapse and petroleum to leak into the sea for months. It was estimated that 476,000 t of petroleum polluted the offshore region in the Gulf of Mexico [49], one of the largest oil spills ever recorded.
- In the 1980s, (February 1983), and during the Iran/Iraq war, an oil tanker collided with the *Nowruz* platform, then Iraqi helicopters attacked it, and the slick caught fire. After several spills and delayed well capping, 260,000 t of petroleum were spilled because of this war [50].
- In the 1990s (January 1991), one of the worst oil spill in history of mankind happened during another war, the *Gulf War*. While retreating from Kuwait after a failed invasion, the Iraqi armed forces opened valves of oil wells and pipelines, causing 1,360,000–1,500,000 t of petroleum to be spilled into the Persian Gulf [51].
- In April 2010, the *Deepwater Horizon* or BP oil spill, the largest accidental marine oil spill in the US history, took place in the Gulf of Mexico 66 km off the coast of Louisiana. The blowout was caused by the expansion of highly pressurized methane from the well up to the oil rig where it ignited and exploded. The platform later sunk and the estimated amount of leaked petroleum was between 492,000 and 627,000 t [52].

In the long run, point sources, although spectacular, remain limited in terms of volume and impact, compared with nonpoint sources where small and recurring amounts of petroleum and its derivatives are being spilled for an extended period of time and from various sources. The non-extensive list contains routine discharges of fuel from huge commercial and leisure ships, airplanes, cars, asphalt-covered roads, and illegal dumping of oil wastes (land and ocean). In addition, the rapid population growth, cities' expansion, and industrial development led to an increased consumption of petroleum and its derivatives with frequent unsafe disposal of the generated wastes. Such products include lubricants, solvents, plastics, paints, etc.

1.5.2 Corruption, Wars, and Geopolitical Instability

While introducing this chapter on the legacy of petroleum-based economy, we have stated that "we were feeding our infinite hunger for wealth from a limited provision." Limited provision refers to petroleum as well as the other fossil fuels.

Let us now develop more this very delicate issue and start by defining the "we." The whole story started in the early twentieth century with small family-owned companies extracting the precious black gold. This story strangely brings in mind another one which happened a century before known as the Gold Rush. The only difference is that cars cannot run on gold.

Soon afterwards, petroleum gained the position of national strategic resource in most parts of the world. Hence, nations became heavily involved in the profitable petroleum-related industries (extraction, transportation, and refining), in order to secure what seemed to be an abundant and easy income to the country. In the meantime, the once small businesses grew bigger and bigger. This spectacular growth became alarming when the end started to justify the means for the "princes" of petroleum.

At this stage, the corporations started to compete with states and governments in terms of involvement in the petroleum production and trading. To satisfy their infinite hunger for wealth, some of those corporations merged together to form giant multinational multi-billion dollar corporations and stand, once and for all, above governments and countries. Democratic or despotic regimes were not a concern for them, but securing a constant flow of petroleum was. This is precisely the source of all petroleum-related problems, or any other strategic resource for that matter, the unholy alliance between money and power.

After identifying the real disease, let us now go deep in the analysis and see how it affected the body of nations. Under normal circumstances, governments in countries, whether exporting or importing petroleum, think primarily about the well-being of their citizens and economic prosperity and environmental safety for future generations. Some big corporations, on the other hand, think only about the well-being of the shareholders and the steady increase of their wealth. A quick look on the daily news for a couple of days will clarify the current world situation (environmental disasters, increasing poverty, political instability, and economic vulnerability in many countries, including oil-rich ones). It leads only to one conclusion: big oil companies won the "battle" against countries.

The question then is how a nation with all its resources, political parties, labor unions, intelligentsia, and military could lose against a corporation, no matter how big it is? Take some time and think about it.

The straightforward answer is corruption. When the subject of petroleum paradox is raised, most of the literature relate it to the corruption in the oil-producing country, forgetting a very important aspect: *corruption is a two-way street paved with greed*. How so? Well, let us start by dividing nations into two major groups and then analyze the disastrous impact of corruption on both of them: despotic regimes and democratic system.

Petroleum-fed corruption in despotic regimes is very simple and direct. One dictator has all the power and owns all the resources of "his" country. Thus, bribing the devil and striking a lucrative deal with him will secure a constant and cheap flow of petroleum and will be a bargain for both of them, but not for the people living under a despotic regime.

Corruption in democratic countries is more subtle yet more dangerous than in dictatorships. It is easy to corrupt an imposed and unelected heads of states or junta. But, to do so for elected officials and even scientists is very difficult. Nonetheless, money (which multi-billion dollars oil corporations have lots of it) seems to ease things by funding election campaigns and research studies of special interest. Another very efficient strategy is to hire skilled people and establish institutions to take care of the corporations' interests through lobbying. The same applies for big pharmaceutical firms, tobacco companies, and military-industrial complex, to name a few.

Petrodollars could also be used to buy newspapers (not from newsstands like all of us), radios, and TV stations, all proven to be very useful to "convince" the masses that, somehow, during the companies' pursuit of wealth, the people will pursue their dreams of happiness. For the better educated people, the petrodollars-financed think tanks will do the convincing.

Overall, it is not just a matter of powerful and influential lobbies capable of bending the laws enabling therefore the petroleum industry to gain more and more money and power. The situation is far more serious. It is about recurrent environmental catastrophes occurring with almost no accountability or impunity and more importantly about their ability to persuade governments to dispatch armies worldwide to secure a constant flow of petroleum causing death and destruction and fueling long-term animosities, nationally and internationally.

Bottom line, if a country finds a valuable resource in its land, ground, or sea, it should fight and eradicate corruption before even thinking extracting it. If it wins the ultimate battle against any kind of political, judicial, administrative and commercial corruption, it will be blessed. In case of defeat, it will be cursed.

1.5.3 Last But Not Least Problem: Consumerism

In the petroleum-based economy, markets became flooded with various affordable commodities from fuels (gasoline, diesel fuel, kerosene...) and plastics, to paraffin wax, lubricants, and many other cheap petrochemicals and commodities.

During this industrial boom, the already existing notion of consumerism shifted from a mere concept synonym of prosperity to a promoted practice believed to increase production and enlarge markets. Thus, in order to make him buy more and consume even more, the BIG production and trading companies told the LITTLE consumer that "he is always right" and that "he is king." Is he really sovereign? Of course not. That's why there are consumers' protection laws and associations worldwide.

Consumer is definitely the weakest element in the economic system. The three others (i.e., exploitation, production, and trading) are lucrative activities that generate wealth, but not consumption. To the contrary, consumers will directly or indirectly generate wealth for exploiters, producers, and traders. Thus, the consumerism concept was built to incite the consumer to buy more so that the trader sells more, the manufactured produces more, and the entrepreneurs or industrials exploits more and more resources.

Overall, at a conceptual level, the current market-based economic model, heavily relying on fossil fuels (petroleum, gas, and coal), is supposed to bring prosperity to nations and welfare to its citizens. In practice, it did it for some, but for most it triggered wars, caused social injustice, provoked famine, and induced environmental disasters as we have seen throughout this chapter.

All those worldwide calamities and the petroleum production were still abundant. Imagine now what will happen when there will be less and less petroleum? Horrible scenarios indeed.

The solution then? Well, gradually free humanity from its severe addiction to fossil fuels and, with a fast but steady pace, shift to bioeconomy as a sustainable production-based economic model, as we shall see in the following chapter.

References

1. Tissot BP, Welte DH. Diagenesis, catagenesis and metagenesis of organic matter. In: Petroleum formation and occurrence—a new approach to oil and gas exploration. Berlin: Springer; 1978. p. 69–73.
2. International Energy Agency (IEA). Key world energy statistics. 2014. p. 6.
3. Hyne N. Dictionary of petroleum exploration, drilling and production. 2nd ed. Tulsa: Pennwell; 2014.
4. Simanzhenkov V, Idem R. Crude oil chemistry. Boca Raton: CRC Press; 2013.
5. American petroleum institute (API). Standard test method for density, relative density, or API gravity of crude petroleum and liquid petroleum products by hydrometer method. 3rd ed. 2012. p. 8.
6. US Energy Information Administration (EIA). International energy statistics. Available online at: http://www.eia.gov/cfapps/ipdbproject/iedindex3.cfm?tid=5&pid=53&aid=1
7. Fahim MA, Al-Sahhaf TA, Elkilani A. Fundamentals of petroleum refining. Amsterdam: Elsevier; 2009.
8. Dincer I, Rosen MA. EXERGY: energy, environment and sustainable development. Amsterdam: Elsevier; 2007.
9. Kent JA. Handbook of industrial chemistry and biotechnology. 12th ed. Berlin: Springer; 2013.
10. Bryan P. The how's and why's of replacing the whole barrel. 2011. Available online at: http://energy.gov/articles/hows-and-whys-replacing-whole-barrel
11. Willeke GP, Weber ER. Advances in photovoltaics: part 2. Waltham, MA: Academic Press; 2013.
12. Schulze S, Pander M, Naumenko K, Altenbach H. Analysis of laminated glass beams for photovoltaic applications. Int J Solids Struct. 2012;49:2027–36.

13. Gross domestic product in US$. World Bank national accounts data 2013. Available online at: http://data.worldbank.org/indicator/NY.GDP.MKTP.CD?order=wbapi_data_value_2013+wbapi_data_value&sort=desc
14. Granger CWJ. Investigating causal relations by econometric models and cross-spectral methods. Econometrica. 1969;37:424–38.
15. Granger CWJ. Testing for causality: a personal viewpoint. J Econ Dyn Control. 1980;2:329–52.
16. Toda HY, Yamamoto T. Statistical inference in vector autoregressions with possibly integrated processes. J Econ. 1995;66:225–50.
17. Chontanawat J, Hunt LC, Pierse R. Does energy consumption cause economic growth? Evidence from a systematic study of over 100 countries. J Policy Model. 2008;30:209–20.
18. Chang T, Gadinabokao OA, Gupta R, Inglesi-Lotz R, Kanniah P, Simo-Kengne BD. Panel granger causality between oil consumption and GDP: evidence from BRICS countries. Int J Sustain Econ. 2015;7:30–41.
19. Zhang W, Yang S. The influence of energy consumption of China on its real GDP from aggregated and disaggregated viewpoints. Energy Policy. 2013;57:76–81.
20. Organization of petroleum exporting countries' statistics. OPEC bulletin 11–12/10. Available online at: http://www.opec.org/opec_web/static_files_project/media/downloads/publications/OB11_122010.pdf
21. Ross ML. The oil curse: how petroleum wealth shapes the development of nations. Princeton: Princeton University Press; 2012.
22. Wijnbergen SV. The Dutch disease: a disease after all? Econ J. 1984;94:41–55.
23. Gary JH, Handwerk GE, Kaiser MJ. Petroleum refining, technology and economics. 5th ed. Boca Raton: CRC Press; 2007.
24. Norway, the rich cousin. The Economist, 2 Feb 2013. Available online at: http://www.economist.com/news/special-report/21570842-oil-makes-norway-different-rest-region-only-up-point-rich
25. Armingeon K, Beyeler M. The OECD and European welfare states. Cheltenham: Edward Elgar; 2004.
26. Statistics Finland. Available online at: http://www.stat.fi/til/index_en.html
27. Sharefkin M, Shechter M, Kneese A. Impacts, costs, and techniques for mitigation of contaminated groundwater: a review. Water Res. 1984;16:1771–84.
28. Harvey RG. Polycyclic aromatic hydrocarbons: chemistry and carcinogenicity. Cambridge: Cambridge University Press; 2011.
29. Pashin YV, Bakhitova LM. Mutagenic and carcinogenic properties of polycyclic aromatic hydrocarbons. Environ Health Perspect. 1979;30:185–9.
30. Riihimaki V, Savolainen K. Human exposure to m-xylene: kinetics and acute effects on the central nervous system. Ann Occup Hyg. 1980;23:411–22.
31. Dudek B, Gralewicz K, Jakubowski M, Kostrzewski P, Sokal J. Neurobehavioral effects of experimental exposure to toluene, xylene and their mixture. Pol J Occup Med. 1990;3:109–16.
32. Malaguarnera G, Cataudella E, Giordano M, Nunnari G, Chisari G, Malaguarnera M. Toxic hepatitis in occupational exposure to solvents. World J Gastroenterol. 2012;18:2756–66.
33. Pies C, Ternes TA, Hofmann T. Identifying sources of polycyclic aromatic hydrocarbons (PAHs) in soils: distinguishing point and non-point sources using an extended PAH spectrum and n-alkanes. J Soil Sediment. 2008;8:312–22.
34. Purushothama RP. Soil mechanics and foundation engineering. London: Pearson; 2007.
35. US Environmental Protection Agency (EPA). What are the six common Air Pollutants? 2015. Available online at: http://www.epa.gov/oaqps001/urbanair/
36. Szopa S, Hauglustaine DA, Vautard R, Menut L. Future global tropospheric ozone changes and impact on European air quality. Geophys Res Lett. 2006;33:1–5.
37. Vahedpour M, Zolfaghari F. Mechanistic study on the atmospheric formation of acid rain base on the sulfur dioxide. Struct Chem. 2011;22:1331–8.

38. Park D, Yoon Y, Kwon SB, Jeong W, Cho Y, Lee K. The effects of operating conditions on particulate matter exhaust from diesel locomotive engines. Sci Total Environ. 2012;419:76–80.
39. EPA report. Addressing air emissions from the petroleum refinery sector. 2011. Available online at: http://www.epa.gov/air/tribal/pdfs/presentationpetroleumrefineries14Dec11.pdf
40. Intergovernmental Panel on Climate Change (IPCC). 4th assessment report: climate change. 2007. Available online at: https://www.ipcc.ch/pdf/assessment-report/ar4/syr/ar4_syr.pdf
41. IPCC. 4th assessment report: projections of future changes in climate. 2007. Available online at: https://www.ipcc.ch/publications_and_data/ar4/wg1/en/spmsspm-projections-of.html
42. Jones G. Global warming, sea level change and the impact on estuaries. Mar Pollut Bull. 1994;28:7–14.
43. Isobe M. Impact of global warming on coastal structures in shallow water. Ocean Eng. 2013;71:51–7.
44. Burningham D, Davies J. Environmental economics. 4th ed. Oxford: Heinemann; 2004.
45. Chiras DD. Environmental science. 10th ed. Burlington, MA: Jones & Bartlett Learning; 2012.
46. Emery KO, Uchupi E. The geology of the Atlantic Ocean. Berlin: Springer; 2012.
47. Wali MK, Evrendilek F, Fennessy MS. The environment: science, issues, and solutions. Boca Raton: CRC Press; 2009.
48. Castillo L. Law of the sea, from Grotius to the international tribunal for the law of the sea. Leiden: Brill; 2015.
49. Payne JR, Phillips CR. Petroleum spills in the marine environment: the chemistry and formation of water-in-oil emulsions and tar balls. Boca Raton: CRC Press; 1985.
50. Libes S. Introduction to marine biogeochemistry. Burlington, MA: Academic Press; 2011.
51. Janardhan V, Fesmire B. Energy explained: conventional energy and alternative. Lanham: Rowman & Littlefield; 2010.
52. Renne JL. Transport beyond oil: policy choices for a multimodal future. Washington, DC: Island Press; 2013.

Chapter 2
Bioeconomy: The Path to Sustainability

Abstract As the global environmental, geopolitical, and socioeconomic situation started to worsen, humanity became aware that the current economic model based on fossil resources is not a viable one and its shortcomings are being sensed all over the world (economic crisis, global warming, accentuated disparities, recurrent pollution incidents, etc.). In response, a general consensus was made about the necessity to reintroduce biomass as the core element for the future economic model allowing a sustainable development, along with dealing with the major issues being faced by humanity nowadays.

In this chapter, the various definitions around the bioeconomy concept are presented, as well as the urgent need elaborate an authoritative definition of this concept in order to synchronize the efforts of all possible contributors (legislators, scientists, industrialists, etc.) for a wider promotion and implementation of this new economic model through a gradual and smooth transition in raw materials from fossil to renewal resources.

The main aim of bioeconomy is primarily to conduct the various agricultural, forestry, and industrial activities in a sustainable manner. Thus, in order to ensure a successful transition to bioeconomy, the key endeavor is to find out different viable schemes to combine both sustainability and profitability. This is definitely the major challenge to face bioeconomy for the next couple of decades. The leading role of science and technology in this vital transition phase towards sustainable bioeconomy is emphasized.

2.1 Introduction

After analyzing the heavy legacy of the petroleum-based economy on mankind and the environment, the ultimate deduction is that the situation has to change, and an alternative economic model has to be proposed to repair mistakes of the past and pave the pathway for a better future, a sustainable and eco-friendly future, for generations ahead.

The utmost important keyword in this very vital endeavor is CHANGE. Indeed, this is the real challenge facing any economic model expected to replace the current fossil fuels-based one. The fear of change is deeply lodged in the psyche of both individuals and societies.

M. Sillanpää, C. Ncibi, *A Sustainable Bioeconomy*,
DOI 10.1007/978-3-319-55637-6_2

No economic model will be able to take over if it is not, at least, as efficient as the current one. This is the only solution to overcome the intrinsic fear of change going all the way from the producers to the consumers. May be a fraction of them is ready to make minor concessions for some time, but no one will be willing to make serious concessions most of the time.

As the geopolitical, socioeconomic, and environmental situation started to deteriorate, humanity suddenly became aware about this "green stuff" around us. We could eat it, cure ourselves with it, feed it to livestock, and produce fuels and many other products from it. Humanity rediscovered then what was always around: BIOMASS.

At this point, we became aware that many commodities came from petroleum, but petroleum too came from biomass. So, what if we replace the petroleum by its source and petroleum-based economy by a biomass-based economy. Thus, instead of "preaching" urgent change to change-fearing population, let us use the comforting "go back to the source" speech.

2.2 What Is Sustainable Bioeconomy?

Defining bioeconomy is a crucial first step. This concept will be used to deal with worldwide population growth, depleting fossil raw materials, climate change, and many environmental problems. Thus, before analyzing bioeconomy and its implementation, the notion itself should be defined.

Although there are many viewpoints about sustainability, the notion itself is quite straightforward. Basically, it is about securing the needs of current generation without compromising the ability of the next one to secure its own needs.

As for bioeconomy or bio-based economy, many scientists, governments, and international institutions presented their own definitions. From those definitions, legislations, strategies, and policies will be developed and later implemented, and this is precisely why defining bioeconomy is crucial. The problem here is that bioeconomy is a multidimensional concept, and its definition basically depends on who's defining. Economists, industrialists, farmers, strategists, and ecologists will have a distinct, sometime contracting definitions of bioeconomy. Imagine decision makers adopting action plans for years ahead based on the "governmental" perception of bioeconomy, but industrials, on the other hand, have another vision of the whole concept. The implementation will be very difficult, especially within an international network (which is the case most of the time).

Here are the widely known (still to be widely accepted) definitions related to bioeconomy. According to the European Union, bioeconomy "*encompasses the production of renewable biological resources and their conversion into food, feed, bio-based products and bioenergy. It includes agriculture, forestry, fisheries, food and pulp and paper production, as well as parts of chemical, biotechnological and energy industries*" [1].

Within the European Union, several countries pursuing their own national strategies aiming at developing bioeconomy and sustainability also adopted other definitions. For instance, the German Federal government defined the concept of bioeconomy in its national research strategy "BioEconomy 2030" as follows: "*The concept of bioeconomy covers the agricultural industry and all manufacturing sectors and their respective service areas, which develop, produce, process, reprocess or use them in any form biological resources such as plants, animals and microorganisms. Thus, it achieves a variety of industries such as agriculture, forestry, horticulture, fisheries and aquaculture, plant and animal breeding, food and beverage, wood, paper, leather, textile, chemical and pharmaceutical industries up to branches of energy industry*" [2].

Finland, one of the leading countries in implementing bioeconomy, also adopted its own definition. According to the Finnish bioeconomy strategy, "*bioeconomy refers to an economy that relies on renewable natural resources to produce food, energy, products and services.*" The objectives of this strategy is to "*reduce dependence on fossil natural resources, to prevent biodiversity loss and to create new economic growth and jobs in line with the principles of sustainable development*" [3].

On the other hand, based on their report entitled "The Bioeconomy to 2030: designing a policy agenda," the OECD countries consider that "*the application of biotechnology to primary production, health and industry could result in an emerging bioeconomy*" which "*is likely to involve three elements: advanced knowledge of genes and complex cell processes, renewable biomass and the integration of biotechnology applications across sectors*" [4].

One concept, one definition. This would be perfect to "harmonize" and "synchronize" the efforts of all possible contributors for a worldwide promotion and implementation of this new economic model. But, this will be very difficult to achieve. First, because the concept in itself depends on the perspectives of the contributors and their involvement context. For instance, what is the worth of a sustainable production if not followed with a sustainable consumption? Second, and more importantly, who has the uncontested authority to define bioeconomy?

For the present book, bioeconomy means the sustainable extraction, exploitation, growth, and production of renewable resources from land and sea and their eco-friendly conversion into food, feed, fuels, fibers, chemicals, and materials, to be consumed and recycled in a sustainable manner.

2.3 The Shift to Sustainable Bioeconomy

2.3.1 Bioeconomy: Necessity or Luxury?

As we've diagnosed in Chap. 1, humanity and the planet are sick; a severe case of intoxication with fossil fuels (and addiction to them). Thus, sustainable bioeconomy is more than a necessity. We have to join force to heal ourselves and the planet. Forget about the moon or Mars, we are born here and we will die here on this beautiful green earth.

Now that the disease has been realistically identified, the next step is to find the cure and quickly start the therapy. The cure has been found. It's called *sustainable bioeconomy*, and it is exclusively made out of biological resources from both land and sea. Nevertheless, despite finding the cure, starting the treatment is taking too much time for two main reasons. First, the disease was falsely diagnosed or at least underestimated right from the start. Second, the initial treatments were localized whereas the therapy should be generalized to the whole body of this sick planet. How should we proceed then to use this cure of sustainable bioeconomy?

We all agree that the treatment of a severe case of combined intoxication and addiction is a long, sometime painful, process. Humanity suffered (and still do) from prolonged exposure to petroleum and fossil fuels, and their consumption became almost compulsive. The healing process is composed of two phases: (1) detoxification and (2) rehabilitation.

The first phase aims at gradually reducing the exposure to fossil fuels. At this advanced stage, it is out of question to abruptly stop using fossil fuels as raw material. The current economies are too dependent and too weak for such drastic approach. Besides, the cure of sustainable bioeconomy is not mature yet and could not be administered except at small and gradual doses. The second process is rehabilitation which aims to cease the recourse to fossil fuels and replace it by the use of renewable bioresources through a wide implementation of bioeconomy. By the time earth regains its "sobriety," little or no petroleum will be left, so no worries about relapse.

In layman's terms, we should continue, while reducing, the use of fossil raw materials and promote, while increasing, the use of renewable raw materials. It is indeed a compromise more than necessary to make sure that all the contributors, whether promoting bioeconomy or benefiting from it, work together in order to attain the same objective: start a green industrial revolution to secure a sustainable future.

2.3.2 Raw Material Change

The term raw material refers to unprocessed organic or inorganic materials or substances used as feedstock for the primary production of energy, fuels, and various intermediates and end products. At a basic level, the change in economic models, national strategies, and international policies could be summarized in one single notion: change in raw material. This change consists of a *gradual shift* from the use of extracted nonrenewable resources to the use of harvested renewable resources.

2.3.2.1 From Fossil to Renewable Raw Material: An Anxious Transition

From the very beginning, securing a constant supply of raw materials at reasonable prices has been the main strategy for any industry to ensure high revenues and great competiveness. But, as the competition became fierce, whether between industrials or between countries, more attention was given to the price, and not the source, of the raw material. From this point, the recourse to relatively cheap fossil nonrenewable materials became a must to be able to compete, either nationally or worldwide. Despite the disastrous consequences related to the extraction, transportation, and use of fossil raw materials, as discussed in Chap. 1, industrial complexes continued to expand and nothing alarmed them until, somehow, they figured that there is a shortage in those precious fossil raw materials.

On the other hand, the steady increase in the world population, the rapid growth of emerging markets like China and India, along with the development of very efficient conversion technologies, strongly deteriorated the raw material supply situation during the previous decade [5]. For many industrialized countries, the dependency on nonrenewable fossil resources as raw material is very advanced and alarming too as most of those developed countries depend on raw material imports [6].

The reaction of most industrialized countries towards the proven shortage in those fossils resources was very symptomatic. Instead of searching for new sources of raw material that are renewable in order to secure a continuous flow of raw supply, they are "advising" to keep relying on other still abundant fossil resources, as if abundance means infinite.

Thus, as the petroleum resources are depleting, a gradual replacement is already planned, but then again with other fossil raw materials. First, they think that it is necessary to increase the lifespan of petroleum exploitation. Thus, the first action to take is to improve the yields of current petroleum deposits and produce more until finding a cheaper resource.

The second action is to start exploiting unconventional fossil resources "resembling" petroleum including tar sand and oil shale. Adopting this strategy will make "*democracies much less dependent on oil flowing from countries like Saudi Arabia, Iran, Iraq, or Venezuela*" [7]. What about their dependency on nonrenewable oil? Is the "democratic" oil better than the "autocratic" one?

In the meanwhile, natural gas will play an increasing role as a raw material for the production of end products. Coal, on the other side, will be considered a long-term replacement, considering "its large reserves and availability in important industrial countries" [8]. The real reason, however, is that the prices of petroleum and natural gas are going to soar, which will make coal cheaper [9]. Thus, considering coal as a replacement knowing its disastrous mining activities [10] and the effects of burning coal [11] on the climate is just wrong.

Proposals to mitigate coal-related environmental problems by carbon dioxide capture and underground storage will have limited impact, considering the massive scale of exploitation and subsequent emissions. Indeed, after immobilizing CO_2, the gas must then be released from the capture medium for sequestration (storage) while limiting the energy consumption. This is where current capture technologies run into problems [12]. As well, it is still unclear how the natural carbon sinks will evolve as climate and atmospheric composition continue to change [13].

All this clearly illustrates how important and delicate is this matter of raw material change. Every decision, no matter how insignificant it could appear, has to be taken after deep discussions, heated ones if necessary. The objective is to reach a consensus between all possible contributors in the field (scientists, industrialists, ecologists, economists, strategists, and decision makers). Even one of us promoting bioeconomy we have to be pragmatic: defend our opinions and visions with cold hard facts and still be ready to make compromises.

Here is one hard fact. Biological resources are only renewable if the rate of regeneration is faster than the rate of exploitation [14], and it is only under this conditions that biomass could be proposed as a renewable raw material. The main distinction is that, at a certain level of utilization, renewable resources can potentially be sustained forever. Climate change, soil quality deterioration, and water pollution are directly affecting biomass production yields in various ecosystems in the world. This will severely disturb or at least delay the worldwide implementation of bioeconomy unless it manages to adapt itself the changing circumstances provoked by the complex climate change phenomenon.

Here is another important and decisive fact; a large-scale breakthrough in the utilization of renewable raw materials is only conceivable in the event of a significant change in the price ratio of fossil to renewable sources.

2.3.2.2 Addressing the Challenges Gradually and Quickly

Bioresources have been exploited by mankind to provide food, energy, and other necessity goods for many millennia. During the last century, a transition to fossil resources occurred in modern societies which changed the status of biomass use. Nowadays, mankind is going back to bioresources as a sustainable alternative to depleting fossil resources. It seems like an easy endeavor but it is not. Indeed, the previous transition from a preindustrial economy to the current fossil based economy took more than 100 years although the preceding model was a sustainable one (*aka*. premodern bioeconomy [15]).

Now, if we take into consideration the disastrous legacy of the fossil-based economic model, on the one hand, and the current worldwide pressure on food, energy, and water supply, on the other, modern bioeconomy will need at least more than a century to take over and be fully operational. The problem is that time do not play in our favor. Serious decisions have to be taken and effective action plans have to be implemented urgently to mitigate climate change, water pollution, and soil degradation.

Thinking about going back to the old sustainable model (premodern bioeconomy) is a waste of time. It is not just a matter of lowering the current state of prosperity and welfare, the old model could not simply feed 7 billion people, let alone provide sufficient energy supply for the industrial and transportation sectors.

Thus, the most practical strategy is to combine the two previous models and "synthesize" a new model: *sustainable* like the old one and *efficient* like the current one. These are the two main objectives for modern bioeconomy. Such approach will benefit from the know-how cumulated by mankind for centuries, especially the experience gained from previous mistakes, thus saving us precious time as we will repair the mistakes of the past while building the better future.

Basically, bioeconomy cannot succeed unless it satisfies the needs of three "big babies": industrials, consumers, and the environment.

1. Industrials have a huge appetite for raw material and energy, and they are not expected to induce any measure of change susceptible to reduce their revenues, at least willingly. Profitability is their main concern and bioeconomy has to take this attitude very seriously because without the industry, it will remain as an abstract notion.
2. Consumers and their growing requirements for food, energy, and clean water. Bioeconomy has to satisfy those basic needs worldwide, but also it has to retain the level of prosperity and welfare for consumers in developed nations and to meet the aspiration of consumers from developing countries to reach the same level of welfare.
3. The environment, the most patient of the big babies, is the easiest to please. It just needs to be left alone. All it asks from us is to limit our wastes and emissions to tolerable levels that could be dealt with using its arsenal for biological and geological resources.

Nonetheless, bioeconomy has to seriously consider the environmental factor because being patient does not mean that it will indefinitely bear our mistakes further. The current level of tolerance is very low after decades of abuse. The "reaction" of the environment, revealed by scientists decades ago, is starting to become visible to all mankind including global warming, severe storms, flooding, desertification, and, last but not least, sea level rise.

So, bioeconomy has to take care of the suffering environment to be able to survive. Indeed, although the needs of our environment are simple, their fulfillment is very difficult mainly because of the attitude of the first big babies: industrials and consumers. We purposely used the term of big babies to connote their lack of maturity, far-sightedness, and self-control, in addition to their reckless attitude all the way from exploiting fossil resources to the unsafe disposal of the generated wastes. More details about this crucial stage of transition are presented later in this chapter where the relation between bioeconomy, on the one hand, and agriculture, industry, and forestry, on the other hand, will be, respectively, analyzed.

2.3.3 *Sustainable Profitability from Bioeconomy*

To be adopted by entrepreneurs and implemented by industrialists, bioeconomy has to, primarily, generate profit. Sustainability is the major aim of bioeconomy. Thus, applying bioeconomy in the industrial sector means adopting sustainable production procedures, which will logically generate "sustainable profitability." How so?

2.3.3.1 From Maximizing Profitability to Sustaining It

All industries are designed and implemented to generate profits. Basically, the whole concept is to invest money, produce or manufacture products, and sell them to generate revenues. After a while (the shorter, the better), the project will compensate the initial investment and starts to generate net benefits. Maximizing those benefits is and will always be the main goal on any industrial activities, from the small artisan shop to the big multinational corporations.

Innovate, produce more, optimize your production process, hire highly skilled personnel, diversify your portfolio, and improve your brand value and reputation are among the adopted strategies to maximize profits and benefits. But, the main strategy to achieve this highly sought objective is and will remain cost reduction. Innovation needs an extensive R&D efforts which, on its turn, needs significant financing to find (or not) "good results." In order to produce more, you have to invest more, to secure a large and, more importantly, a continuous supply of raw materials. As well, to optimize and diversity you portfolio, you have to hire experienced executives, dedicated employees, and skilled workers. All those requirements, combined, are achievable only for big companies.

Most industries in the world endeavor to maximize their profits via the easiest and proven to be efficient strategy: cost reduction. The main targets in this context are raw material, energy, and labor. Thus, searching and using cheap fossil resources and nonrenewable energy supply led us to the previously discussed heavy dependency on fossil raw materials. The same strategy to cut costs in order to maximize benefits seriously affected the labor sector. Three worldwide phenomena are sufficient to illustrate the gravity of this inclination: low wages, outsourcing, and massive layoffs.

Overall, during the industries' quest to maximize profitability, two major factors were seriously affected, or at least neglected, namely, the social and environmental factors. Now, considering the disastrous legacy of this hunger for wealth (detailed in Chap. 1), it is no longer tolerable for an industry to generate profit without considering the impacts on both society and the environment. One of the results of such inevitable standpoint is the introduction of the notion of corporate social responsibility in the mid-1970s [16]. The idea is that companies will start to address and manage the impact of their industrial activities on the economic, environmental, and social levels, altogether. How so? Some rely on the companies' self-conscious, others on the inside pressure from some activist stakeholders, and many of restricting laws.

First of all, relying on big multinational corporations to become all of a sudden conscious about the environment and the well-being of societies is just idealistic. Even counting on activism pressures (few from inside or more from the outside) in order to "induce" this self-regulating sprit has no to little impact. Indeed, adopting this notion implies that companies are admitting their responsibilities for the great damages done to the environments and the society. More importantly, how is it possible to induce a change of this magnitude that will shake the core of the industry outside the strategic decision-making process. This is the real target to make any shift in policies and strategies.

To make a real and significant breakthrough in this very important matter, we have to think like entrepreneurs and industrials. How do big corporations perceive the notion of corporate social responsibility? They cannot neglect the urgent pressures calling for actions from the industries to seriously consider the social and environmental factors in their strategic planning. In most big corporations, this issue is dealt with, not by the executives or shareholders, but by the division of public affairs, and this alone summarizes how delicate is this subject of profitability.

In practice, the discussed subject is far removed for ideological opinions. It is a matter of balancing the social and environmental responsibilities of companies with their legitimate economic goal to generate benefit and maximize profit. The outcome is called sustainable profitability.

2.3.3.2 Ensuring a Sustainable Profitability

Basically, industrial complexes are assembled to manufacture products with economic value in the marketplace. The major aim of each industry is to maximize its profit gain by producing more (quantity) and/or better (quality), at the lowest possible cost. With this strategy, profitability can only be maximized if the industry is constantly able to adjust the produced amount of each product with the ever fluctuating raw materials cost, on the one hand, and the equally fluctuating costumers' demand, on the other hand [17].

Now, what if industries rely on a renewable, thus continuous, flow of feedstock and energy supply (instead of the current reliance on fossil resources), such shift would have two major repercussions:

1. Maximizing the production value by minimizing the costs of raw material and energy since their sources under this scheme are stable (i.e., less fluctuation considering the renewable character of those resources which will block the path to the damaging monopoly and speculation activities).
2. Sustaining the generated profitability on the long run. Indeed, profitability from renewable resources will exceed than that from nonrenewable resources when the cost related to the exploration and exploitation of fossil resources becomes

higher than the cost related for the industrial conversion of renewable resources into the desired products.

Let us take the case of a petroleum-extracting company. The process will start with a heavy exploration cost, added to the high initial investment (heavy equipment, large sites, and highly skilled staff). Once it starts to extract and sell crude oil, it will have to manage low variable costs (maintenance, wages, emergency management, etc.), as long as the supply flow is constant (or appearing so). With time, however, the amount of extracted crude oil will start to decrease, thus gradually increasing the related variable costs until the extraction costs become too high and the company starts to lose money.

On the other hand, in the same energy sector, the production of fuel from renewable resources (bioethanol for instance) will need larger initial investments as the technology is not yet mature for large-scale worldwide production systems and the workforce is less skilled (for now) compared with that involved in the field of fossil fuels. These costs are mainly related to building biorefineries, buying equipments, acquiring renewable raw materials and managing the co-products.

Nonetheless, it is precisely the renewable character of the raw material which will give the profitability its sustainable dimension. Indeed, under this scheme, the feedstock flow will be continuous and the production procedure will be optimized, so that the output will remain constant and could even be increased as some variable costs are expected to decline with each technological breakthrough.

Currently, during the exploitation phase, the "well established" nonrenewable system requires less investment, compared with the "still developing" system based on renewable resources. Then, during the production phase, the nonrenewable system is able to generate profit quicker and higher than the competing renewable system, but for a certain period of time. However, on the long run, the fossil resources will gradually start to deplete, thus increasing the exploitation cost until the whole nonrenewable system becomes unprofitable (even before the complete depletion of the resource). On the other hand, the system based of the exploitation of renewable resources, although having a slow start, will sustain profitability almost indefinitely (pending an exploitation rate lower than the bioresources' regeneration rate).

In addition, as an emerging exploitation and production system, bioeconomy possesses a very important asset compared to the still highly performing fossil-based economy: a wide margin for progress. Thus, each R&D breakthrough and each technological innovation will considerably impact the profitability of bioeconomy in terms of decreasing the initial investment, quickening the profit generation phase, and stabilizing the profitability for long-lasting period of time [18].

Now, in practice, how to change the exploitation, production, and conversion systems from using nonrenewable resources to using renewable ones? In other terms, how to make the shift to bioeconomy financially profitable? Well, two major strategies could be implemented.

1. Industries which are able to financially support the slow and costly implementation of sustainable production units until becoming productive and thus

profitable. Those high initial investment will permit installing highly efficient operational units from the start, which will lead, on the medium to long term, to decrease the costs and increase the production yields, therefore increasing the profitably and stabilizing it when the production process reaches "maturity." This strategy could be adopted and implemented by big corporations, whether committed to bioeconomy and sustainable development or just planning to invest in sustainability in order to diversify their portfolio, thus enhancing the returns, as well as lowering the risks and overall volatility.

2. Most industries committed to the transition to bioeconomy will adopt a gradual approach. Basically, this strategy deals with already operational units for which small and gradual changes are applied to the established conversion and production procedures. Such approach will avoid the high initial investment, thus bringing about substantial cost savings, which could be channeled to afford costly, but more efficient, technologies and skilled workforce. Combined, both the technological and the human factors will enable a more efficient exploitation, conversion, and production of renewable raw materials, thus generating a gradual increase in profitability, on the short and medium run, and stable returns, on the long run. For instance, already functional industrial complexes like fossil-fuel power plants, petroleum refineries, and wood-based industries could be "upgraded" to the sustainability concept, thus benefiting from their proven efficiency to ease and quicken the transition to bioeconomy.

From this analysis, the availability and renewability of the biological resources will attract both governmental and private investment capitals. In addition, the "green" processes involved in biomass conversion will generate less pollution and less carbon footprint than the processes involved during the extraction and transformation of fossil resources. In this context, environmental taxation policies are expected to become more severe [19] which will make investment in renewable resources more profitable. As well, bioeconomy will become more attractive for entrepreneurs and industrialist as the technologically advanced exploitation and conversion procedures of renewable resources will decrease the initial investment, hasten the profit generation, and increase the net profitably.

2.3.4 Leading Role of Science and Technology in the Transition

Bioeconomy is a knowledge-based concept developed to generate sustainable growth and to bring better life conditions for mankind and safer conditions for the environment. Many of us promoting sustainable bioeconomy are "selling" it as a vision for the future. But even so, we have to agree that in order to build a better future we have to start now.

2.3.4.1 A Multidisciplinary Effort

Research and Development (R&D) are among the main drivers of growth for bioeconomy [20, 21]. Indeed inventions and innovations are the only way to enhance productivity, induce economic growth, and justify big conceptual and structural changes. R&D should and will enable the expansion of sustainable bioeconomy, guarantee its efficiency and competitiveness, and later ensure its worldwide implementation. Therefore, in order to reach the objective of worldwide implementation of sustainable bioeconomy, we should promote R&D efforts with all its multidisciplinary dimension. Indeed, integrating high-level scientific expertise from various scientific disciplines is a key requirement to set up the knowledge-based bioeconomy.

If bioeconomy is the ship that will carry us safely to the shores of "sustainable world," R&D is the engine of that ship. However, to make this engine work and propel the ship to the desired destination we need fuel, and money is the fuel of that engine. Indeed, with money you can acquire the two indispensable factors to ensure reliable and efficient R&D efforts: high-tech equipment and highly skilled researchers.

To achieve the transition towards a sustainable, low carbon, resource-efficient bioeconomy, we need to invest more in science and technology (S&T). Proper investment, good management of those financial resources, and a multidisciplinary R&D effort will quicken the process of building an efficient, robust, and viable economic paradigm. Indeed, the commitment of the S&T community, as a whole, in this major human endeavor is crucial to quickly attain the goal.

The involvement of scientists and researchers from fields as diverse as chemistry, biology, biochemistry, economics, agronomics, engineering, medical, and social studies (to name a few) is a prerequisite for success, since the issues related to the concept of sustainable bioeconomy are not defined by specific disciplines but rather by complex problems that need to be dealt with and solved in an multidimensional manner. It starts with debates and interactions among scientists, development and/or optimization of procedures and technologies, and later assessing their impacts on human society and the environment. The amount and quality of the produced knowledge will be decisive to properly (i.e., quickly and efficiently) address urgent problems of sustainability, economic development, and environmental preservation.

One of the most visible illustrations of this cooperation is the growing collaboration between academia and industry, especially when the latter started to recognize the importance of the free thinking spirit embodied by universities. Such collaboration is and will remain delicate to manage as conflicts of interest could easily erupt because academia need the money from the industry to boost its R&D potential, and the industry needs brains from academia to optimize what is working and fix what is not.

The Energy Biosciences Institute (EBI) is a good example of this kind of cooperation. This self-proclaimed "largest public–private partnership of its kind in the world" involves a consortium made of the University of California at Berkeley, Lawrence Berkeley National Laboratory, the University of Illinois at Urbana-Champaign, and the energy company British Petroleum (BP). This interesting, yet delicate, cooperation was based on a couple of important guiding principles including that inventions made during the research investigations conducted by EBI are owned by the involved academic institutions according to U.S. patent law. The private partner BP will receive an automatic, but nonexclusive, license in return for its research funding. As well, all four partners have representation in the EBI's governing board of directors, with none having a majority or veto power, thus encouraging consensus in all decisions [22].

Overall, the model of university research backed by industry will open the door for new innovations and therefore will be a major strategy for the development of sustainable bioeconomy.

2.3.4.2 Managing the Transition to Sustainability

Basically, the transition management is a coordinated transdisciplinary effort aiming to ease and speed up large-scale changes based on the concept of sustainable development. This vital transitory phase will help breaking up with conventional way of thinking and doing, thus opening more windows for visionary thinking, which is the key to invent and innovate.

This coordinated effort of scientists led to concrete transdisciplinary interactions, best illustrated by the emergence of the field of sustainable science. The founding principle of this science is visibly sustainability. As for the objectives, Kates et al. [23], in their paper in Science magazine, defined three major targets:

1. Understanding the fundamental interactions between nature and society
2. Guiding these interactions along sustainable trajectories
3. Promoting the social learning necessary to navigate the transition to sustainability

Nowadays, after more than two decades of research, development and innovation, sustainability science has emerged as a vibrant field of study. Today, it is profoundly contributing in the transition phase towards bioeconomy based on the increasing numbers of involvement of research centers and laboratories from various scientific backgrounds and, equally important, based on the increasing number of universities committed to teaching sustainable science [24].

The generated theoretical and practical results from extensive studies and debates on how to manage this transition phase paved the path for promoting sustainable development. The whole process was developed by large network of scientists, in collaboration with industrialists and decision makers. Nonetheless, some researchers are urging for more progress down this path of sustainability as they think that socio-technical change remains underappreciated and relatively unexplored in sustainability research, which is key for society and its institutions to articulate visions of sustainability [25].

At this point, it has to be noticed that neither top-down government policies nor bottom-up market forces or scientific findings can alone support the heavy weight of introducing changes in various strategic sectors, not only related to the well-being of mankind, but even with its survival. Such major changes have to be managed, especially during the transition phase, through combinations of R&D efforts, government policies, market forces, and constructive initiatives from civil society [26].

In short, sustainability science should set up a new research agenda for scientists from different backgrounds and convert all the resulting R&D effort towards the single most important objective for decades ahead: reforming mankind–nature interactions.

2.4 Bioeconomy and Agriculture

Agriculture will always be the pillar of any economic model. Its place is even vital in the bioeconomy concept. As an efficient biomass-producing system, agriculture should have a privileged position in bioeconomy. As for the dilemma of using agriculture to produce food and feed or fuel and chemicals, the response is quite simple. This is a rich world dilemma. The question itself is an aberration in the poor world, which happens to constitute most of the inhabitants of planet earth.

Thus, as a global economic model, bioeconomy should focus on agriculture to secure the food requirements of the still growing seven billion people. Agriculture in bioeconomy should have a clear directive: fill the stomach first, then the fuel tank.

2.4.1 Why Sustainable Agriculture

Basically, sustainability in agriculture means satisfying mankind's needs for food, feed, fibers, and fuel, while enhancing (or at least maintaining) the quality of environment and conserving natural resources for the next generations to fulfill their needs.

For the previous and current economic systems, the objective is very clear: use all the available resources (land, water, and energy) to feed the growing population. Such approach led to the expansion of unsustainable agricultural practices, which deteriorated the quality and quantity of the exploited natural resources. The main repercussions are the increasing shortage in arable land, soil, and water pollution with herbicides and pesticides, deforestation, and soil erosion. Such conditions are not only seriously affecting the environment but started to seriously threaten the viability of future agricultural production activities.

In this gloomy context, sustainable agriculture was founded in order to meet three major objectives:

1. Reform our "unsustainable" practices by carefully managing our natural resources, while ensuring a highly efficient production system.
2. Connect sustainability and profitability
3. Reconsider the social dimension, intrinsically related to agriculture

2.4.2 How to Make Agriculture Sustainable

The supply of agricultural commodities is vital to human existence. The big dilemma is how can sustainable agriculture reconcile the needs of mankind and that of the environment? In other terms, how can we demand more from Nature while trying to preserve it?

The agricultural practices carried out during the last decades have greatly increased global food supply. For instance, the worldwide cereal production has doubled in the past 40 years to meet the need of 7 billion people. This remarkable achievement was mainly due to the increased yields resulting from the exploitation of more lands, greater inputs of fertilizers, water, pesticides, new crop strains, and other technologies of the so-called "Green Revolution" [27].

Now that scientists are looking back into the unsustainable practices used in modern agriculture to increase the crop yields; they are concluding that agriculture is pushing itself towards stagnation with severe damage to ecosystems. The best illustration is the unbalanced cycles of nitrogen and phosphorus and the subsequent nutrients deficiencies which seriously affected the food production and contributed to land degradation in some parts of the world [28].

Therefore, urgent calls for action are being made by the scientific community regarding the need to apply more sustainable agricultural methods. The main target is to ensure the continuous production of sufficient amount of good quality food, with minimum harm to the environment.

In practice, and in order to become sustainable, agriculture has to deal with those key aspects: soil, water, energy, and cultivation practices.

2.4.2.1 Sustainable Land Use

Sustainable land management is primarily based on maintaining and restoring (if necessary) the soil fertility in order to allow the production of food supplies and other renewable natural resources on a long-term basis. This implies that the natural ecosystems should be managed in such a way that the nutrients cycles and energy fluxes among soil, water, and atmosphere are preserved [29].

In agriculture, soil fertility is the main characteristic to enable high crop yields, a very crucial feature to meet the needs of an increasing world population, as well as the limitations in terms of availability of new arable lands. The sustainable dimension in agriculture aims at preserving soil fertility for improved production yields with less harm to the environment [30].

In this regard, scientists are advising that improving soil fertility should not only consider enhancing crops yields but also maintaining the balance between nutrients requirement on the one hand and nutrients supplies on the other [31]. The reason is that either excessive or deficient nutrients in soils have harmful repercussions not only on agricultural productivity but also the living organism, especially microbes, which are the key players in various nutrients cycles in soil [32]. Studies showed that excessive nitrogen amendments to the soil during fertilization could lead to serious consequences on the agricultural ecosystem and the climate, respectively, through nitrate (NO_3^-) leaching and the emissions of nitrous oxide (N_2O), a greenhouse gas [33, 34].

In the same context, another study revealed that N_2O emissions from agriculture are responsible for more than 75% of total anthropogenic emissions [35], which constitute 40% of the global N_2O emissions (natural and anthropogenic sources combined) [36].

Considering all the previously mentioned facts, soil "health" is definitely one of the most important constituents of efficient agricultural ecosystems. Overall, healthy soil means vigorous crops and dynamic microbes. If the soil is subjected to unsustainable practices, it will require frequent fertilization campaigns. Then, to increase the production yields, highly productive, but often vulnerable, crop varieties will be cultivated. Thus, more pesticides and herbicides have to be used to protect the crops. Overall, a "sick soil" will induce more chemical contamination, thus further sickening the soil and polluting water resources and the atmosphere. To face those environmental threats, sustainable agriculture has to work on two different fronts regarding soil quality:

1. Maintain the fertility of healthy soils.
2. And enhance the fertility of « sick » soils.

Another important point regarding sustainable agriculture regards its interaction with its immediate environment. A symbiotic relationship between agriculture and the ecosystem could benefit both. The productivity could be improved while preserving the integrity of the natural habitat and biodiversity of the region where the exploitation is occurring. This is the essence of sustainable agriculture. For instance, local crops are more in harmony with the ecosystem as they have adapted themselves to its pedoclimatic conditions. As well, maintaining the biodiversity of local wildlife can help increase the production yields via enhanced pollination through bees or quick and efficient pest management through natural predators.

2.4.2.2 Sustainable Water Management

Although irrigation consumes approximately 70% of the world's freshwater supply per year [37], the demands for water from the strategic agricultural sector are increasing, but the amount and quality of the water supply is decreasing. Numerous factors intervene in this decrease including the urban use from an increasing world population and the competition from the industrial sector. As well, the pollution of surface and groundwater resources further limits the availability of water for agricultural use, in addition to the public health and environmental contamination issues. To reclaim those contaminated waters, costly treatment procedures had to be applied which contributes in increasing the cost of clean water.

Currently, new kinds of pressures on water resources are coming from the repercussions of climate change, which are likely to alter both water availability and agricultural water demands. Therefore, scientists are warning that the agricultural sector is particularly vulnerable to climate change as it affects both water resources and land [38]. Indeed, even minor changes in average temperatures, precipitation patterns, or the frequency of extreme weather conditions will cause serious damages to the agricultural activities [39].

Considering all those threats, sustainable management of water resources is a key endeavor in sustainable agriculture by maintaining good water quality and sufficient supply. Thus, pollution prevention will help keeping harmful contaminants such as pesticides and chemicals fertilizers away from surface and groundwater reserves, as well as preserving soils from contamination and therefore ensuring a continuous and sustainable exploitation of valuable arable lands.

Furthermore, managing water consumption is an equally important endeavor in sustainable agriculture, especially in arid climates were high evaporation rates and frequent droughts could significantly decrease the water reserves, if coupled with unsustainable irrigation practices. Thus, in order to sustain their agricultural activities with respect to water consumption, farmers working in arid regions or with limited access to water resources should manage well the available supply of water by adopting water-saving technologies such as micro-irrigation or drip systems, in addition to cultivating drought-tolerant crops.

2.4.2.3 Sustainable Cultivation Practices

The great challenge facing modern agriculture is to increase food and feed production in a sustainable manner without overexploiting natural ecosystems and compromising public health. This has to be echoed not only in the management of land and water but also on the cultivation techniques. Indeed, sustainable agricultural practices are developed and applied in order to limit and replace chemical-based agriculture by minimizing the systematic recourse to pesticides and chemical fertilizers. If efficient in terms of productivity, sustainable agriculture would both save money and reduce the impact on the environment to tolerable levels.

Most of the sustainable practices in agriculture aim to preserve the quality of soil and avoid erosion and enhance the efficiency of natural pest and weed control. In practice, those objectives could be reached via different old and innovative methods such as:

- *The use of cover crops* which are plants grown between two cropping periods (instead of leaving the land uncultivated), mainly to restore fertility to the exploited soil, to prevent erosion [40], and to control invasive weeds [41]. As well, using nitrogen-fixing cover crops help reducing the leaching of nitrogen from cultivated fields [42], thus reducing the threat of groundwater contamination.

- *Amendments of biochars* is currently being promoted by researchers in order to improve soil health. Indeed, numerous studies reported several agronomic and environmental benefits for the addition of biochars into soils including enhancing soil fertility [43], improving carbon sequestration (thus mitigating climate change), and reducing the bioavailability and leachability of heavy metals and organic pollutants in soils [44].
- *Adopting an eco-friendly pest management approach* through the use of natural predators, natural biopesticides [45] (instead of chemical pesticides), and natural bioherbicides to protect crops from invasive weeds [46]. Two strategies could be adopted in this context. The first relies exclusively on biological measures like in organic farming. The second adopts the integrated pest management approach where the initial interventions are biological, with the possibility to use chemical pesticides as a last resort solution.
- *Crop rotation* is an ancient method still being practiced in order to maintain the fertility of soil. It is based on alternating different crops so that the nutrients absorbed by this year's crop would be replenished by cultivating another species next year. For instance, research studies proved that including grain legumes in the rotation helped increasing cereal crop yield and improving soil fertility without adding nitrogen fertilizers [47], as leguminous plants are able to fix nitrous oxide and biological nitrogen [48].
- *Recycling* is also a common practice in sustainable farming. It consists of collecting and reusing crop wastes or animal manure as biological fertilizers. As well, leaving crop residues in the field after harvest or composting them could also be an organic source for nutrients to replace chemical fertilizers. Reclamation and utilization of treated wastewater or the collection of rainwater for irrigation are other examples of sustainable recycling practices.

2.4.3 Bioeconomy and Food Security

Unfortunately, despite the worldwide dimension of bioeconomy, most countries lean towards implementing it individually. Thus, each country will adapt bioeconomy and its agricultural activities to feed its population and livestock. As a consequence, the rich and developed countries will easily reach this strategic goal of food security and also end up with a production surplus which will be put on the world market or used to produce fuel and chemicals. For the poor and even the developing countries, they will struggle securing the food requirements of their populations, let alone reserving bits of arable land to produce energy crops.

Thus, implementing bioeconomy independently will for sure benefit developed and industrialized countries, for a while, but it will be an epic mistake since mankind will miss a rare (may be the last) opportunity to join forces to combat world hunger and poverty (among many other challenges facing bioeconomy, notably climate change).

Let us consider this fact. Agriculture is and will remain the major source for food. If there is not enough food, hunger will occur. According to the "Hunger Statistics" [49] from the World Food Programme, the vast majority of the world's hungry people live in developing countries. Asia is the continent with the highest count of hungry people (two-thirds of the total), and Sub-Saharan Africa is the region with the highest prevalence of hunger. Why is that? Because 20% of the world's wealthiest population is consuming 76.6% of the natural resources, while 80% of humanity gets the remainder [50].

In Europe, the wealthiest continent in the world, the perception of food is completely different. While other continents are suffering from hunger, European experts in their white paper "*En Route to the Knowledge-Based Bio-Economy*" are expecting that food will be designed for special consumer groups. One priority is the development of foods for groups with defined risk factors or diseases like diabetes, obesity, and cardiovascular problems [51]. Paradoxically, all these diseases are caused or accentuated by the overconsumption of food. In this context, a Canadian study showed that, on a population-wide level, wars, economic crises, and the widespread food shortages are "the only interventions that have dramatically reduced the prevalence of obesity and cardiovascular disease in modern times" [52].

During the previous century or so, we have lost many battles during the fossil-based economy, especially in the environmental and social fronts. As well, international inequalities were accentuated, which fueled more animosity to the already tense relationships between poor developing countries and rich developed one (military invasions, economic sanctions, illegal immigration, etc.).

If bioeconomy could narrow this gap between countries, at least for the most basic human needs of water and food, we would have learned valuable lesson from our previous lost battles and then alone we could stand together to win the war on hunger and poverty with the sustainable economic model.

2.5 Bioeconomy and Industry

The main concern of industrialists is to secure a continuous flow of feedstock to their production units at the lowest cost possible. They could be flexible about the production procedure itself, and even about the end products, but the raw material is another matter. Let us now reflect on how industrials think about this important matter of raw material. It is a very important analysis because industry is the pillar for bioeconomy implementation on the ground, and a healthy transition towards using renewable raw materials is a keystone in building bioeconomy and preparing for a sustainable future.

2.5.1 Bioeconomy and the Energy Industry

For an industrialist, fossil fuels (petroleum, coal, and natural gas) are the backbone for many industries as feedstocks for many lucrative industrial products including fuels, heating oils, and fine chemicals. Nowadays, despite the shortage in fossil supplies, the reliance on those nonrenewable resources will continue until the alternative renewable supplies become affordable and available for large-scale exploitation.

Based on the current consumption pattern, on the one hand, and the amount of proven reserves on the other, petroleum is likely to be exploited and depleted first. What to do next? With all the disastrous legacy of petroleum-based economy, one should look for a new type of raw material, a renewable and eco-friendly one. But, as long as there is enough petroleum cheaper than the closest alternatives, the reliance will continue to be on the former. The infrastructure is already there; the know-how was acquired over decades of extraction and refining.

Thus, in order to have the strategic energy sector "onboard," sustainable bioeconomy has to provide reliable alternative fuel supplies. In this context, several aspects have to be considered and carefully planned. A detailed description and discussion on the role and implication of the energy industry in the development and implementation of sustainable bioeconomy will be presented in the coming Chaps. 4, 7, and 8.

2.5.2 Bioeconomy and the Chemical Industry

In the chemical industry, a gradual replacement of fossil resources by renewable raw materials is expected in the future but definitely not before long. Thus, crude oil will remain the dominant raw material in the chemical industry in the foreseeable future, at least as long as the extraction and transportation of fossil fuels remains cost effective.

The transition towards sustainable industries needs to move forward as fast as possible in order to reduce the time for this transitory phase and quicken the full implementation of bioeconomy and benefit (economically, socially, and environmentally) from its sustainable exploitation, conversion, production, and recycling procedures.

Such crucial objective could be attained mainly through (1) enhanced R&D efforts, (2) bold initiatives from the industry especially in terms of investments, and (3) legislations promoting and easing this transition from parliaments and governments. As this process is advancing, renewable raw materials will gradually become more economically attractive for the production of various chemicals compounds, mainly the ones produced via improved large-scale conversion technologies. As well, in the event of a significant shift in the price ratio of fossil fuels to renewable sources, an increase in the share of renewable raw materials in the production of basic chemicals is conceivable [53].

2.5.3 *Bioeconomy and the Forest Industry*

Compared with the other biomass-producing ecosystems, forestry has a tremendous advantage to become one of the pillars of bioeconomy. Three major assets could be pointed out:

1. Forests have large biomass production potential.
2. They do not compete with the strategic agricultural sector.
3. They contribute in climate change mitigation through carbon capture and sequestration.

In Europe, for instance, forests cover about 40% of the total land area (equivalent to 157 million ha), and those areas have increased by about 3 million ha in the last 10 years. Production wise, the total amount of wood is estimated at 24.1 billion m^3, corresponding to an increase of 4.9 billion m^3 over the past 20 years. In addition, the European forest sector involves around 3.5 million workers, mainly located in rural areas. All those considerations give a strategic importance to forestry within the bioeconomy model as it is already dealing with the industrial, environmental, and societal challenges.

Currently, with the technological developments in biomass conversion procedures, on the one hand, and the decreasing demand for paper, on the other, the role of the forest industry and its contribution in strengthening the sustainable dimension of bioeconomy are increasing [54]. Indeed, for many decades, when spoken of the forest industry, two main products come in mind: pulp and paper and timber. With the various R&D breakthroughs in the field of biomass valorization, especially lignocellulosic wood, modern forest industry became a high-tech industry providing world markets with value-added bioproducts including biofuels, biopolymers, and chemicals.

In addition, the serious challenges facing humanity regarding energy supply and climate change are putting a great deal of pressure on researchers and policymakers to find sustainable solutions which create new incentives for bioenergy and biofuel production. In this context, using wood from sustainable forest or industrial residues such as wood chips and sawdust to produce biofuels and bioenergy provides an option that provides increasing profits along with tightening energy and fossil fuel policies [55].

Equally important is the fact that the contribution of forests in the bioeconomic model is not restricted to its biomass (wood) production. Indeed, sustainability-managed forests play a very important role in three major environmental issues: (1) biodiversity conservation, (2) water and soil protection, and (3) climate change mitigation and adaptation. Although those "services" by forests are important to the well-being of our environment, they remain not properly appreciated. Indeed, scientists are confirming that this forest contribution (wood production aside) are excluded from the market, and they are suggesting that introducing payments for them would encourage private landowners to manage their forests sustainably [56].

2.6 Challenges Facing the Transition to Bioeconomy

During the transition phase, and in order to replace the current fossil fuels-based economy, bioeconomy will face various obstacles, especially that the expectations from this new economic model are very high, both quantitatively and qualitatively. Indeed, bioeconomy is expected to replace old industries, products, and practices by new eco-friendly industries, bioproducts, and procedures, thus enhancing the sustainable dimension in the production and consumption processes, and meeting market feedstock demand and price [57].

Another major expectation from bioeconomy is ensuring a continuous supply of feedstock for the various industrial sectors (food/feed, energy, chemicals, pharmaceuticals, construction, etc.) while mitigating climate change and preventing further degradation of clean water supply and biodiversity [58].

Overall, the big challenge facing bioeconomy is ensuring an economic growth using renewable resources without harming the environment and responding to the global issues of energy and food security, climate change, poverty, and the ever increasing shortages of clean water supply and productive land.

Now, to give more insights about possible barriers for the transition to sustainable bioeconomy, the Technical University of Denmark conducted an interesting survey on behalf of the International Energy Agency (IEA Bioenergy—Task 42 Biorefinery), in which diverse factors affecting this transition to bioeconomy were analyzed [53]. The involved countries were Australia, Austria, Canada, Denmark, Italy, Japan, the Netherlands, New Zealand, and the United States. Overall, the participants to this survey were from the industrial sector (64%), academia (16%), governmental institutions (12%), and public organization (8%).

The main finding reported in this study is that profitability is the most important obstacle for the transition to bioeconomy. The second limiting factor was governmental policies, followed by finding suitable markets for the bioproducts, in addition to securing the supply of biomass resources.

Another very important point regarding this transition phase is how should we proceed? Unfortunately, despite the worldwide dimension of bioeconomy, countries are implementing it individually. For instance, each country will adapt bioeconomy and its agricultural sector to feed its population. As a consequence, the rich and developed countries will easily reach this strategic goal of food security and also end up with a production surplus which will be put on the world market or used to produce fuel and chemicals. For most of the developing countries, they will struggle securing the food requirements of their growing populations, let alone reserving bits of arable land to produce energy crops.

After studying the legacy of the petroleum-based economy in the first chapter and the transition to the sustainable bioeconomy in the second chapter, let us now start analyzing bioeconomy, study its various constituents, and discuss the impact of this sustainable economic model on both mankind and the environment, as a green industrial revolution. In the next chapter, we will start with the most important sustainable pillar of bioeconomy: BIOMASS.

References

1. European Commission – Press release data base (February 2012). Commission adopts its strategy for a sustainable bioeconomy to ensure smart green growth in Europe. MEMO/12/97. http://europa.eu/rapid/press-release_MEMO-12-97_en.htm?locale=en
2. Federal Ministry of Education and Research (BMBF). National Research Strategy BioEconomy 2030 – Our Route towards a biobased economy. Berlin; 2011. p. 56. Available online at: http://www.bmbf.de/en/1024.php?hilite=bioeconomy
3. The Finnish bioeconomy strategy. http://biotalous.fi/wp-content/uploads/2014/08/The_Finnish_Bioeconomy_Strategy_110620141.pdf
4. Organisation for Economic Co-operation and Development (OECD). The Bioeconomy to 2030: Designing a Policy Agenda. OECD Publishing, Paris; 2009. p. 322. http://www.oecd.org/futures/bioeconomy/2030
5. Rosenau-Tornow D, Buchholz P, Riemann A, Wagner M. Assessing the long-term supply risks for mineral raw materials – a combined evaluation of past and future trends. Resour Policy. 2009;34:161–175l.
6. Behrens A, Giljum S, Kovanda J, Niza S. The material basis of the global economy: worldwide patterns of natural resource extraction and their implications for sustainable resource use policies. Ecol Econ. 2007;64:444–53.
7. Jansen RA. Second generation biofuels and biomass: essential guide for investors, scientists and decision makers. Weinheim: Wiley; 2013.
8. Keim W, Röper M. Use of renewable raw materials in the chemical industry. Position paper of DECHEMA, GDCh, VCI, DGMK, Frankfurt; 2008.
9. Coal: The fuel of the future, unfortunately. The Economist, April 16th 2014.
10. Betz MR, Partridge MD, Farren M, Lobao L. Coal mining, economic development, and the natural resources curse. Energy Econ. 2015;50:105–16.
11. Castleden WM, Shearman D, Crisp G, Finch P. The mining and burning of coal: effects on health and the environment. Med J Aust. 2011;195:333–5.
12. Cooper AI. Materials chemistry: cooperative carbon capture. Nature. 2015;519:294–5.
13. Brienen RJW, Phillips OL, Feldpausch TR, et al. Long-term decline of the Amazon carbon sink. Nature. 2015;519:344–8.
14. Manahan SE. Environmental science and technology: a sustainable approach to green science and technology. 2nd ed. Boca Raton, FL: CRC Press; 2006.
15. Kircher M. The transition to a bio-economy: national perspectives. Biofuels Bioprod Biorefin. 2012;6:240–5.
16. Wood DJ. Corporate social performance revisited. Acad Manag Rev. 1991;16:691–718.
17. Martin PG. Sustainable profitability. Invensys systems white paper. 2011. http://iom.invensys.com/EN/pdfLibrary/WhitePaper_Invensys_SustainableProfitability_04-11.pdf
18. Zilberman D, Kim E, Kirschner S, Kaplan S, Reeves J. Technology and the future bioeconomy. Agric Econ. 2013;44:95–102.
19. Spratt S. Environmental taxation and development: a scoping study. IDS Working Papers. 2013;433:1–52.

20. Cichocka D, Claxton J, Economidis I, Högel J, Venturi P, Aguilar A. European Union research and innovation perspectives on biotechnology. J Biotechnol. 2011;156:382–91.
21. Pfau SF, Hagens JE, Dankbaar B, Smits AJM. Visions of sustainability in bioeconomy research. Sustainability. 2014;6:1222–49.
22. The Energy Biosciences Institute. Annual report. 2014. http://www.energybiosciencesinstitute.org/sites/default/files/publications/2014_EBI_AR_0.pdf
23. Kates RW, Clark WC, Corell R, et al. Sustainability science. Science. 2001;292:641–2.
24. Clark WC. Sustainability science: a room of its own. Curr Issue. 2007;104:1737–8.
25. Miller TR, Wiek A, Sarewitz D, et al. The future of sustainability science: a solutions-oriented research agenda. Sustain Sci. 2014;9:239–46.
26. Loorbach D. Governance for sustainability. Sustain Sci Pract Policy. 2007;3:1–5.
27. Tilman D, Fargione J, Wolff B, et al. Forecasting agriculturally driven global environmental change. Science. 2001;292:281–4.
28. Sutton MA, Bleeker A, Howard CM, et al. Our nutrient world: the challenge to produce more food and energy with less pollution. Global overview of nutrient management. Edinburgh: Centre for Ecology and Hydrology; 2013.
29. De Wrachien D. Land use planning: a key to sustainable agriculture. In: García-Torres L, Benites J, Martínez-Vilela A, Holgado-Cabrera A, editors. Conservation agriculture: environment, farmers experiences, innovations, socio-economy, policy. Netherlands: Springer; 2003.
30. Tilman D, Cassman KG, Matson PA, Naylor R, Polasky S. Agricultural sustainability and intensive production practices. Nature. 2002;418:671–7.
31. Tanaka H, Katsuta A, Toyota K, Sawada K. Soil fertility and soil microorganisms. In: Tojo S, Hirasawa T, editors. Research approaches to sustainable biomass systems. Oxford: Academic Press; 2014.
32. Ramirez KS, Lauber CL, Knight R, Bradford MA, Fierer N. Consistent effects of N fertilization on soil bacterial communities in contrasting systems. Ecology. 2010;91:3463–70.
33. Liu C, Yao Z, Wang K, Zheng X. Three-year measurements of nitrous oxide emissions from cotton and wheat–maize rotational cropping systems. Atmos Environ. 2014;96:201–8.
34. Gagnon B, Ziadi N, Rochette P, Chantigny MH, Angers DA. Fertilizer source influenced nitrous oxide emissions from a clay soil under corn. Soil Sci Soc Am J. 2011;75:595–604.
35. Benoit M, Garnier J, Billen G, Tournebize J, Gréhan E, Mary B. Nitrous oxide emissions and nitrate leaching in an organic and a conventional cropping system (Seine basin, France). Agric Ecosyst Environ. 2015;213:131–41.
36. U.S. Environmental Protection Agency, EPA Methane and nitrous oxide emissions from natural sources. Washington, DC. 2010. http://www.epa.gov/outreach/pdfs/Methane-and-Nitrous-Oxide-Emissions-From-Natural-Sources.pdf
37. UNESCO Facts and figures – managing water under uncertainty and risk. 2012.http://www.unesco.org/new/fileadmin/MULTIMEDIA/HQ/SC/pdf/WWAP_WWDR4%20Facts%20and%20Figures.pdf
38. Parry ML, Canziani OF, Palutikof JP, van der Linden PJ, Hanson CE. Contribution of working group II to the fourth assessment report of the Intergovernmental Panel on Climate Change (IPCC). Cambridge: Cambridge University Press; 2007.
39. Pimentel D, Berger B, Filiberto D, et al. Water resources: agricultural and environmental issues. BioScience. 2004;54:909–18.
40. Haruna SI, Nkongolo NV. Cover crop management effects on soil physical and biological properties. Prog Environ Sci. 2015;29:13–4.
41. Abdin OA, Zhou XM, Cloutier D, Coulman D, Favis MA, Smith DL. Cover crops and interrow tillage for weed control in short season maize (*Zea mays* L.). Eur J Agron. 2000;12:93–102.
42. Plaza-Bonilla D, Nolot JM, Raffaillac D, Justes E. Cover crops mitigate nitrate leaching in cropping systems including grain legumes: field evidence and model simulations. Agric Ecosyst Environ. 2015;212:1–12.
43. Solaiman ZM, Anawar HM. Application of biochars for soil constraints: challenges and solutions. Pedosphere. 2015;25:631–8.

44. Zhang X, Wang H, He L, et al. Using biochar for remediation of soils contaminated with heavy metals and organic pollutants. Environ Sci Pollut Res Int. 2013;20:8472–83.
45. Wratten SD. Conservation biological control and biopesticides in agricultural. In: Jorgensen SE, Fath B, editors. Encyclopedia of ecology. Amsterdam: Elsevier; 2008. p. 744–7.
46. Ash GJ. The science, art and business of successful bioherbicides. Biol Control. 2010;52:230–40.
47. Díaz-Ambrona CH, Mínguez MI. Cereal–legume rotations in a Mediterranean environment: biomass and yield production. Field Crop Res. 2001;70:139–51.
48. Zhong Z, Lemke RL, Nelson LM. Nitrous oxide emissions associated with nitrogen fixation by grain legumes. Soil Biol Biochem. 2009;41:2283–91.
49. World Food Programme Hunger statistics. 2015. https://www.wfp.org/hunger/stats
50. Shah A. Poverty around the world. 2011. http://www.globalissues.org/article/4/poverty-around-the-world
51. En route to the knowledge-based bio-economy. 2007. Koln conference paper. http://www.bio-economy.net/reports/files/koln_paper.pdf
52. Padwal RS, Sharma AM. Prevention of cardiovascular disease: obesity, diabetes and the metabolic syndrome. Can J Cardiol. 2010;26:18–20.
53. Jørgensen H. The role of industry in a transition towards the bioeconomy in relation to biorefinery. IEA Bioenergy report, Task 42 Biorefinery; 2015. p. 1–17.
54. Ollikainen M. Forestry in bioeconomy – smart green growth for the humankind. Scand J Forest Res. 2014;29:360–6.
55. Lal P, Alavalapati JRR. Economic of forest biomass-based energy. In: Kant S, Alavalapati J, editors. Handbook of forest resource economics. London: Routledge; 2014. p. 275–89.
56. Fares S, Mugnozza GS, Corona P, Palahi M. Sustainability: five steps for managing Europe's forests. Nature. 2015;519:407–9.
57. Sheppard AW, Gillespie I, Hirsch M, Begley C. Biosecurity and sustainability within the growing global bioeconomy. Curr Opin Environ Sustain. 2011;3:4–10.
58. Robertson GP, Dale VH, Doering OC, et al. Sustainable biofuels reflux. Science. 2008;322:49–50.

Chapter 3
Biomass: The Sustainable Core of Bioeconomy

Abstract Bioeconomy, through its industrial and agricultural activities, as well as forestry and fishery, is aiming at providing markets with various bio-based commodities at competitive prices. Thus, the use and conversion of bioresources is at the core of bioeconomy, and in order to produce a wide range of end products, a multitude of bio-based production and manufacturing processes are applied or being upgraded, all depending on the availability and biochemical composition of biomass.

In this chapter, a detailed characterization of many bioresources is presented as the primary step for its industrial or agricultural valorization. This includes woody, herbaceous, and aquatic biomass as well as agro-industrial and municipal wastes. The main components of bioresources valued by several industrial activities are cellulose, hemicellulose, and lignin (i.e., lignocellulosic biomass), along with starch, lipids, proteins, chitin, and chitosan. Based on the data presented in this third chapter, three main characteristics could be given to biomass: renewability, availability, and versatility, which are among the main prerequisite to implement bioeconomy on a sustainable ground.

3.1 Introduction

The primary purpose of bioeconomy is to provide markets with various bio-based products able to compete with fossil-based ones and over time replace them. A substantial fraction of bioeconomy is based on renewable biomass as feedstock for the different industrial activities. Indeed, the diversity of the end products, the multitude of production and manufacturing processes, and the involved engineering and technological aspects are all dependent upon the one single element: *Biomass*.

Therefore, the composition and availability of biomass are the real instigators of the worldwide efforts, in both academia and industry, to develop new sustainable production schemes and the subsequent breakthroughs in converting and refining those bioresources. But, before we get started, we should first know what we are dealing with and ask this following simple question.

M. Sillanpää, C. Ncibi, *A Sustainable Bioeconomy*,
DOI 10.1007/978-3-319-55637-6_3

3.2 What Is Biomass?

Basically, biomass is a generic term for all plants or to be more specific all materials derived from growing plants.

Over the last decades, biologists and ecologists lost the monopole of defining biomass and engineers and industrialists adopted and extended it to include such diverse sources as algae, municipal solid waste, food wastes, and agro-industrial by-products. Such shift was promoted by the valuable breakthroughs in R&D that facilitated the production of various commodities from biomass, especially biofuels after the oil crisis in the 1970s. For instance, the French national institute for statistics and economic studies proposed this energy-oriented concept to define biomass: "*The biomass is all the organic matters which can become sources of energy. They can be used either directly (wood energy) or after a methanation of the organic matter (biogas) or the new chemical transformations (biofuel). They can be also used for the composting*" [1].

The most influential of those definitions are the ones given by legislative bodies. Indeed, laws and regulations will be based on those definitions which will impact all the involved industrial sectors including food, energy, and chemicals, to name a few, along with the repercussions on the environment. For instance, the European parliament, in its Renewable Energy Directive [2], states that biomass is "*the biodegradable fraction of products, waste and residues from biological origin from agriculture (including vegetal and animal substances), forestry and related industries including fisheries and aquaculture, as well as the biodegradable fraction of industrial and municipal waste.*"

In the United States, the Energy Security Act [3] tried to give an exhaustive definition. Thus, the term biomass becomes referring to "*any organic material that is available on a renewable or recurring basis, including agricultural crops, trees grown for energy production, wood waste and wood residues, plants (including aquatic plants and grasses), residues, fibers, animal wastes and other waste materials, fats, oils, and greases (including recycled fats, oils, and greases), and does not include paper that is commonly recycled or unsegregated solid waste.*"

Overall, it seems that the increase in the economic value of biomass pushed towards widening its definition so that it encompasses any new potential feedstock for the industry. Based on this assumption, one would argue that biomass will be defined as every renewable resource that is widely available and could be acquired at minimal cost. If in addition, less greenhouse gases were emitted, compared with fossil resources, we are talking about sustainable biomass.

3.3 Biomass Classification

Based on the definitions presented earlier, the term biomass encompasses a variety of plants, fractions of plants, or transformation residues and wastes, hence the practical need to group them into categories. Several attempts have been made to develop a consistent grouping and classification of biomass resources. As a result, biomass was divided into several categories based on a number of factors including its origin (land and sea or rural and urban), production (forest, agricultural, or municipal sources), nature (feedstock or residue), characteristics (moisture content or biochemical composition), and its end-use purpose (energy, food, chemicals, or materials).

The diagram (Fig. 3.1) illustrates the classification of biomass that is going to be used in the present book:

1. Natural resources that have to be harvested
2. The ones requiring cultivation
3. Wastes and residues produced after the transformation of those natural or cultivated resources.

3.3.1 Woody Biomass

Woody biomass is basically the combined biological mass of roots, wood, bark, and leaves of trees and shrubs from the forestry or agriculture sectors, whether in its natural form or after transformation (wood residues and by-products). This is how the authors of this book define woody biomass. Indeed, in the literature, this term has different definitions based on standpoints of the defining authority.

In the United States for instance, The Forest Service defined woody biomass as "*trees and woody plants, including limbs, tops, needles, leaves, and other woody parts, grown in a forest, woodland, or rangeland environment, that are the by-products of forest management*" [4]. Other institutions including USDOE are using the term forest biomass instead of woody biomass. Overall, they are referring to the same bioresources, which are wood materials in the forests or wood materials derived from manufacturing and processing factories or urban waste [5].

In practice, woody biomass could be procured from various sources. Forest harvesting is a major source whether through thinning young stands and cutting older ones for lumber or pulp or the management harvesting consisting in the removal of dead trees and ones of small diameter from overpopulated stands for wildfire hazard fuel reduction [6]. As well, forest plantations of fast growing trees (3–15 years), destined mainly for the energy sector, are also a significant provider of woody biomass to global biomass markets [7].

On the other hand, and based on the composition of wood (mainly cellulose, hemicellulose, and lignin), the terms woody biomass and lignocellulosic biomass

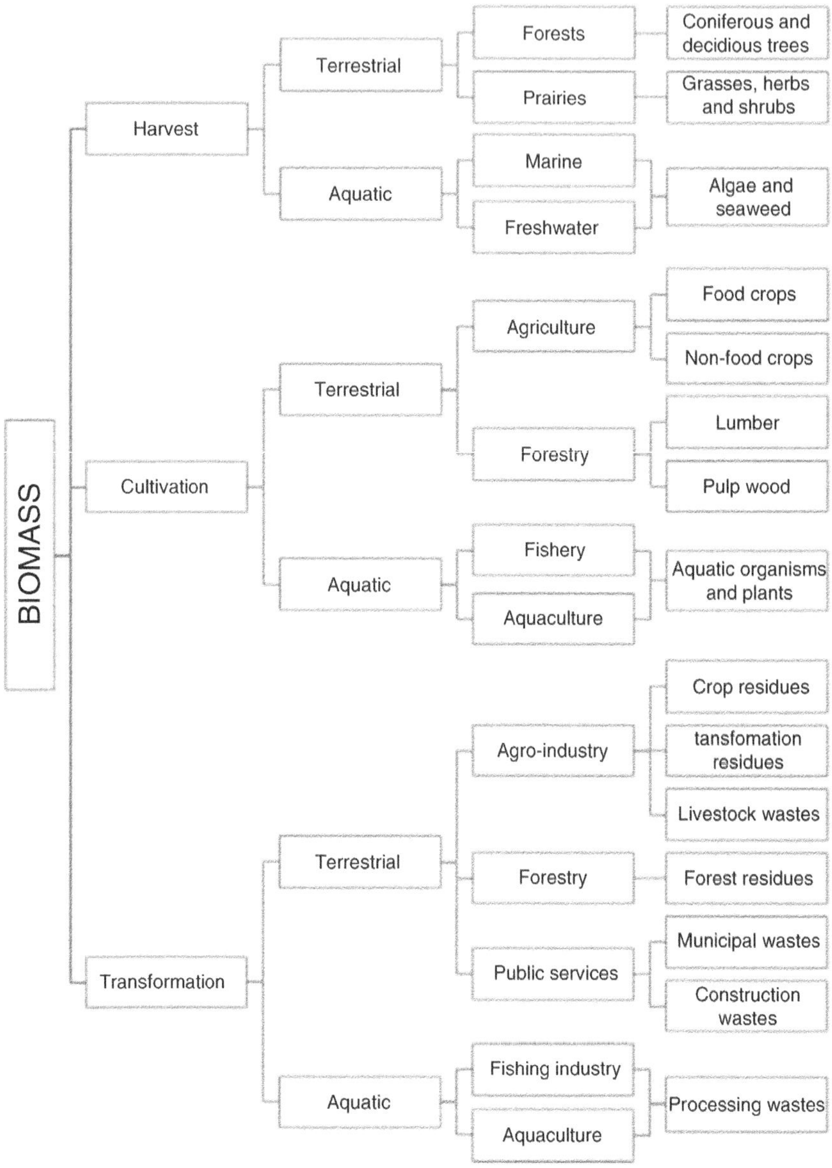

Fig. 3.1 Diagram of biomass classification

are sometimes interchangeably used, which is not accurate. Indeed, the term lignocellulosic biomass is more wide-ranging as it encompasses the term woody biomass in addition to other plants composed of cellulose and lignin (at various ratios) such as herbaceous biomass. In this context of classification based on the bioresource composition, two main subdivisions could be emphasized:

1. Cellulosic biomass: referring to plants and derived solid biological materials having cellulose as the major biochemical component. In practice, they are mostly linked to agricultural crops and crop residues (corn cobs, cotton stalks, cereals straws, etc.) and grasses (Miscanthus, Switchgrass, Bermuda grass, etc.)
2. Lignocellulosic biomass: comprising plants and derived solid biological materials containing significant amounts of both cellulose and lignin polymers, including but not limited to forest or woody biomass.

More details on the biomass composition will be presented in the following Sect. 3.4.

3.3.2 *Herbaceous Biomass*

Based upon the European Standard EN 14961-1, herbaceous biomass is "*from plants that have a non-woody stem and which die back at the end of the growing season. It includes grains or seeds crops from food processing industry and their by-products such as cereal straw.*"

Generally, herbaceous biomass resources could be classified into three major groups: natural herbaceous vegetation resources, agricultural crops, and agro-industrial residues.

1. *Natural herbaceous vegetation* are defined as plants without persistent stem or shoots above ground and lacking definite firm structure [8]. From an ecological perspective [9], there are main two categories of those graminoids: grasses and forbs. In our context, a natural herbaceous vegetation is every nonwoody (i.e., could be lignocellulosic), annual and perennial ($\geq$2 years), plant spontaneously growing in prairies or other ecological system.
2. *Agricultural crops* are herbaceous plants cultivated and harvested for consumption (food crops) or for the industrial utilization in the energy sector (energy crops), and also for the production of feed, materials, and chemical industries. The group includes plants like corn, wheat, triticum, rice, and flax [10].
3. *Agro-industrial residues* refer to the fraction of the crop left on the field after the harvest or other standard agricultural operations. It includes straws from cereals (wheat, barley, rye, and rice); stalks of corn, cotton, and sunflower; and many other by-products of the industrial production of food, feed, and fiber.

   One of the main characteristics of most herbaceous biomass is that they contain more holocelluose (cellulose and hemicellulose) than lignin. Knowing that holocellulose are relatively easier to digest and/or ferment compared to lignin, herbaceous biomass is a better feedstock for certain industrial activities, mainly liquid and gaseous biofuels production. In addition, and due to their rapid growth rate, when compared to woody biomass, herbaceous plants could secure a supply for valuable bioresources on a short-term basis. But on the other side, scientists revealed that this kind of biomass tend to have high chlorine and ash contents, which could limit its application in certain activities involving thermal processes [11].

3.3.3 Aquatic Biomass

Aquatic biomass is all micro/macroalgae, seagrasses, and other plants growing or being cultivated in aquatics environments including seas, oceans, lakes, ponds, and rivers. Those marine and freshwater bioresources have considerable potential to meet part of the rising demand for biomass from various industrial sectors. One of the major assets related to aquatic biomass is that they could be procured from their native ecosystems and even cultivated without competition with food crops as no agricultural lands are required, with the possible cultivation of those bioresources using marginal waters such as municipal wastewaters [12]. Furthermore, aquatic plants are noted for their high levels of productivity, wide range of synthesized biochemicals, and significant contribution in carbon capture and global warming mitigation [13, 14].

Overall, aquatic biomass could be classified into two major groups:

1. *Algae* are microscopic or macroscopic photosynthetic organisms that can live and grow in freshwater and marine ecosystems. Basically, their photosynthetic process is similar to terrestrial plants, but due to a simple cellular structure (unicellular for microalgae and multicellular for macroalgae, *aka*. Seaweed), and the fact that they are submerged in an aquatic environment, they have substantial supply of CO_2, water, and nutrients. Besides, some microalgae have the ability to acclimate to severe fluctuations in environmental conditions including temperature, pH, light exposure, carbon dioxide, and nutrients supply. This high adaptation faculty is due to the capability to modify their chemical composition in response to environmental variability [15]. In terms of biomass production, algae yield more dry matter per hectare than normal agricultural products. For instance, certain microalgae are able to produce 15–300 times more lipids for biodiesel production than traditional crops, on an area basis [16], which make it a prime choice for various industrials activities, especially food, feed, and energy.
2. *Seagrasses:* Contrary to the literal meaning of their name, seagrass are not grasses that grow in the sea. Rather, they are flowering photosynthesizing marine plants mostly present in coastal habitats. Seagrasses tend to form large meadows, either made of a single species or mixed beds where several species coexist including Zostera, Posidonia, and Cymodocea. Commonly, seagrass meadows grow in shallow, sheltered soft-bottomed marine coastlines and estuaries, as the amount and quality of light available for photosynthesis is the determining factor controlling the depth at which seagrass could grow and flourish [17]. In addition to its ecological role in providing food and habitats for other marine organisms such as algae, prawns, and various fish species, seagrass meadows depict by their high biomass production. Indeed, their primary productivity (i.e., rate at which energy is converted by photosynthesis to organic substances) can reach levels

similar to or greater than many cultivated terrestrial systems [18]. Other studies revealed that although seagrass meadow occupy only 0.15% of the ocean surface, they contribute nearly 1% of the net primary production of the global ocean [19].

3.3.4 Wastes and Residues

In addition to natural resources from both land and seas, biomass also includes residues, wastes, and by-products from:

1. The agricultural sector. Indeed, each year, agricultural practices leave substantial amounts of residues on the fields such as cereal plant straws, corn stovers, and rice husks, making therefore *agricultural residues* a renewable resource of biomass. Generally, a fraction of those bioresources is recycled back into the agricultural production systems as feed, litter, or fertilizer. However, a large fraction remains unused, thus creating a disposal problem most of the time dealt with burning those wastes. Such practice, still frequent in some countries, is a serious waste of certain extra income, at least in terms of energy supply.
2. Several industrial activities using organic matter as feedstock (agriculture or forest-derived resources) generate wastes and by-products that could also be recycled either in the same industrial process as heat and power source [20] or as new feedstock for other industries [21, 22]. In real case scenario, still substantial amounts of *industrial by-products* are being discarded, thus creating serious environmental threats to soil and water resources. This includes wastes from the pulp and paper mills (sludge and black liquor), olive oil mills, and many food-processing industries (jam, meat, cheese, etc.).
3. *Municipal wastes* refer to any non-liquid waste generated by households, small businesses, construction sites, and any other institution linked to the waste management system of a municipality. In general, this kind of wastes, commonly called trash or garbage, is composed of items as diverse as rubber, plastic debris, broken glass, metals, and textiles. But, in our context of biomass resources, the term municipal wastes will be reserved for every organic matter containing cellulose, lipids, or proteins. For instance, this includes for food wastes, cooking oils, yard trimmings, newspapers, cardboard for packaging materials, etc. [23].

3.4 Biomass Composition

A detailed characterization of biomass is the primary step for its industrial or agricultural valorization. Indeed, knowing the chemistry of the various biomass components will help anticipating their reaction and interaction behaviors during the conversion process.

As we have seen in the previous section, there are several varieties of biomass. This has a direct impact on the variation in their biochemical composition. Nonetheless, there are some typical components based on which groups of biomass could be categorized. Primarily, this includes cellulose, hemicellulose, and lignin (forming lignocellulosic biomass), as well as starch, lipids, and proteins.

3.4.1 Cellulose

Basically, cellulose is a long linear chain of β(1–4) linked D-glucose building blocks (Fig. 3.2) with degrees of polymerization varying from 1000 to 10000. These chains could interact with each other via hydrogen bonds by the hydroxyl groups on these linear cellulose chains [24], thus promoting their agglomeration into a crystalline cluster.

This results in intermediate degrees of cellulose crystallinity and the subsequent formation of a multitude of amorphous, partially crystalline, and crystalline fibrous structures [25]. In this context, analyzing the supramolecular structure of cellulose reveals that both the amorphous and crystalline states intertwine to form the natural cellulose [26]. Another study showed the amorphous (noncrystalline) phase mainly due to the fact that most hydroxyl groups on glucose are amorphous [27].

In terms of natural production, cellulose is believed to be the most common organic polymer on Earth, representing about 1.5×10^{12} tons of the total annual biomass production [28]. This natural polymer is the major component of all plant cell walls [29], but other organisms could also produce cellulose including algae [30] and bacteria [31].

3.4.2 Hemicellulose

Hemicellulose is the second important component in the plants' cell walls. It contributes to strengthening the cell wall, primarily along with cellulose and also by interacting with lignin. It is formed by short-branched chains giving an amorphous structure with less strength than the unbranched and crystalline cellulose [32]. Structurally, it is a polysaccharidic polymer primarily made of pentoses (five-carbon sugars) like xylose and arabinose, but also some hexoses (six-carbon sugars) including glucose, galactose, and mannose, linked together by β-1,4-glycosidic bonds in main chains and β-1.2-, β-1.3-, β-1.6-glycosidic linkages in the side chains.

Hemicelluloses include xyloglucans, xylans, mannans, and glucomannans [33]. The presence of certain types of hemicellulose and their abundance in the cell walls depend on the species itself, and even within the same species, it depends on the cell functionality (Fig. 3.3).

Because of the low degree of polymerization (80–200), hemicelluloses are easily hydrolyzed (chemically or enzymatically) than cellulose [34]. The results of

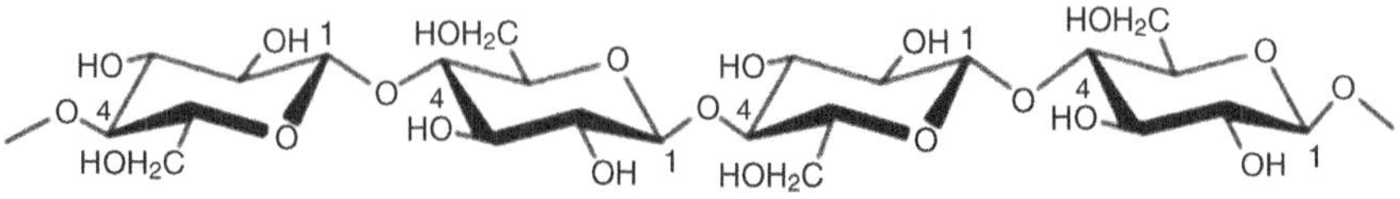

Fig. 3.2 Molecular structure of cellulose

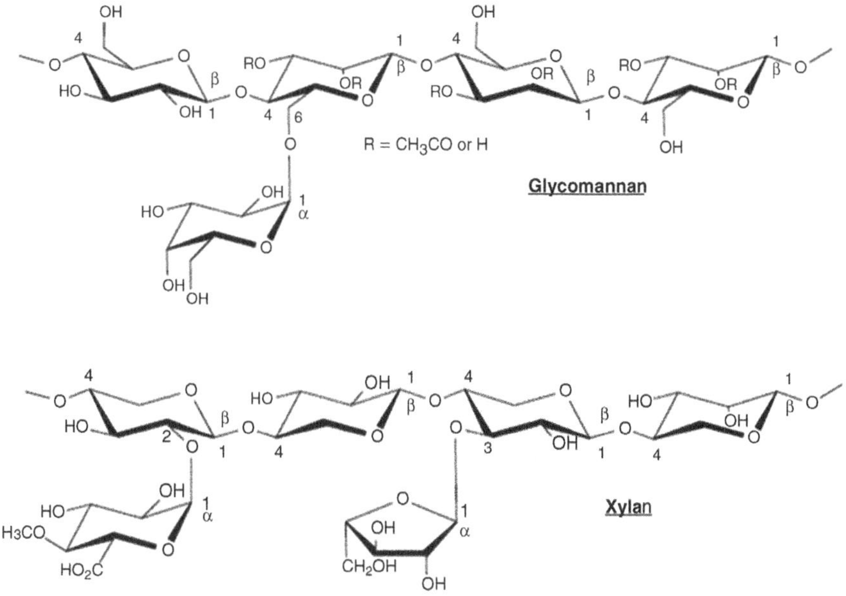

Fig. 3.3 Molecular structure of two hemicelluloses: Glycomannan and Xylan

hydrolysis are hexoses like D-glucose, D-fructose, D-galactose, and D-mannose, as well as pentoses such as D-xylose, L-arabinose, and D-ribose [35]. The other interesting feature is that sugars from hemicelluloses could be easily separated from the lignocellulosic biomass hydrolysate since they are usually in a noncrystalline state [36].

3.4.3 *Lignin*

Lignin is the most abundant aromatic polymer on earth and second most abundant organic polymer after cellulose. It could be found in most terrestrial plants at various dry weight percentages and is responsible of providing structural integrity and strength to different parts of those plants [37].

Lignin is a complex polymer composed of phenylpropane building blocks, nonlinearly and randomly cross-linked into a three-dimensional network. The

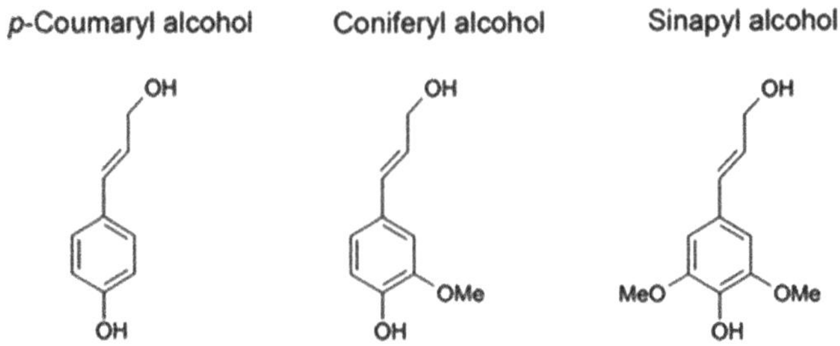

Fig. 3.4 The three building blocks of lignin

three main monomer units are *p*-coumaryl alcohol, coniferyl alcohol, and sinapyl alcohol (Fig. 3.4).

The lignin fraction interacts with both cellulose and hemicellulose through chemical linkages such as ethers and esters [38]. A strong lignin–carbohydrate network is therefore formed in the cell wall and the separation of those three components becomes a major challenge as we will see later in this book when the subjects of pretreatment and conversion of lignocellulosic biomass are going to be discussed.

The following table illustrates the cellulose, hemicellulose, and lignin contents of various lignocellulosic bioresources in % weight (dry basis) categorized into *woody*, *herbaceous*, and *aquatic* biomass, along with *biowastes* (Table 3.1).

3.4.4 Starch

Starch is the main carbohydrate reserve in plants and the most important source of digestible carbohydrate for human nutrition. In fact, contrary to the cellulose, starch present in dietary fibers is digested by humans (broken down by enzymes known as amylases) and can represent up to 80% of the carbohydrates present in the staple foods of people worldwide. Starch therefore constitutes a major source of glucose and significantly contributes to total food energy intake [56].

Starch is also a valuable molecule for various nonfood industries including papermaking, starch-based glues, and adhesives. Starch is also the main precursor to produce various biodegradable films in the form of thermoplastic starch [57].

Basically, a starch molecule is a polysaccharide made of glucose units (from five hundred to several hundred thousand units [58]), chemically joined by covalent bonds forming a single polysaccharidic structure, whether linear amylose (up 20–30% of native starch) or branched amylopectin (70–80%).

As shown in Fig. 3.5, amylose is composed of glucose residues (α-D-glucopyranosyl moieties) linearly connected through α-(1 → 4)-glycosidic linkages. Amylopectin has a highly branched structure composed of α-(1 → 4)-glucans

Table 3.1 Chemical composition of various biomass (% weight on a dry basis)

Classification	Biomass	Cellulose	Hemicellulose	Lignin	Reference
Woody	Eucalyptus (*Eucalyptus camaldulensis*)	45	18.2	31.3	[39]
	Spruce (*Picea glauca*)	39.5	30.6	27.5	
	Pine (*Pinus sylvestris*)	40	28.5	27.7	
	Black Locust (*Robinia pseudoacacia*)	39.2–42.6	16.6–18.8	23.8–28.5	[40]
	Hybrid Poplar DN-34	39.2–45.7	16.6–22.6	23.3–25.1	
Herbaceous	Switchgrass	30–35	20–30	20–25	[41]
	Flax	~65	~16	2.5	[42]
	Hemp	53.8	10.6	8.7	[43]
	Amur silver grass (*Miscanthus sacchariflorus*)	42	30.1	7	
	Rice straw	39	20.9	5.7	[44]
	Wheat straw	37.6	28.8	14.5	[45]
	Barley straw	42.4	27.8	6.8	
	Corn stover	31.3	21.1	3.1	
Aquatic	Seagrass *Posidonia oceanica* fibers	38	21	27	[46]
	Posidonia oceanica leaves	31.4	25.7	24.7	[47]
	Green alga *Ulva lactuca*	9.1	20.6	1.5	[48]
	Seagrass *Zostera marina*	57.3	28.5	5.1	[49]
	Brown seaweed *Sargassum sp.*	22	19.6	0	[50]
	Algal biomass from sewage sludge	7.1	16.3	1.5	[51]
Wastes and residues	Sugarcane bagasse	46.4	25.9	23.6	[52]
	Agave bagasse	24	20	15	[53]
	Nut shells	25–30	25–30	30–40	[54]
	Corn cobs	45	35	15	
	Lemon peel	12.7	5.3	1.7	[51]
	Orange peel	13.6	6.1	2.1	
	Corn stover	30.6–38.1	19.1–25.2	17.1–21.2	[40]
	Oil palm empty fruit bunch fiber	44.2	33.5	20.4	[55]

with inter-linked α-(1 $\rightarrow$ 6)-glycosidic linkages at intervals of approximately 20 units [59].

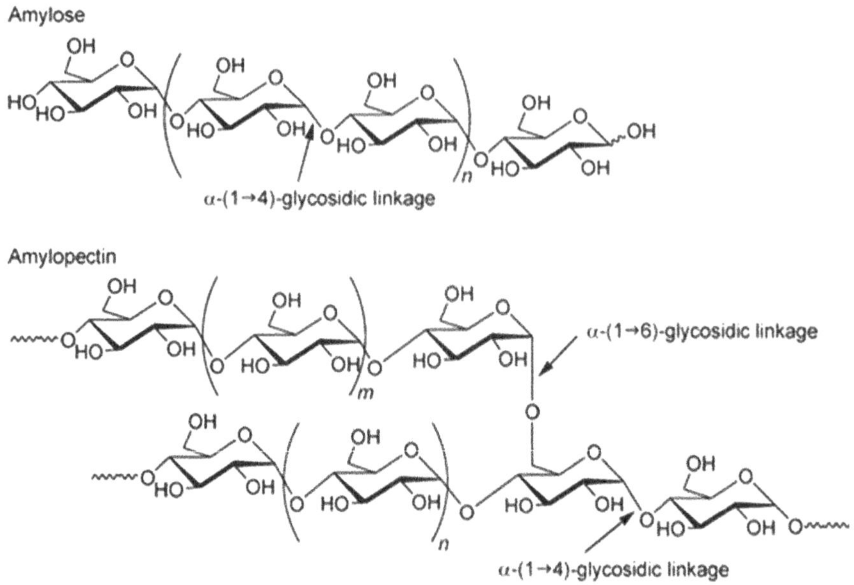

Fig. 3.5 Amylose and amylopectin structures [60]

3.4.5 Proteins

Proteins are a major and vital component in the human diet. In the last decades, a growing interest has been seen in recovering proteins from nonconventional feedstocks include forage crops, cattle by-products, and some microorganisms. Thus, the valorization and utilization of biomass to produce natural, renewable, and biodegradable proteins became an area of extensive study and research.

Proteins are naturally occurring polymers produced by animals (casein, collagen ...), plants (soy and what proteins ...), and bacteria (mainly enzymes such as chymotrypsin, fumarase ...). Basically, proteins are highly sophisticated 3-D macromolecules containing 20 different amino acids [61]. Amino acids are organic compounds synthesized by cells through a set of biochemical processes. They contain a basic amino group and an acidic carboxyl group, hence their name. Chemically, this dual functionality allows individual amino acids to interact and form long chains via peptide bonds (–CO–NH–) when the amino group (–NH2) of one amino acid reacts with the carboxyl group (–COOH) of another amino acid, with the release of a molecule of water.

Sequence amino acids up to 10 units are oligopeptides and chains of 10 or more amino acids are termed polypeptides, the backbone structures of proteins. Commonly, the transition from polypeptide to protein is not well defined, but insulin with its 50 amino acid residues is regarded as the smallest protein [62, 63].

Structurally, proteins can be configured at four consecutive levels from the primary to the quaternary structure. The *primary structure* is a linear sequence of amino acids in a protein, resulting in polypeptide chains, each one characterized by a specific arrangement of amino acids [64].

Typically, there are several intramolecular and intermolecular interactions within a protein molecule, mainly via covalent and hydrogen bonds. As a result, we encounter specific regions in protein where chains are organized into regular structures, thus constituting the second level of protein structure, namely the *secondary structure*. The most common secondary structures are the *α-helix* (present in keratin, protein in hair [65]) and *β-sheet* (present in fibroin, protein in silk [66]).

As for the *tertiary structure*, it describes the folding (i.e., three-dimensional arrangement) of the secondary structures into a compact shape such as in globular proteins (enzymes and immunoglobins) where the complex structure is held together by non-covalent interactions including hydrogen bonds, disulfide bridges, and ionic bonds. This folding pattern confers specific biological functions to the protein, as it is in the case for enzymes and other proteins acting as molecular receptors recognizing specific compounds based on their shapes [67].

Finally, a *quaternary structure* is also possible when some proteins (tertiary-structured subunits) interact with each other via non-covalent bonds to form agglomerates or units. One illustrative example is human hemoglobin which is composed of four subunits: two α-chains of 141 amino acid residues each and two β-chains of 146 residues each [68].

3.4.6 Lipids

Lipids from biomass are oil and fats synthesized by plants, animals, and microorganisms from renewable carbon substrates. The main producers of vegetable oils are in Asia (Malaysia and Indonesia) for suitable climatic conditions and intensive agricultural practices. Animal fats on the other hand are mainly produced in the United States and Europe.

Those "bio-lipids" possess various chemical properties based on the length of the fatty acids chain and the unsaturation degree. Indeed, lipids, consisting of fatty acids, are classified based on the presence or absence of double bonds (DB) as follows:

- No DB → Saturated fatty acids
- One DB → Monounsaturated fatty acids
- Two or more DBs → Polyunsaturated fatty acids

The main sources for lipids are:

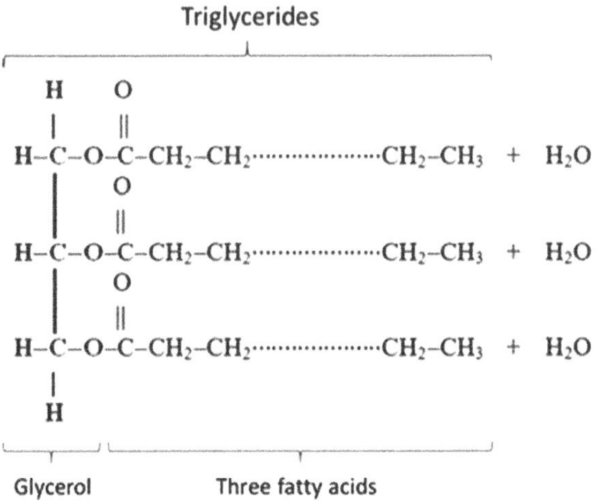

Fig. 3.6 Esterification of triglycerides

3.4.6.1 Vegetable Oils

The oil is derived from various species producing oleaginous seeds (grape, sunflower, canola. . .) or fruits (olives, palms, nuts...) and also from heat-processed woody biomass (pyrolysis). It is estimated that 80% of the produced vegetable oils are for human consumption and the remaining 20% for industrial applications [69], especially the growing biodiesel sector, but also in the production of other oil-based products including lubricants, surfactants, and various polymeric materials. Thus, the rapid increase in demand for vegetable oils is still sparked off by the food market rather than the industrial or biodiesel sectors.

Vegetable oils are mainly constituted of triglycerides synthesized through the esterification of one molecule of glycerol with three fatty acids, as depicted in Fig. 3.6. At room temperature, this chemical compound could be either solid (butter or fat) or liquid (oil).

Regarding the fatty acids, their chain length and degree of condensation determine the oil characteristics. For instance, as the number of double bonds (i.e., unsaturation) increases, the melting point of the oil decreases. As well, due to the presence of double bonds, unsaturated fatty acids are more reactive than the statured ones. Table 3.2 gives the fatty acids content in several vegetable oils.

3.4.6.2 Animal Fats

Animal fats are the by-products or wastes from slaughtered animals that are bred to produce a variety of products such as meat, milk, eggs, leather, etc. Animal fats can be categorized into two main groups: edible and inedible fats. The former is used in the human consumption and also in the feed and oleo-chemical industries. The

Table 3.2 Main fatty acids in various vegetable oils [70–72]

Oil	Fatty acids (% of the total content)				
	Palmitic (16:0)[a]	Stearic (18:0)	Oleic (18:1)	Linoleic (18:2)	Linolenic (18:3)
Soybean	11	4	23.4	53.3	7.8
Sunflower	6	4	42	47	1
Canola	4.1	1.8	60.9	21	8.8
Palm	39	5	45	9	–
Corn	10.9	2	25.4	59.6	1.2
Olive	13.7	2.5	71.1	10	0.6
Grape	6.6	3.5	14.3	74.7	0.15
Sesame	9.7	6.5	41.5	40.9	0.21
Peanut	7.5	2.1	71.1	18.2	–
Coconut	–	2.7	6.2	1.6	–
Cottonseed	21.6	2.6	18.6	54.4	0.7
Rapeseed	4	2	56	26	10
Linseed	5.5	3.5	19.1	15.3	56.6

[a]Chain length: Number of DB

latter, on the other hand, are not allowed in human consumption and can only be used in feed, oleo-chemistry, and biofuels (mainly biodiesel and also biogas).

In general, the wastes from animal-processing industries are subjected to various rendering processes to produce other added-value products. Such strategy allows the generation of valuable extra income and contributes to the preservation of the environment and public health. Basically, rendering refers to the processes (physical, chemical, and thermal) used to separate the various components in those raw animal wastes including fats, bones, feathers, hoofs, horns, blood, etc.

As far as fat is concerned, the rendering process targets the entire fatty tissues in an animal and purifies it into fats like tallow, greases, or lard. Figure 3.7 illustrates the rendering process for poultry wastes.

Table 3.3 gives the fatty acids content in several animal fats. Such analysis is crucial to determine which products would be the suitable, as their quality will depend on the nature and amount of specific fatty acids in the fat. For instance, biodiesel production is one of the promising field in which animal fats could be incorporated as a low cost feedstock [74].

In this sector, the use of animal fat wastes produced biodiesel with low NOx emissions and high Cetane number at high ratios of saturated fatty acids [75].

3.4.6.3 Microbial Lipids

Some microorganisms, including fungi, bacteria, and microalgae, are capable of synthesizing and storing lipids and fatty acids as membrane components, storage products, and metabolites. To be termed "oleaginous microorganism," the

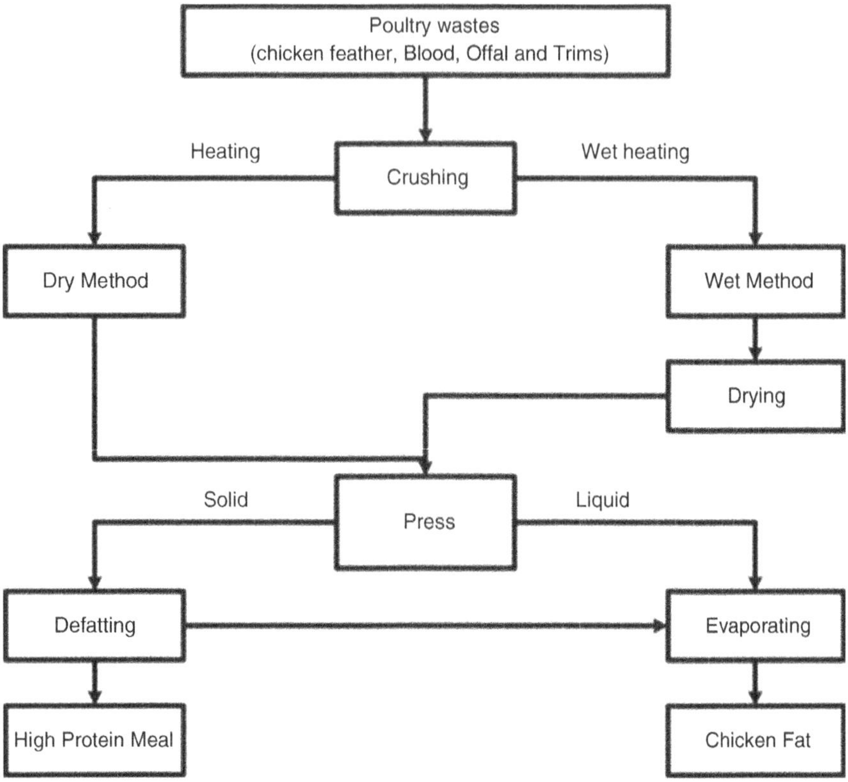

Fig. 3.7 Flow diagram of poultry wastes rendering process [73]

Table 3.3 Main fatty acids in various animal fats [76–81]

	Fatty acids (% of the total content)				
Animal fat	Palmitic (16:0)	Stearic (18:0)	Oleic (18:1)	Linoleic (18:2)	Linolenic (18:3)
Mutton tallow	27	24.1	40.7	2	–
Duck tallow	17	4	59.4	19.6	–
Chicken fat	24	5.8	38.2	23.8	1.9
Yellow grease	23.2	12.9	44.3	6.9	0.6
Brown grease	22.8	12.5	42.3	12.1	0.8
Choice white grease	25.5	15.3	35.8	12.6	0.9
Lard	23.2	10.4	42.8	19.1	64.7

microbial species has to have a lipid content of at least 20%, mostly in form of triacylglycerol stored in cytosolic lipid bodies [82].

Among those oleaginous microbial species, some are able to accumulate around 20% of lipid in mass. Others, including certain oleaginous yeasts, can synthesize

Table 3.4 Lipid content in selected oleaginous microorganisms [83–85]

Classification	Microorganism	Lipid content (%) on a dry matter basis
Bacteria	*Arthrobacter* sp.	>40
	Acinetobacter calcoaceticus	27–38
	Rhodococcus opacus	24–25
	Bacillus alcalophilus	18–24
	Agrobacterium tumefaciens	23
Fungi (including yeasts and molds)	*Mortierella isabellina*	86
	Humicola lanuginosa	75
	Rhodotorula glutinis	72
	Aspergillus oryzae	57
	Gibberella fujikuroi	41
	Cryptococcus adeliensis	33
Microalgae	*Schizochytrium* sp.	50–77
	Botryococcus braunii	25–75
	Nannochloropsis sp.	31–68
	Neochloris oleoabundans	35–54
	Nitzschia sp.	45–47
	Cylindrotheca sp.	16–37
	Chlorella sp.	28–32

and accumulate up to 70% of lipids. Noting that the chemical characteristics of those lipids vary from species to species and from strain to strain, thus producing lipids with various carbon chain length and unsaturation degree. Furthermore, for the same stain, the lipid content could significantly fluctuate according to specific culture conditions.

Table 3.4 gives an account on the lipid content in a selection of bacterial, fungal, and algal biomass.

Compared to lipids from vegetables and animals, microbial lipids offer several advantages related to the short life cycles on microorganisms, continuous harvesting, more homogeneous physical structure, and low requirements (space, water, and nutrients). In this context, it was reported that photosynthetic microorganisms can produce 100 times more lipids per hectare than plants [86]. Furthermore, scientists proved that microbial lipids are valuable feedstock for biodiesel production because their fatty acid composition is similar to that of vegetable oils [87]. Also, lipids from oleaginous microorganisms, rich in polyunsaturated fatty acids oils, are added to infant formulas and nutritional supplements [88].

3.4.7 Chitin and Chitosan

Chitin is the second most abundant polysaccharide on earth after cellulose. It consists of 2-acetamido-2-deoxy-β-D-glucose through a β(1–4) linkage. It is a white, hard, inelastic, and nitrogenous compound, occurring in form of ordered crystalline microfibrils in the exoskeleton of arthropods including insects, arachnids, and crustaceans.

Chitin could also be found in the cell walls of fungi and yeast and other organisms, mainly for reinforcement purposes [89].

Chemically, it resembles cellulose in its insolubility and low reactivity due to great structural similarity. In fact, cellulose is a linear β(1–4)-linked polymer of D-gluco-pyranose units, and chitin is therefore a cellulose derivative where the 2-hydroxyl group has been replaced with an acetamide group (*Cf.* Fig. 3.8). In this context, chitin produced by fungi was reported to have interesting advantages compared to animal chitin such as more homogeneity in composition, renewability, and absence of inorganic salts in its matrix [90].

On the other side, however, fungal chitin is tightly linked to other polysaccharidic components (cellulose, glucan, manna…) via covalent bonds, which make its chemical isolation difficult, with the possibility to enhance this process enzymatically [91].

Chitosan is a water-soluble chitin derivative formed by removing varying fractions of the *N*-acetyl groups (deacetylation). The solubilization occurs by protonation of the amine function on the C-2 position of the D-glucosamine unit. Thus, being soluble in aqueous solutions, chitosan and chemically modified chitin are industrially valorized in various applications including food processing, paper making, cosmetics, textile fibers, and many biomedical products such as wound dressings and artificial skins [92].

Table 3.5 illustrates the chitin content in various marine and terrestrial bioresources.

3.5 Concluding Remarks

Based on what we have seen in this chapter, three main attributes could be given to biomass: *renewability*, *availability*, and *versatility*. The renewability is the most obvious of the three. That's why both raw and residual biomass, from both land and sea, could secure a constant and indefinite flow of feedstock materials, on condition that we let those bioresources regenerate themselves in peace.

To have an idea about the availability of biomass, we just have to know that the total amount of biomass on earth (based on dry weight) is about 1.8 trillion tons on land and 4 billion tons in the ocean [96].

The following table gives an indicative idea about the magnitude of biomass availability for selected bioresources (Table 3.6).

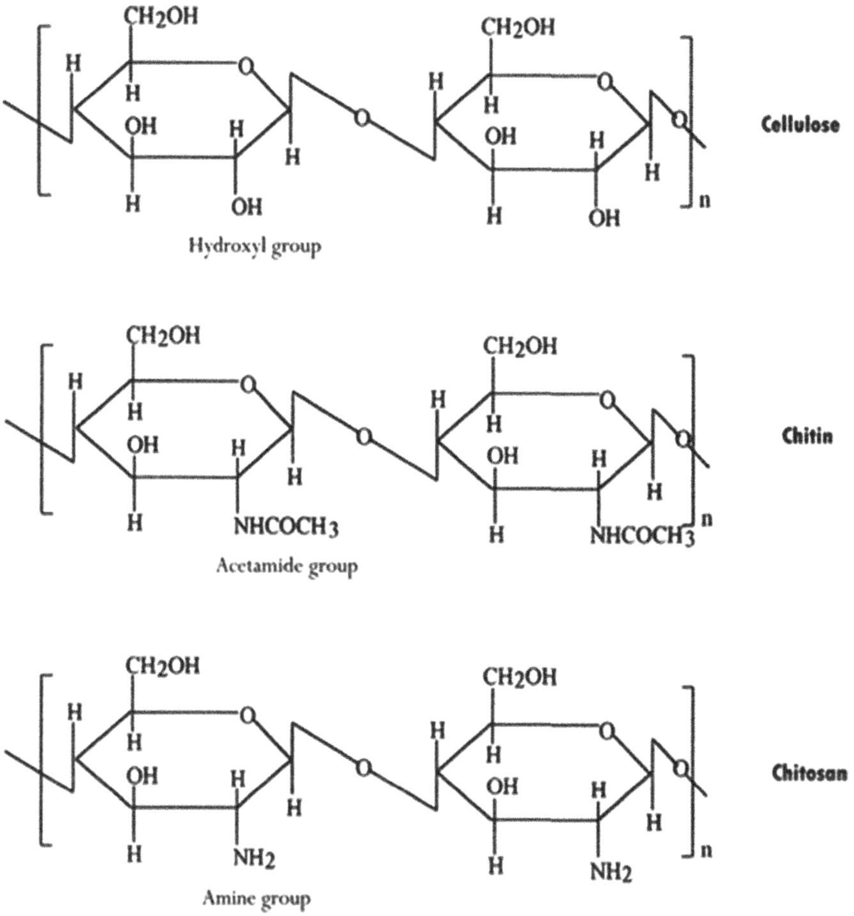

Fig. 3.8 Chemical structures of cellulose, chitin, and chitosan: Similarities and differences

Thus, as long as you have a bit of land or sea, you possess a valuable biomass supply to be valorized. Just this one advantage should justify the shift from petroleum (benefiting some countries) towards biomass (benefiting all countries). Think about the geopolitical stability the biomass-based economy would bring to the troublesome international affairs, providing that biomass itself does not trigger a new era of struggle over valuable resources. After all, petroleum is from biomass.

As for its versatility, the rich biochemical composition and various structural properties of biomass, whether from plants, animals or microorganisms, along with the possibly to extract valuable compounds (cellulose, hemicellulose, lignin, proteins, lipids, chitin. . .) make them suitable to be used in large scale in the food, feed, energy, textile, and pharmaceutical industries, to name a few.

As a result, many biomass-based commodities will be produced (biofuels, biomaterials, and biochemicals), and with more breakthroughs in R&D, the

Table 3.5 Chitin content for selected biomass (% on a dry matter basis) [93–95]

Classification	Biomass	Chitin content (%)
Crustaceans	*Penaeus monodon* (shrimp)	40.4
	Penaeus sp. (prawn)	33.0
	Chinoecetes opilio (carb)	26.6
	Euphosia superba (krill)	24.0
	Procamborus clarkia (crawfish)	13.2
Insects	*Pieris* (butterfly)	64.0
	Diptera (fly)	54.8
	Bombyx mori (silkworm)	44.2
	Galleria mellonella (wax moth larva)	33.7
	Blatella (cockroach)	18.4
Fungi	*Aspergillus niger*	42.0
	Histoplasma capsulatum	26.4
	Aspergillus phoenicis	23.7
	Penicillium notatum	18.5
	Blastomyces dermatidis	13.0
Plants	*Grass hay*	6.2
	Wheat straw	5.2

Table 3.6 Worldwide data on area, production, and yields of selected crops [97, 98]

Biomass	World production (Million metric tons)	Cultivated area (Million ha)	Average yield (metric ton/ha)
Sugarcane	1877.11	26.52	70.78
Corn	1008.99	178.69	5.65
Wheat	725.91	222.52	3.26
Rice	478.14	159.98	4.46
Soybean	318.80	118.14	2.70
Cassava	276.72	20.73	13.34
Sugar beet	250.19	4.48	55.84
Rapeseed	72.12	35.91	2.01
Sorghum	63.49	42.85	1.48
Cottonseed	44.35	33.01	1.34
Sunflower seed	39.57	23.17	1.71

competiveness of those bioproducts will increase until they become the ultimate market choice (high quality and low price). This is why biomass is the sustainable core of bioeconomy.

References

1. National Institute of Statistics and Economic Studies. Definitions and methods. http://www.insee.fr/en/methodes/default.asp?page=definitions/biomasse.htm
2. Directive 2009/28/EC of the European Parliament and of the Council of 23 April 2009 on the promotion of the use of energy from renewable sources and amending and subsequently repealing Directives 2001/77/EC and 2003/30/EC. http://eur-lex.europa.eu/legal-content/EN/TXT/PDF/?uri=CELEX:32009L0028&from=EN
3. Energy Independence and Security Act of 2007. Title XII section 1203(e)(z)(4)(A) P.L. 110–140. https://www.gpo.gov/fdsys/pkg/PLAW-110publ140/pdf/PLAW-110publ140.pdf
4. Memorandum of understanding on policy principles for woody biomass utilization for restoration and fuel treatments in forests, woodlands, and rangelands. http://www.fs.fed.us/woodybiomass/documents/BiomassMOU_060303_final_web.pdf
5. Perlack RD, Wright LL, Graham RL, Turhollow A, Stokes B, Erbach D. Biomass as feedstock for a bioenergy and bioproducts industry: the technical feasibility of a billion-ton annual supply. ORNL/TM-2005/66. U.S. Department of Energy. 2005.
6. Shelly JR. Woody biomass: what is it – what do we do with it? http://www.pelletheat.org/assets/docs/industry-data/infoguides43284.pdf
7. Hall JP. Sustainable production of forest biomass for energy. For Chron. 2002;78:1–6.
8. Scoggan HJ. The flora of Canada. National Museums of Canada: Ottawa; 1978.
9. Kuechler AW. Zonneveld I.S. Handbook of vegetation science. Dordrecht: Kluwer Academic; 1988.
10. Buranov AU, Mazza G. Lignin in straw of herbaceous crops. Ind Crops Prod. 2008;28:237–59.
11. The bioenergy system planners handbook – BISYPLAN. Herbaceous biomass – resources. http://bisyplan.bioenarea.eu/html-files-en/02-02.html
12. Şirin S, Sillanpää M. Cultivating and harvesting of marine alga *Nannochloropsis oculata* in local municipal wastewater for biodiesel. Bioresour Technol. 2015;191:79–87.
13. Irigoien X, Huisman J, Harris RP. Global biodiversity patterns of marine phytoplankton and zooplankton. Nature. 2004;429:863–7.
14. Brown LM, Zeiler KG. Aquatic biomass and carbon dioxide trapping. Energy Conver Manage. 1993;34:1005–13.
15. Bonachela J, Raghib M, Levin S. Dynamic model of flexible phytoplankton nutrient uptake. Proc Natl Acad Sci USA. 2012;108:20633–8.
16. Schenk PM, Thomas-Hall SR, Stephens E, et al. Second generation biofuels: high-efficiency microalgae for biodiesel production. Bioenerg Res. 2008;1:20–43.
17. Collier CJ, Waycotta M, McKenzie LJ. Light thresholds derived from seagrass loss in the coastal zone of the northern great barrier reef. Aust Ecol Indic. 2012;23:211–9.
18. Duarte CM, Chiscano CL. Seagrass biomass and production: a reassessment. Aquat Bot. 1999;1334:1–16.
19. Duarte CM, Cebrian J. The fate of marine autotrophic production. Limnol Oceanogr. 1996;41:1758–66.
20. Sharma A, Shinde Y, Pareek V, Zhang D. Process modelling of biomass conversion to biofuels with combined heat and power. Bioresour Technol. 2015;198:309–15.
21. Sila A, Bougatef A. Antioxidant peptides from marine by-products: isolation, identification and application in food systems: a review. J Funct Foods. 2016;21:10–26.
22. Dreschke G, Probst M, Walter A, Pümpel T, Walde J, Insam H. Lactic acid and methane: improved exploitation of biowaste potential. Bioresour Technol. 2015;176:47–55.
23. Sokka L, Antikainen R, Kauppi P. Municipal solid waste production and composition in Finland. Resour Conserv Recycl. 2007;50:475–88.
24. Somerville C. Cellulose synthesis in higher plants. Annu Rev Cell Dev Biol. 2006;22:53–78.
25. Hall M, Bansal P, Lee JH, Realff MJ, Bommarius AS. Cellulose crystallinity – a key predictor of the enzymatic hydrolysis rate. FEBS J. 2010;277:1571–82.

26. Chen H. Chemical composition and structure of natural lignocellulose. In: Biotechnology of lignocellulose. Dordrecht: Springer; 2014.
27. Zhang JQ, Lin L, Sun Y, Mitchell G, Liu SJ. Advance of studies on structure and decrystallization of cellulose. Chem Ind For Prod. 2008;28:109–14.
28. Klemm D, Heublein B, Fink HP, Bohn A. Cellulose: fascinating biopolymer and sustainable raw material. Angew Chem Int Ed. 2005;44:3358–93.
29. Keegstra K. Plant cell walls. Plant Physiol. 2010;154:483–6.
30. Mihranyan A. Cellulose from cladophorales green algae: from environmental problem to high-tech composite materials. J Appl Polym Sci. 2011;119:2449–60.
31. Esa F, Tasirin SM, Rahman NA. Overview of bacterial cellulose production and application. Agric Agric Sci Proc. 2014;2:113–9.
32. Rowell RM. Handbook of wood chemistry and wood composites. Boca Raton: CRC Press; 2005.
33. Scheller HV, Ulvskov P. Hemicelluloses. Annu Rev Plant Biol. 2010;61:263–89.
34. Thakur VK, Thakur MK. Handbook of sustainable polymers: processing and applications. Boca Raton: CRC Press; 2015.
35. Mäki-Arvela P, Salmi T, Holmbom B, Willför S, Murzin DY. Synthesis of sugars by hydrolysis of hemicelluloses – a review. Chem Rev. 2011;111:5638–66.
36. Aguilar R, Ramirez JA, Garrote G, Vazquez M. Kinetic study of the acid hydrolysis of sugar cane bagasse. J Food Eng. 2002;55:309–18.
37. Ragauskas AJ, Beckham GT, Biddy MJ. Lignin valorization: improving lignin processing in the biorefinery. Science. 2014;344:665–772.
38. Gellerstedt G. Softwood kraft lignin: raw material for the future. Ind Crops Prod. 2015;77:845–54.
39. Sjostrom E. Wood chemistry. Fundamentals and applications. 2nd ed. San Diego: Academic Press; 1993.
40. Biomass program. US Energy Department. 2004. http://www.afdc.energy.gov/biomass/progs/search3.cgi?25312
41. Keshwani DR. Microwave pretreatment of switchgrass for bioethanol production. Raleigh, NC: North Carolina State University; 2009.
42. Bismarck A, Aranberri-Askargorta I, Springer J, et al. Surface characterization of flax, hemp and cellulose fibers; surface properties and the water uptake behavior. Polym Compos. 2002;23:872–94.
43. Kikas T, Tutt M, Raud M, et al. Basis of energy crop selection for biofuel production: cellulose vs. lignin. Int J Green Energy. 2016;13:49–54.
44. Huang C, Han L, Liu X, Ma L. The rapid estimation of cellulose, hemicellulose, and lignin contents in rice straw by near infrared spectroscopy. Energy Sources A. 2010;33:114–20.
45. Mani S, Tabil LG, Sokhansanj S. Effects of compressive force, particle size and moisture content on mechanical properties of biomass pellets from grasses. Biomass Bioenergy. 2006;30:648–54.
46. Ncibi MC, Jeanne-Rose V, Mahjoub B, et al. Preparation and characterisation of raw chars and physically activated carbons derived from marine *Posidonia oceanica* (L.) fibres. J Hazard Mater. 2009;165:240–9.
47. Bettaieb F, Khiari R, Dufresne A, et al. Nanofibrillar cellulose from *Posidonia oceanica*: properties and morphological features. Ind Crops Prod. 2015;72:97–106.
48. Yaich H, Garna H, Besbes S, et al. Chemical composition and functional properties of *Ulva lactuca* seaweed collected in Tunisia. Food Chem. 2011;128:895–901.
49. Davies P, Morvan C, Sire O, Baley C. Structure and properties of fibres from sea-grass (*Zostera marina*). J Mater Sci. 2007;42:4850–7.
50. Tamayo JP, Rosario EJD. Chemical analysis and utilization of *Sargassum* sp. as substrate for ethanol production. Iran J Energy Environ. 2014;5:202–8.

51. Ververis C, Georghiou K, Danielidis D, et al. Cellulose, hemicelluloses, lignin and ash content of some organic materials and their suitability for use as paper pulp supplements. Bioresour Technol. 2007;98:296–301.
52. Abril D, Medina M, Abril A. Sugar cane bagasse prehydrolysis using hot water. Braz J Chem Eng. 2012;29:31–8.
53. Saucedo-Luna J, Castro-Montoya AJ, Rico JL, Campos-García J. Optimization of acid hydrolysis of bagasse from *Agave tequilana* Weber. Rev Mex Ing Quím. 2010;9:91–7.
54. Sun Y, Cheng J. Hydrolysis of lignocellulosic materials for ethanol production: a review. Bioresour Technol. 2002;83:1–11.
55. Khalid M, Ratnam CT, Chuah TG, Ali S, Choong TSY. Comparative study of polypropylene composites reinforced with oil palm empty fruit bunch fiber and oil palm derived cellulose. Mater Design. 2008;29:173–8.
56. Slaughter SL, Ellis PR, Butterworth PJ. An investigation of the action of porcine pancreatic a-amylase on native and gelatinised starches. Biochim Biophys Acta (BBA) Gen Sub. 2001;1525:29–36.
57. Shirai MA, Grossmann MVE, Mali S, Yamashita F, Garcia PS, Müller CMO. Development of biodegradable flexible films of starch and poly(lactic acid) plasticized with adipate or citrate esters. Carbohydr Polym. 2013;92:19–22.
58. Ambily C, Sunny K, Tessymol M. Synthesis, characterization and thermal studies on natural polymers modified with 2-(5-(4-dimethylamino-benzylidin)-4-oxo-2-thioxothiazolidin-3-yl) acetic acid. Res J Chem Sci. 2012;2:37–45.
59. Cheetham N, Tao L. Variation in crystalline type with amylose content in maize starch granules: an X-ray powder diffraction study. Carbohydr Polym. 1998;36:277–84.
60. Kadokawa J. Preparation and applications of amylose supramolecules by means of phosphorylase-catalyzed enzymatic polymerization. Polymers. 2012;4:116–33.
61. Zhang L, Zeng M. Proteins as sources of materials. In: Belgacem MN, Gandini A, editors. Monomers, polymers and composites from renewable resources. Oxford: Elsevier; 2011. p. 479–91.
62. Langel U, Cravatt BF, Graslund A, et al. Introduction to peptides and proteins. Boca Raton: CRC Press; 2009.
63. National Institutes of Health. National Human Genome Research Institute. http://www.genome.gov/glossary/
64. Protein structure – boundless biology. https://www.boundless.com/biology/textbooks/boundless-biology-textbook/biological-macromolecules-3/proteins-56/protein-structure-304-11437/
65. Feughelman M, Lyman D, Menefee E, Willis B. The orientation of the alpha-helices in alpha-keratin fibres. Int J Biol Macromol. 2003;33:149–52.
66. Hu Y, Zhang Q, You R, Wang L, Li M. The relationship between secondary structure and biodegradation behavior of silk fibroin scaffolds. Adv Mater Sci Eng. 2012;2012:1–5.
67. Saha R, Rakshit S, Pal SK. Molecular recognition of a model globular protein apomyoglobin by synthetic receptor cyclodextrin: effect of fluorescence modification of the protein and cavity size of the receptor in the interaction. J Mol Recogn. 2013;26:568–77.
68. Lukin JA, Kontaxis G, Simplaceanu V, Yuan Y, Bax A, Ho C. Quaternary structure of hemoglobin in solution. Proc Natl Acad Sci USA. 2003;100:517–20.
69. Rosillo-Calle F, Pelkmans L, Walter A. A global overview of vegetable oils, with reference to biodiesel. A report for the IEA Bioenergy Task 40; 2009. 89 p.
70. Guner FS, Yagci Y, Erciyes AT. Polymers from triglyceride oils. Prog Polym Sci. 2006;31:633–70.
71. Khot SN, Lascala JJ. Development and application of triglyceride-based polymers and composites. J Appl Polym Sci. 2001;82:703–23.
72. Orsavova J, Misurcova L, Ambrozova JV, Vicha R, Mlcek J. Fatty acids composition of vegetable oils and its contribution to dietary energy intake and dependence of cardiovascular mortality on dietary intake of fatty acids. Int J Mol Sci. 2015;16:12871–90.

73. Adewale P, Dumont MJ, Ngadi M. Recent trends of biodiesel production from animal fat wastes and associated production techniques. Renew Sustain Energy Rev. 2015;45:574–88.
74. Alptekin E, Canakci M, Sanli H. Evaluation of leather industry wastes as a feedstock for biodiesel production. Fuel. 2012;95:214–20.
75. Adewale P, Dumont MJ, Ngadi M. Rheological, thermal and physicochemical characterization of animal fat wastes destined for biodiesel production. Energy Technol. 2014;2:634–42.
76. Panneerselvam SI, Miranda LR. Biodiesel production from mutton tallow. Int J Renew Energy Res. 2011;1:45–9.
77. Chung KH, Kim J, Lee KY. Biodiesel production by transesterification of duck tallow with methanol on alkali catalysts. Biomass Bioenergy. 2009;33:155–8.
78. Arnaud E, Trystram R, Relkin P, Collignan A. Thermal characterization of chicken fat dry fractionation process. J Food Eng. 2006;72:390–7.
79. Canakci M, Van Gerpen J. Biodiesel production from oils and fats with high free fatty acids. Trans ASAE. 2001;44:1429–36.
80. Benz JM, Tokach MD, Dritz SS. Effects of choice white grease and soybean oil on growth performance, carcass characteristics, and carcass fat quality of growing-finishing pigs. J Anim Sci. 2011;89:404–13.
81. Lee KT, Foglia TA, Chang KS. Production of alkyl ester as biodiesel from fractionated lard and restaurant grease. J Am Oil Chem Soc. 2002;79:191–5.
82. Hu Q, Sommerfeld M, Jarvis E, et al. Microalgal triacylglycerols as feedstocks for biofuel production: perspectives and advances. Plant J. 2008;54:621–39.
83. Chisti Y. Biodiesel from microalgae. Biotechnol Adv. 2007;25:294–306.
84. Meng X, Yang J, Xu X, Zhang L, Nie Q, Xian M. Biodiesel production from oleaginous microorganisms. Renew Energy. 2009;34:1–5.
85. Li SL, Lin Q, Li XR, et al. Biodiversity of the oleaginous microorganisms in Tibetan plateau. Braz J Microbiol. 2012;43:627–34.
86. Huber GW, Iborra S, Corma A. Synthesis of transportation fuels from biomass: chemistry, catalysis, and engineering. Chem Rev. 2006;106:4044–98.
87. Huang C, Zong M, Wu H, Liu Q. Microbial oil production from rice straw hydrolysate by *Trichosporon fermentans*. Bioresour Technol. 2009;100:4535–8.
88. Spolaore P, Cassan CJ, Duran E, Isambert A. Commercial applications of microalgae. J Biosci Bioeng. 2006;101:87–96.
89. Rinaudo M. Chitin and chitosan: properties and applications. Prog Polym Sci. 2006;31:603–32.
90. Peter MG. Chitin and chitosan in fungi. In: Peter MR, editor. Biopolymers online. Germany: Wiley; 2005.
91. Hudson SM, Smith C. Polysaccharides: chitin and chitosan: chemistry and technology of their use as structural materials. In: Kaplan DL, editor. Biopolymers from renewable resources. Dordrecht: Springer; 2013.
92. Dutta PK, Dutta J, Tripathi VS. Chitin and chitosan: chemistry, properties and applications. J Sci Ind Res. 2004;63:20–31.
93. Synowiecki J, Al-Khateeb NA. Production, properties, and some new applications of chitin and its derivatives. Crit Rev Food Sci Nutr. 2003;43:145–71.
94. Kaur S, Dhillon GS. Recent trends in biological extraction of chitin from marine shell wastes: a review. Crit Rev Biotechnol. 2015;35:44–61.
95. Tshinyangu KK, Hennebert GL. Protein and chitin nitrogen contents and protein content in *Pleurotus ostreatus* var. *columbinus*. Food Chem. 1996;57:223–7.
96. Nomiyama T, Aihara N, Chitose A, Yamada M, Tojo S. Biomass as local resource. In: Hirasawa T, Tojo S, editors. Research approaches to sustainable biomass systems. Oxford: Academic Press; 2014.
97. US Department of Agriculture. World agricultural production. Circular series WAP 2–16. 2016. http://apps.fas.usda.gov/psdonline/circulars/production.pdf
98. FAO Statistics Division. http://faostat.fao.org/site/567/default.aspx#ancor

Chapter 4
Biofuels and Bioenergy

Abstract The depletion of fossil fuels and the global environmental awareness along, with several economic concerns, are the major driving forces behind the worldwide orientation towards renewable bioresources and agro-industrial wastes for the production of alternative fuels in a sustainable manner. Consequently, the development of more efficient biomass-producing systems and biomass-processing technologies are becoming serious challenges for industrialists and researchers in order to provide markets with eco-friendly fuels at competitive prices and contribute to the reduction of CO_2 emissions.

In this chapter, multiple opportunities to valorize biomass as feedstock for fuel and energy production are highlighted. This includes various woody, herbaceous, agro-industrial, and aquatic bioresources, as well as animals, and microorganisms, rich in cellulose, hemicellulose, starch, chitin, and lipids for the production of bioethanol, biodiesels, and biogas. The conversion of those bioresources to the desired biofuels involves a variety of technologies and processes, which are presented and compared in this chapter, including biomass pretreatments, thermochemical and biological conversion procedures, as well as separation, purification, and upgrading technologies. The need for more R&D breakthroughs enabling the production of biofuels at more competitive prices is also highlighted as a major step to accelerate the shift towards bioeconomy.

4.1 Introduction

Energy is essential to ensure economic production and, consequently, economic growth. To fulfill the requirements of the industrial, transportation, and residential end-use sectors, a heavy resort to fossils fuels characterized the last century or so, with severe repercussions on the environmental, socioeconomic, and geopolitical fronts.

Thus, considering the strategic importance of the energy sector, and in order to face serious challenges related to the rarefaction of the fossil fuels and the urgent need to decrease the amounts of greenhouse gas emissions, scientists from around the world made (and still are making) tremendous contributions to find out sustainable procedures of producing alternative energy sources. Such effort led to a worldwide orientation, or shall we say reorientation, towards natural resources.

M. Sillanpää, C. Ncibi, *A Sustainable Bioeconomy*,
DOI 10.1007/978-3-319-55637-6_4

With its eco-friendly approach, bioenergy and biofuels production provided an opportunity to:

- Address the energy security issues by using locally available bioresources and wastes to produce solid, liquid, and gaseous fuels.
- Gradually lower the emission of greenhouse gases and therefore contribute in the effort to mitigate global warming.
- Induce more geopolitical stability by reducing countries' dependence on foreign energy resources.
- Create new green jobs.

In this chapter, we will present and discuss every opportunity to valorize biomass as a valuable energy source. In clearer terms, we think that we have to move past the food/nonfood dilemma. Why? Well, we could cultivate food crops for energy in marginal land and with marginal waters reclaimed from industrial or municipal wastewaters. The ethical issue is limited to the scenario where nonfood crop, with high energy output, are cultivated in arable lands and irrigatted with fresh waters. Nowadays, with the threatening climate change impacts on agriculture [1], the serious shortages of water in many countries [2], and the recurring starvations and undernourishments in many others [3], it would be common sense (not to say humane) to prioritize feeding the population. Nonetheless, if food is a vital issue for people, energy is a vital one for countries, which are supposed to "take care" of those people (food, education, medication, transportation, jobs, etc.). Thus, between food and fuel, it is not really a dilemma; it is just a matter of priority and common sense. The keyword in this regard is wisdom. We have to be wise in choosing the right feedstock, in managing our lands and water resources, in allocating governmental subsidies, and especially in promoting R&D effort.

As far as the authors of this book are concerned, the priority is very clear: fill the stomachs and then fill the fuel tanks.

Taking all these factors into consideration, in this chapter, every biomass is susceptible to be valorized for bioenergy production that provided an eco-friendly and conflict-free acquisition process whether by growing crops, harvesting naturally occurring resources, or collecting biowastes.

We are going to start by showcasing various feedstocks for each biofuel, then the different steps to produce it:

1. Pretreatment/conversion procedures
2. Separation/purification technologies
3. Upgrading/refining (when applicable) before the end use in the industrial, transportation, residential, or commercial sectors

4.2 Bioethanol

Currently, the transportation sector is still heavily relying on petroleum-based fuels including gasoline, aviation kerosene, and diesel, all liquid fuels. In order to face the coming shortage in fossil resources and provide renewable and eco-friendly alternatives, scientists from around the world and from multidisciplinary backgrounds are searching, developing, and optimizing processes to generate various automotive biofuels [4–7].

Despite the use of other renewable energy sources in the transportation sector (electric, hydrogen, nuclear, and solar powered vehicles and vessels), liquid fuels remain the main source to power the engines. Thus, researchers focused on producing various liquid biofuels from renewable, abundantly available, and low-cost bioresources. The main aim was to develop efficient, large-scale, and economically competitive processes to produce alternative biofuels.

Among the various liquid biofuels, we will start by discussing the production of bioethanol, based on the rapid expansion of its industrial production and end uses in transportation.

4.2.1 Bioethanol Feedstocks

Sugars are fermented into alcohol. This is basically the main mission of the ethanol industry. Hence, since all sugar-containing raw and residual bioresources (mono, di, or polysaccharides) are suitable to be converted into bioethanol, then all plants could be considered as possible feedstock since cellulose is the basic structural component of their cell walls.

In nature, there are four main kinds of sugar-loaded bioresources: (1) *simple sugars-based biomass* such as sugarcane and sugar beet, (2) *starch-based biomass* such as corn and cassava, (3) *lignocellulose-based biomass* which includes woody and herbaceous plants, and (4) some marine species.

As well, the marine world comprises valuable feedstock for bioethanol production, especially the alginate and chitin-based biomass such as macroalgae and crustaceans, respectively. Noting that simple sugars and starch constitute only a small fraction of the plant matter, as the main polysaccharidic compounds of most plants are cellulose, and to a lesser extent, hemicellulose.

Based on the climatic conditions and the expertise in agricultural practices, the bioethanol production industries, although trying to diversify it, tend to rely on specific crops. For instance, in tropical areas such as India, Brazil, and the Caribbean, sugarcane is the first choice. In other parts of the world, including the United States, Europe, and China, grains such as corn and wheat are the main feedstock [8]. According to some estimates, bioethanol production from starch-based crops accounts for about 60% and from sugar-based crops (mainly sugarcane and sugar beet) for nearly 40% [9].

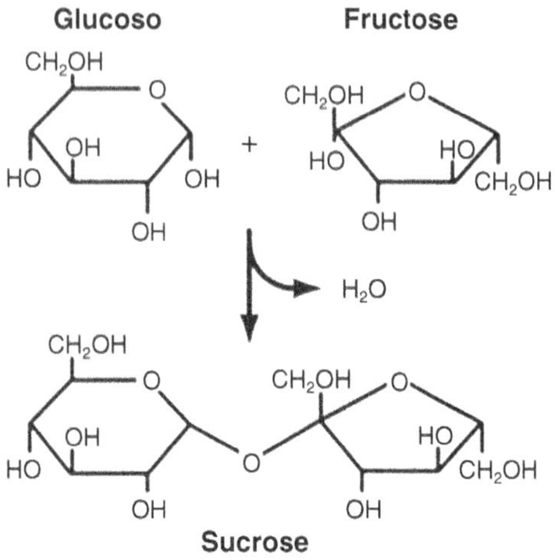

Fig. 4.1 Glucose, fructose, and the result of their combination, sucrose

4.2.1.1 Sugar-Containing Biomass

Certain plants such as sugarcane, sugar beet, and sweet sorghum can be the direct source of fermentable sugars, mainly fructose, glucose, and sucrose (Fig. 4.1).

Their main assets are high production of simple sugars per cultivated surface and production yield, in addition to a straightforward fermentation procedure and therefore low conversion costs. Nonetheless, their seasonal availability tends to limit their use as economically viable feedstock. Later, when we will discuss the production of bioethanol from wastes and residues, we will see how sugarcane or many other sugar-based feedstocks could have their by-products and residues further converted into valuable biofuels and fine chemicals, thus increasing the overall profitability.

Table 4.1 gives the sugar content of selected simple sugar-containing bioresources.

Sugarcane (*Saccharum officinarum*)

It is a tropical, perennial grass (C4 carbohydrate metabolism belonging to the Poaceae family that includes also sorghum and maize). The currently grown sugarcane cultivars are *Saccharum officinarum hybrids*, a domesticated sugar-producing and strong-growing species. Those sugarcane hybrids were selected over the years for enhanced biomass and sugar yields, improved resistance to pests and drought, and better adaptation to intensive cultivation practices. Under favorable conditions, sugarcane has the potential to yield up to 100 tons of dry biomass per hectare annually [11].

The stems grow into cane stalk of 5 m in height. Composition wise, a mature stalk of sugarcane contains on average 70% water, 16% sugars, and 14% fibers,

Table 4.1 Sugar composition and content in various biomasses (% weight on a dry basis) [8, 10]

Biomass	Glucose	Fructose	Sucrose
Sugarcane	0.2–1.0	0.2–1.0	11–16
Sugar beet	0.1–0.5	0.1–0.5	16–17
Sweet sorghum	0.6–1.6	0.3–1.0	9.6–17.6

along with proteins, ash, and extractives. After harvesting, the sugarcane stalks are washed, cut into pieces by rotating knives and pressed to produce a juice loaded with sugars, mostly sucrose. This juice is dehydrated and refined to produce sugar and could also be directly fermented to produce bioethanol. The big question here is not could we produce biofuel from sugarcane but how much could we use without compromising the necessary requirement of the food industry. To answer this question, locally centered economic and environmental assessments have to be planned and discussed in order to develop sustainable strategies to valorize sugarcane, or any valuable feedstock, efficiently.

The process of extracting sugar from sugarcane also generates residues including bagasse, molasses, vinasse, and filter cakes that could be further valorized into biofuels [12], biomaterials [13], and organic fertilizers [14].

Sugar Beet (*Beta vulgaris*)

It is a biennial plant, belonging to the Chenopodiaceae family, grown in more than 50 countries in Europe, Asia, North Africa, and the Americas [15]. It is mainly cultivated for its conical and fleshy white taproot rich in sucrose. The average yield is 50 tons/ha. Nonetheless, depending on climatic conditions and cultivation practices, specific yields can fluctuate between 30 and 70 tons/ha. In the same context, research studies revealed in addition to the increase in biomass yields, the concentration of sucrose in the sugar beet roots could also be increased using selective cultivars and/or intensive agricultural practices [16].

The industrial processing of beets to produce sugar generates interesting residues including pulp (suitable for animal feed), press cake (organic fertilizer), and molasses containing a residual fraction of sugars that could be fermented into alcohol. A typical sugar beet root consists of 76% water, 18% sugar, and 5.5% pulp. From the total sugar content, roughly 83% of crystalline sucrose (*aka.* table sugar) is recovered along with 12.5% of molasses [17].

Among the sugar-based bioresources, sugar beet is the feedstock with the highest bioethanol production potential. Indeed, in France, sugar beets produce 7000 L of ethanol per hectare compared with the 5000 L/ha produced from sugarcane in Brazil [18].

Sweet Sorghum (*Sorghum bicolor* var. *saccharatum*)

It is a drought-tolerant perennial crop cultivated for the fermentable sugars located in the main stalk and also for its vegetative biomass. A mature sweet sorghum stalk is composed of pith (the spongy core) and bark (external layer), with the former accounting for 65% of the fresh biomass matter. Regarding the sucrose distribution in the stem, the pith and bark are composed of 67.4 and 32.2% of dry weight, respectively [19]. These sugars are extracted via several crushing or squeezing techniques to produce the stalk juice (similar to the sugarcane processing).

One of the main advantages of using sweet sorghum production to produce bioethanol is that its juice provides an aqueous source of simple sugars (mono and disaccharides) that could be directly fermented to alcohol (in some cases, a pH adjustment for optimum bacterial activity). In this context, a scientific investigation showed that by simply adding yeast to freshly pressed juice, the fermentation was complete in 24 h with 90% of sugar content converted into bioethanol [20]. Nonetheless, to benefit from this straightforward availability, the sugar-loaded juice has to be processed quickly. Indeed, considering the high sugar content in the juice, spontaneous fermentation by naturally occurring microorganisms could consume part of the sugars or generate compounds that will inhibit the yeasts with highest conversion capabilities.

In terms of production yields, sweet sorghum, with its approximate 50–85 tons/ha of stalks, produces a juice yield of 39–42 tons/ha. After fermentation of the juice (84% water and 14.2% sugars including sucrose, glucose, and fructose), one hectare of sweet sorghum could produce between 3450 and 4132 L of bioethanol [21].

4.2.1.2 Starch-Containing Biomass

As we have seen in the previous chapter (Sect. 3.4.4), starch is a polysaccharidic molecule made of glucose units (from five hundred to several hundred thousand units) that could be hydrolyzed (thermo-chemically or enzymatically) into simple sugars, *aka*. Saccharification. The sugar content could therefore be fermented in the same way as those from sugar-based bioresources. Thus, right off the bat, the bioethanol production process from starch-based biomass is more complex compared to sugar-based one, as it requires the extra step of saccharification to depolymerize the amylose and amylopectin structures and release the fermentable sugars.

In a plant, starch is synthesized and stored in the cells as granules that differ in size and shape. Overall, all starch granules have a very dense and semi-crystalline structure insoluble in cold water (i.e., temperatures below gelatinization [22]. They act as a carbon reserve for various periods of time starting with one day in leaves up to many years in dormant seeds. Within those granules, the amylopectin content is estimated at 75% and that of amylose at 25% [23].

Table 4.2 Starch content in various biomass (% weight on a dry basis) [24, 25]

Storage organ	Plant	Starch content (%)
Grains	Corn	60–68
	Wheat	60–65
	Barely	55–65
	Sorghum	60–65
	Oats	50–53
	Rye	60–65
	Rice	70–72
Root tubers	Cassava	25–30
	Yam	8–18
	Sweet potato	10–29
Stem tubers	Potato	22–24

Several starch-containing sources could be identified such as cereal grains, tubers, and also some nuts and legumes. Table 4.2 provides a general guide to the content of starch in selected plants.

It has to be noted that the starch content of grains and tubers can fluctuate considerably within the same species, according to the maturity level, genetic traits, and environmental factors during the cultivation and the subsequent storage phase.

In the following section, we will briefly describe the most used grains in the bioethanol industry, namely, corn and wheat. Also, we will consider a representative from the tubers' group, Cassava.

Corn (*Zea mays*)

Also called maize, corn is a grain plant from the Poaceae family which also includes wheat, rice, and barley. The corn plant has a simple stem reaching at maturity 2–3 m in height. The stem contains nodes and internodes from which develops pairs of large leaves (8–21 per plant) [26].

One of its main assets is that corn can be cultivated over a wide range of agro-climatic zones. Indeed, it can be cultivated from 58°N to 40°S, from below sea level to altitudes higher than 3000 m, and in areas with 250 mm to more than 5000 mm of rainfall per year [27, 28]. Nonetheless, the major corn production areas are located in temperate regions, and the United States, China, Brazil, and Mexico account for 70% of the worldwide corn production.

Corn constitutes a staple food in many countries. As well, it could be cultivated for other industrial purposes, based on its starch content, such as the bioethanol sector, but also to produce other products such as plastics, fabrics, and adhesives.

The corn-to-ethanol industry is the most mature industry in this sector, benefiting from the age-long expertise in corn cultivation and brewing. In the United States, the largest producer of bioethanol in the world with an annual production of 14,370 gallons in 2014 ($\approx$ 54.4 billion liters) [29], corn is the primary feedstock. Indeed according to the US Energy Information Administration "*nearly all ethanol produced in the United States is derived from corn.*" To secure this production, between 35 and 40% of the US corn production was used to produce bioethanol and other derived products.

Later in this chapter, we will discuss the conversion of corn into bioethanol starting with the basic milling processes (dry and wet), the advanced saccharification and fermentation procedures, and the latest R&D in the field.

Wheat (*Triticum* spp.)

It is a cereal grain belonging to the Poaceae family (also called Gramineae). Although the global production of wheat is lower than corn, it has a wider distribution throughout the temperate zone from Scandinavia to Argentina and at higher elevations in the tropics [30]. Currently, largest wheat producers in the world are China, India, the United States, Russia, and France.

In order to properly assess the predisposition of wheat as a feedstock for the energy sector, we have to take these facts into serious consideration:

- Wheat is the most widely grown cereal grain, occupying 17% of the total cultivated land in the world.
- Wheat is the staple food for 35% of the world's population.
- Wheat provides more calories and protein in the world's diet than any other crop [31].

Overall, the rich biochemical composition of wheat makes it a valuable source for human nutrition and also bioethanol production. Indeed, wheat grains comprise three main components: carbohydrates (59.4%), protein (11.3%), and dietary fibers (13.2%) [32]. Other studies reported that the starch content in wheat could vary between 60 and 75% of the total dry weight of the grain [33].

In Europe, for instance, wheat is the main crop grown for bioethanol production, roughly accounting for 0.7% of EU agricultural land and 2% of the continent's grain supply [34]. The process of bioethanol production from wheat grains is based on its predominant starch content in the grains (60 to 65% of the kernel), and, therefore, like all grainy feedstock, it could be summarized in the following steps:

- Milling (dry or wet)
- Liquefaction/saccharification (chemical or enzymatic)
- Fermentation (Fungi or bacteria)
- Distillation and dehydration

To give a comparative estimate on the production yields between sugar and starch-based biomass, one report quantified that one ton of wheat grains could produce 336 L of ethanol (roughly as much as corn). As for the sugar-based feedstock, the yields were 70–90 L/t for both sugarcane and sugar beet.

Nonetheless, based on the huge difference in biomass production yields (8 tons/ha for wheat, compared to 53 and 70 tons/ha for sugar beet and sugarcane, respectively), the annual bioethanol yield per cultivated hectare is the highest for sugarcane (4900–6600 tons/ha/year), followed by sugar beet (3700–5000 tons/ha/year) and wheat (2700 tons/ha/year) [35]. Further details about the grain-to-ethanol process will be provided in the next section.

Cassava (*Manihot esculenta*)

It is a shrubby perennial plant belonging to the Euphorbiaceae family. It is a staple crop in many tropical countries, especially with little rainfall. Indeed, framers in those countries cultivate cassava from its nutritional starch content and its tolerance to prolonged drought in the tropics [36]. Typically, a cassava root is composed of 70% moisture, 24% starch, and other substances including minerals (3%), fibers (2%), and protein (1%) [25].

Based on the starch content, estimated at 25% in fresh roots [37], its high rate of CO_2 fixation, high water-use efficiency, and superior starch conversion ratio into ethanol compared to other crops [38], cassava is a suitable feedstock for bioethanol production.

This was the case in Brazil and many Asian countries, especially China and Thailand [39]. The production of bioethanol from cassava in sub-Saharan Africa is also under investigations [40, 41]. In India, it was reported that 60% of the produced cassava was for industrial purpose, 30% for human consumption, and 10% for animal feed [42].

In practice, fresh cassava roots pose serious storage challenges. Indeed, the root starts to spoil just after 48 h of harvest due to physiological changes and microbial activity [43], hence the resource to the drying process to lower the moisture content at around 14%. The produced material is called cassava chips. Some estimates reported that it takes 2–2.5 kg of fresh roots to produce 1 kg of chips [44].

Therefore, the cassava-to-ethanol industry is based on two main feedstocks: fresh roots and dried chips. The production process varies in the beginning, i.e., how to recuperate the starch content. Then, the starch to ethanol procedures is basically the same with the other starch-based feedstocks.

Definitely, the conversion ratio of any feedstock to simple sugars and then to ethanol is a very important factor to be taken into consideration, along with the biomass productivity of that feedstock. For example, sugarcane annual biomass production yield is higher than cassava. But, when it comes to conversion ratios, cassava is converted into sugars and ethanol than sugarcane better (roughly the double as shown in Table 4.3).

Table 4.3 gives a comparative perspective relating both the feedstock conversion ratios (to sugar and ethanol) on the one hand and the related biomass production yields on the other.

Table 4.3 Potentialities of various bioresources as feedstock for bioethanol production [45]

Biomass	Annual yield (ton/ha)	Conversion rate to sugar or starch (%)	Conversion rate to ethanol (L/ton)	Annual ethanol yield (kg/ha)
Sugarcane	70	12.5	70	4900
Cassava	40	25	150	6000
Sweet sorghum	35	14	80	2800
Corn	5	69	410	2050
Wheat	4	66	390	1560

4.2.1.3 Lignocellulosic Biomass

Avoiding the competition with food crops and/or valorizing local bioresources where the major driving forces behind the worldwide orientation towards producing bioethanol from lignocellulosic materials, loaded with polysaccharidic polymers (cellulose and hemicellulose). The product is now commonly known as second-generation bioethanol.

In general, lignocellulosic feedstocks could be grouped into three main groups: woody and herbaceous bioresources and agro-industrial residues. The biochemical composition of several bioresources in terms of cellulose, hemicellulose, and lignin contents are already given in the previous chapter (Table 3.1).

Worldwide, mastering and optimizing the production of biofuels from lignocellulosic bioresources was (and still is) the concern of many researchers and inventors. Indeed, although renewable and highly available, lignocellulosic biomass poses many technological challenges related to the removal of the complex and recalcitrant lignin [46] and the depolymerization of the rigid polysaccharides to recover fermentable sugars [47].

Woody Feedstock

Woods could be grouped into two main categories: softwood and hardwood. Softwoods (Coniferous with needle-like leaves) are the dominant source of lignocellulosic biomass in the northern hemisphere including the United States, Europe, Canada, Russia, and Japan. It comprises species such as pine, spruce, and cedar trees. Hardwoods, on the other hand, are angiosperms (seed-producing and flowering plants) with broad leaves including wood species such as poplar, willow, Eucalyptus, locust, and oak trees.

Some reports estimate that hardwood is a better feedstock for ethanol production than softwood due to more suitable biochemical composition [48]. First, because generally hardwood has significantly less lignin than softwood. Second, the hemicellulose content of hardwood is mainly composed of hexoses (C6 sugars), while that of softwood is mainly composed of pentoses (C5 sugars). Such difference in biochemical composition benefits hardwood as hexoses (e.g., glucose), are much

Table 4.4 Proportion of the main components in various parts of Scots pine [50]

Scots pine	Cellulose (%)	Hemicellulose (%)	Lignin (%)
Stem wood	43.9	28.9	20.2
Bark	10.7	11.2	14.7
Branches	33.3	23.4	20.8
Leaves	–	–	11.1
Stump	29.5	19.4	13.4
Roots	26.0	17.1	27.1

easier to be fermented to ethanol than pentoses (e.g., xylose) [49], using the current technologies.

Many wood species were proven to be valuable feedstocks for bioethanol production. The following shortlist illustrates the potentialities of 2 softwoods and 2 hardwoods, based on related research studies.

Pine Trees (*Pinus* sp.) belonging to the Pinaceae family are trees ubiquitous to most of the Northern Hemisphere. The components of various pine tree species were investigated for bioethanol production based on their polysaccharidic content. Table 4.4 gives an overview about the biochemical composition of the various parts of Scots pine (Pinus sylvestris) which is a useful tool to apply the adequate conversion procedures based on the proportion of those chemical compounds.

In the United States, loblolly pine (*Pinus taeda*) was tested for bioethanol production [51]. The first step of transforming the wood to fermentable sugars resulted in conversion ratios between 75 and 95% of the carbohydrate content. After fermentation and distillation, the produced bioethanol amount was, respectively, estimated between 302 and 386 L/ton of dry biomass. The same study predicted the cost of producing bioethanol from loblolly pine wood. Based on the initial conversion ratios (75 and 95%), the prize was estimated at 0.40 and 0.34 US $/L, respectively.

Lodgepole pine (*Pinus contorta*) was also tested to recover fermentable sugars from its polysaccharidic content and ferment them into bioethanol using *Saccharomyces cerevisiae* yeast [52]. Under optimum operating conditions (35 °C and pH 5.5), a maximum bioethanol concentration of 47.4 g/L was reported, corresponding to a theoretical yield of 285 L/ton of Lodgepole pine wood.

Cedar Trees (*Cedrus* spp.) are coniferous softwoods. Like pine trees, they belong to the Pinaceae family and are mostly located in the Northern hemisphere. In Japan, for instance, 66% of the total land area is forest with the local cedar specie *Cryptomeria japonica* (*aka.* Japanese cedar) accounting for nearly 21% [53]. The same team reported that the biomass produced from forest thinning, corresponding to roughly 20 million m^3 per year, was kept in forests as logging residues with valorization.

Thus, several research teams focused on this matter and investigated the use of cedar trees as a potential feedstock for bioethanol production. One of those studies tried to enhance the bioethanol production yield from Japanese cedar by increasing the recovery of fermentable sugars through various pretreatment techniques [54]. The related results showed that the maximum amounts of recovered glucose and reducing sugars (408 and 462 mg/g of initial dry matter, respectively) were attainted using consecutive steam explosion and ionic liquid pretreatments.

Poplar Trees (*Populus* sp.) are natives to the Northern hemisphere, belonging to the Salicaceae family which comprises thirty to forty species [55]. The possibility of applying intensive growing practices on hybrid poplar tress resulted in a substantial increase in productivity up to nearly 24.5 ton/ha annually [56]. Populus species are believed to be the most plentiful fast-growing trees suitable for bioethanol production [57]. On the environmental preservation front, those species make a valuable contribution, considering their rapid growth, in carbon sequestration and soil phytoremediation.

Several procedures were applied to produce bioethanol for poplar species. Like any lignocellulosic resource, pretreatment of poplar wood prior to hydrolysis and fermentation was revealed to be a decisive factor to optimize bioethanol production yields. In this context, the application of various pretreatment techniques on four poplar species resulted in conversion ratios to simple sugars between 47 and 55%. The subsequent fermentation produced different bioethanol yields, the highest (0.17–0.20 L/kg) was attainted for aspen wood [58].

Eucalyptus Trees (*Eucalyptus* sp.), belonging to the Myrtaceae family, are native to Australia. In addition to their fast growth, the wood from Eucalyptus is characterized by a high cellulose content attaining around 45% on a dry matter basis [59]. 15% of the total cellulose content in an Eucalyptus tree is located in its bark (outermost layers of the trunk) [60]. In Australia, a research study proposed eucalyptus (*Eucalyptus dunnii*) forest thinnings as a local feedstock to produce bioethanol [61]. Under optimum experimental conditions, a conversion ratio to monomeric sugars of 74% was attained. After fermentation using *Saccharomyces cerevisiae*, the ethanol production yield was 18 g/L of hydrolysate, corresponding to a sugar to ethanol conversion ratio of 92%.

In Spain, where Eucalyptus is present in an area approximating half a million hectares, many studies were conducted to valorize this woody biomass in the bioenergy sector. For instance, the wood of *Eucalyptus globulus* was tested for bioethanol production [62]. The aim of the study was to solubilize the hemicellulose content and produce treated feedstock with an increased cellulose content and therefore easier enzyme digestibility. With this procedure, the research team reported the production of 291 L of bioethanol per metric ton of oven-dry Eucalyptus wood.

Herbaceous Feedstock

Lignocellulosic herbaceous plants are among the high-yielding biomass in the world [8], which made them suitable feedstocks for biofuel production in general and bioethanol in particular. Commonly termed as energy crops, lignocellulosic herbaceous bioresources include a wide range of woody, biennial or perennial, rhizomatous grasses such as Miscanthus, Switchgrass, Reed canary grass, Bermudagrass, and Alfalafa.

Miscanthus (*Miscanthus* sp.) is also a high-yielding perennial grass, belonging to the Poaceae family. It is native to Africa and South Asia and was introduced into Europe as an ornamental garden grass. Based on their high biomass yields (*approx.* as much as Switchgrass), low cultivation inputs, and substantial polysaccharidic content, several Miscanthus species attracted the attention of scientists, farmers, and industrials to be tested, cultivated, and converted into bioethanol. In Europe for instance, 15 different genotypes of Miscanthus were cultivated with biomass production yields varying between 38 and 41 ton/ha for *Miscanthus x giganteus* and other Miscanthus hybrids, respectively [63]. As for the biochemical composition, the holocellulose content (cellulose + hemicellulose) ranges typically between 76.2 and 82.7%, with a lignin content of 9.2 to 12.6% [64]. Those variations are mainly related to genetic factors, soil and irrigation conditions, and also cultivations practices.

A research team in Korea assessed the suitability of 12 Miscanthus genotypes (8 lines of *Miscanthus sinensis*, 1 line of *Miscanthus x giganteus*, and 3 lines of *Miscanthus sacchariflorus*) to be converted into bioethanol [65]. Based on the findings, two genotypes of *Miscanthus sacchariflorus* were found to be most suitable for biofuel production, based on high polysaccharide content (~65%) and low lignin content (<20%).

In the other side of the world, several comparative studies conducted in the United States tried to assess the potentialities of Miscanthus to produce ethanol, compared with other feedstocks. One of those studies made a comparison between Miscanthus, the challenger, and corn, the energy crop already in use [66]. The investigation proved that, in order to produce enough ethanol to secure 20% of US gasoline consumption, the corn-to-ethanol option would divert 25% of US cropland currently in production. On the other hand, to secure the same demand for gasoline, the Miscanthus-to-ethanol strategy would divert only 9.3% of the land. Years later, the same team reported that if their projections can be proven in the market place, Miscanthus could be the solution to the meet the US objective of replacing 30% of the gasoline currently in use with biofuels by 2030 [67].

Switchgrass (*Panicum virgatum*) is warm-season, high-yielding perennial grass belonging to the Poaceae family. At maturity, Switchgrass could reach more than 3 m of height and have a root depth of more than 3.5 m. It has the aptitude to grow in various types of soils including marginal lands with limited requirement for fertilizers. Based on the pedo-climatic conditions, the annual biomass production yields oscillate between 15 and 37 dry tons per hectare. Switchgrass could be exploited in various ways, mainly as fodder but also in preserving soils from erosion. A growing interest to use this grass as feedstock for potential production biofuel is taking place in the United States and Europe [68]. Regarding the conversion of Switchgrass into bioethanol, many studies around the world proved that it is a valuable feedstock. In this context, several research investigations estimated the bioethanol production from switchgrass. The first projections estimated a production yield fluctuating between 1711 and 2000 L of ethanol/ha [69, 70]. Another estimate, based on a more productive farm, reported the higher yield of 3000 L of ethanol/ha [71].

Alfalfa (*Sativa medicago*) is a perennial flowering plant belonging to the Fabaceae family. It can be grown under a wide range of climates providing an average daily temperature during the growing period above 5 °C. Thus, alfalfa is the one of the most cultivated forage crops in the world, especially in North America and Europe.

In addition to its adaptation faculties and low cultivation inputs (no need for nitrogen amendments), alfalfa has a very interesting biochemical composition, considering the high protein content in the leaves (28.5%, dry matter basis) and high polysaccharide content in the stalks (48.5% cellulose and 6.5% hemicellulose) [72].

In order to benefit from those characteristics, alfalfa was proven to be a suitable feedstock to both produce protein from its leaves (primary product) for animal feed and bioethanol (coproduct) from its stalks [73]. A study reported the production yields of 232 to 278 L/ton of alfalfa [74].

Overall, researchers involved in this integrated valorization strategy (feed and fuel) are promoting it as essential to increase the profitability of the production and conversion systems using alfalfa [75], or any other food or feed crop for that matter. Such orientation generated a worldwide R&D effort to further valorize by-products and residues for bioethanol production, along with other biofuels, biomaterials, and biochemicals. The following section presents a selection of those agro-industrial residues.

Agro-Industrial Feedstocks

The industrial processing of many crops to produce various commodities (grains from cereals, sugar from sugarcane and sugar beet, oil from oilseeds, jam from fruits, and so on) generates a substantial amount of residues and wastes, still containing valuable components such as sugars, starch, and holocellulose. The biochemical composition of some of those "residual" bioresources is given in Table 3.1.

Sugarcane Bagasse like we have seen for alfalfa where the protein-containing leaves are used as animal feed and the cellulose-containing stalks for ethanol production, sugarcane allows the same dual valorization. Indeed, once processed, the sugary juice is processed and refined into table sugars, leaving a significant amount of solid waste, sugarcane bagasse, and also liquid waste, vinasse (viscous to be accurate). In Brazil, the largest producer of sugarcane in the world, 1 ton of this crop generates 280 kg of bagasse [76] and 700 to 900 L of vinasse [77].

In the same country, and based on their national ethanol program "Proalcool," roughly half of the sugarcane yield is used to produce sugar (38.7 million tons) and the other half to produce bioethanol fuel (~ 27 billion liters) [78].

Considering its polysaccharidic content (cellulose and hemicellulose) exceeding 70% of the dry matter [79] and its large worldwide production yield estimated at 1900 million tons [80], sugarcane bagasse is one of the most suitable feedstock for bioethanol production in the world.

Research wise, several experiments were conducted to maximize the conversion ratio of bagasse into bioethanol by optimizing the pretreatment, hydrolysis, and fermentation procedures. From those studies, various volumetric productivity of bioethanol from sugarcane bagasse were reported such as 0.25 and 0.1 g of ethanol per liter of hydrolyzate per hour in one study [81] and 1.22 g/L/h in another one, corresponding to 95% sugar conversion efficiency and a bioethanol yield of 96% of its theoretical value [82].

Corn Stover is the residual biomass from corn farms left in the field after the harvest of kernels or the industrial processing to harvest the grains. It includes stalks, leaves, and husks cobs. On average, corn stover accounts for 3 to 4.5 dry tons per acre of corn fields (*equiv. 7.4 to 11.1 tons/ha*). Thus, based on its biomass production yields and a polysaccharidic content of 58.3% [83], corn stover is considered to be one of the most suitable feedstock for the second generation biofuels.

In the United States, the largest producer of corn in the world, annual corn stover yields are estimated between 80 and 100 million dry tons of per year [84]. The same study estimated that if, on the long term, the demand for corn stover for nonenergy applications would be around 20 million dry tons per year. Thus, 60 to 80 million dry ton per year of this by-product could be used as energy feedstock. Assuming that 40% of the available corn stovers are valorized in producing bioethanol, a prospective production yield approximating 11 billion liters per year could then be achieved.

In another study conducted in China, corn stovers were subjected to chemical treatment and resulting in the removal of about 90% of lignin and 80% of hemicellulose. The remaining cellulose fraction was enzymatically hydrolyzed and more than 90% of its content was converted to fermentable sugars. The resulting production yield was 1kg of bioethanol from 6.2 kg of raw corn stover [85].

Wheat Straw is the residual biomass left in the field after the harvesting campaign of wheat grain. It basically includes rest of the plant, consisting of the stems, leaves, chaff, and the roots. Under intensive cultivation practices, 1 to 3 tons per acre of wheat straw could be generated, which in the United States would give an annual production of around 82 million dry tons [86]

A detailed analysis of the carbohydrates in wheat straw was conducted by the National Renewable Energy Laboratory in Colorado [87]. The results revealed the presence, on a dry matter basis, of 30.2% cellulose and 22.3% hemicellulose, with 18.7% xylan, 2.8% arabinan, and 0.8% galactan, which makes this wheat farming residue a suitable feedstock for bioethanol production.

Many research studies worked on converting wheat straw into bioethanol and others on R&R optimizing the conversion ratios and production yields while using integrated green process. In one of those studies, wheat straws were subjected to dilute acid pretreatment followed by enzymatic hydrolysis of the polysaccharides into simple sugars. The highest yield of monomeric sugars from wheat straw was about 565 mg/g. After fermentation, the registered bioethanol production yield was between 13 and 17 g per liter of hydrolysate [88].

In another investigation, and after the successive pretreatment and hydrolysis of wheat straw, the generated hydrolysate was co-fermented (both pentose and hexose) using *Saccharomyces cerevisiae* and *Fusarium oxysporum*. The outcome was a bioethanol production yield of 58 to 62 g/L [89].

Rice Straw is a residue generated from rice harvesting. It includes stems, leaves, blades, leaf sheaths, and the remains of the panicle after threshing [90]. The global rice production is estimated around 741 million tons per year, 203 million tons (27.4%) in China alone, the world leading rice producer [91]. Each kilogram of produced rice grain was assessed to be associated with an average production of 1.325 kg of straw [92]. Thus, an annual production yield of 981 million tons per year of rice straw could be projected.

As for the biochemical composition, it varies with respect to soil quality and water supply during the growth as well as the implemented agricultural practices. On average, rice straw contains 33–47% of cellulose, 19–27% of hemicellulose, and 5–24% of lignin [93].

Several studies investigated the potentialities of producing bioethanol from rice straw. It was unanimously promoted as a suitable feedstock, providing an efficient pretreatment procedure, as it is the case for any lignocellulosic biomass. Among the numerous research efforts, one interesting study detailed the mass balance for ethanol production from rice straw [94]. It was reported that the pretreatment process caused a 32% loss in mass and the recovery of 88.3% of polysaccharides. The high-solid enzymatic hydrolysis of pretreated rice straw allowed the conversion of 76% of the polysaccharidic content, corresponding a production yield of fermentable sugars of 510 mg/g. Fermentation of the sugar-loaded hydrolysate under optimum conditions was estimated to generate 220 L of ethanol from a ton of raw rice straw.

4.2.1.4 Marine Biomass

Marine biomass refers to various kinds of aquatic organisms, able, in most cases, to synthesize chlorophyll, including single cell and multicells *algae* and the more evolved flowering plants, *seagrassses*.

Based on the high area productivity of marine biomass, their rich content in polysaccharides, ability to capture CO_2, and the fact that they do not compete with conventional crops over land and fresh water [94, 95], bioethanol production from those marine bioresources constitutes a promising alternative not only to fossil fuels but also to some biofuels constrained by technical or ethical issues or their low competitive potential.

Algae/Seaweeds

Algae, aka. seaweeds, could be grouped into two main classes:

- Microalgae, which are microscopic unicellular organisms ranging in size from few micrometers to several hundred micrometers. It includes diatoms, green algae (Class Chlorophyceae), golden brown algae (Class Chrysophyceae), prymnesiophytes (Class Prymnesiophyceae), and the eustigmatophytes (Class Eustigmatophyceae) [96].
- Macroalgae, *aka.* seaweeds, are macroscopic multicellular algae living near the seabed (benthic specie). It includes green algae (Chlorophyta) such as *Caulerpa* and *Ulva* species, brown algae (Heterokontophyta) such as *Sargassum* species and large kelps, and red algae (Rhodophyta), which is the most diverse group of all [97]. In some classifications, blue-green algae, or cyanobacteria, (Class Cyanophyceae) were also considered as algae (photosynthesizing bacterial specie in others).

Basically, the production of bioethanol from algae is founded on converting the carbohydrates content to fermentable sugars and then to ethanol. The nature of those carbohydrates varies from an algal group to another. In general, the following compounds could be found:

- Microalgae → cellulose, starch, rhamnose, and mannose
- Macroalgae →
  - Green algae: cellulose, starch
  - Brown algae: alginate, cellulose, laminarin, fucoidan, and mannitol
  - Red algae: Carrageen, agar, and cellulose [98].

Many worldwide research studies were performed to investigate the potentialities of many algae as feedstock for bioethanol production. In one of those studies, several algae, namely *Ulva lactuca*, *Gelidium amansii*, *Laminaria japonica*, and *Sargassum fulvellum* were tested [99]. The marine biomass was first subjected to mild acid treatment, followed by enzymatic hydrolysis. Afterwards, the sugars in the hydrolysates (mainly glucose, mannose, galactose, and mannitol) were fermented using *Escherichia coli* KO11. For the case of *L. japonica* for example, the bacterial strain was able to convert both mannitol (30%) and glucose (7%), and the result was a bioethanol production yield of 0.4 g per g of substrate.

In another research work, the brown macroalga, *Saccharina japonica*, was used as substrate to produce bioethanol [100]. After a low acid pretreatment process, a glucan content of 29% and an enzymatic digestibility of 84% were achieved. The production yield obtained after saccharification and fermentation was 6.65 g per liter of hydrolysate.

Table 4.5 gives an account on the bioethanol production potentialities of selected microalgae, along with their carbohydrate content.

Table 4.5 Carbohydrates content in selected micro- and macroalgae [101–105]

	Algal species		Carbohydrates content (%)	Ethanol yield (g/g of substrate)
Microalgae	*Chlorococcum humicola*		32.5	0.52
	Chlorella vulgaris		47.5	0.40
	Chlamydomonas reinhardtii		60	0.29
	Chlorococcum infusionum		43.9	0.26
	Scenedesmus obliquus		51.8	0.21
	Dunaliella tertiolecta[a]		51.9	0.14
Macroalgae	Green algae	*Ulva lactuca*	54.3	0.40
		Ulva fasciata	53.3	0.08
		Ulva pertusa	65.2	0.12
	Brown algae	*Sargassum* spp.	0.3	0.13
		Laminaria japonica	54.4	0.28
	Red algae	*Gelidium amansii*	75.2	0.39
		Eucheuma cottonii	50.5	0.42
		Gracilaria verrucosa	74.1	0.15

[a]Lipid-extracted microalgae

Seagrasses

Not be confused with macroalgae, seagrasses are fully evolved marine plants having roots and vascular tissues and producing flowers and pollen. Seagrass meadows constitute a highly productive ecosystem and are vitally important in the coastal environment as source of food and shelter for various marine organisms including large fish, crabs, turtles, and marine mammals [106]. They also contribute in water oxygenation and sediment accumulation and stabilization.

Numerically, the studies on the use of algae for biofuels production are more abundant than the ones on seagrasses, probably due to the easier availability of algae or their predisposition to aquaculture [107]. Nonetheless, many seagrass species possess valuable carbohydrates content. For instance, Mediterranean seagrass *Posidonia oceanica* have an interesting holocellulosic content (i.e., cellulose + hemicellulose) in both its leaves (57.1%) and leaf sheaths, aka. spheroids or aegagropila (59%, on a dry matter basis) [108, 109].

One research study investigated the potentialities of *P. oceanica* biomass to produce bioethanol [110]. Acid and enzymatic hydrolyses were applied on the marine feedstock in order to optimize the ethanol conversion efficiency. The results showed that the separate use of acid and enzymatic hydrolyses gave the maximum reducing sugar concentrations of 27.6 g and 28.4 g/100 g of dry biomass, respectively. On the other hand, the consecutive enzymatic and acid hydrolysis enhanced the saccharification yield to 52.3 g of reducing sugar/100 g of dry biomass. The reported bioethanol production yield was 44.1%, corresponding to a productivity of 0.46 kg/m^3/h.

Zostera marina is another seagrass with interesting biochemical composition (27.4% cellulose, 16.9% hemicellulose, and 22.4% lignin). The availability of this marine biomass in Turkish shores led a local research team to study the possible utilization of *Z. marina* as feedstock for bioethanol production. Under optimized hydrolysis conditions, 58.2 g of fermentable sugars were extracted from 100 g of dry biomass. The subsequent fermentation process produced a bioethanol yield of 8.72% (v/v) [111].

4.2.2 Biomass-to-Ethanol Conversion Processes

Some estimates stated that around 95% of the worldwide bioethanol production is generated by agricultural products [112]. It was also reported that the bioethanol production from sugar-containing crops such as sugarcane and sugar beets accounted for nearly 60% of the global bioethanol production, with the remaining 40% mainly from starch and other lignocellulosic bioresources [113, 114].

At this point, it has to be noted that the share of bioethanol from lignocellulosic biomass is still low as most of the work is still confined in research laboratories or pilot plants. The current situation could be resumed as follow: It is still easier and cheaper to produce bioethanol from the juice of sugar-containing biomass rather than starch and, to a greater extent, lignocellulosic biomass, mainly due to technical challenges and the incumbent extra cost related to the pretreatment and/or saccharification stages.

Depending on the feedstock's biochemical properties, several conversion technologies were proven to be efficient in transforming biomass into bioethanol. From a simple juice extraction to the more elaborate polysaccharide hydrolysis, the first step is to maximize the extraction of fermentable sugars and then proceed to the fermentation and purification stages

Figure 4.2 summarizes the conversion processes to produce bioethanol from sugar-containing (e.g., Sugarcane), starch-containing (e.g., Corn), and lignocellulosic biomasses.

In this section, we will proceed step by step. Thus, first we will see how to extract simple fermentable sugars from biomass. Then, how to convert those sugars into ethanol and, at the end, how to recuperate this liquid biofuel and make it available for the transportation sector and other applications

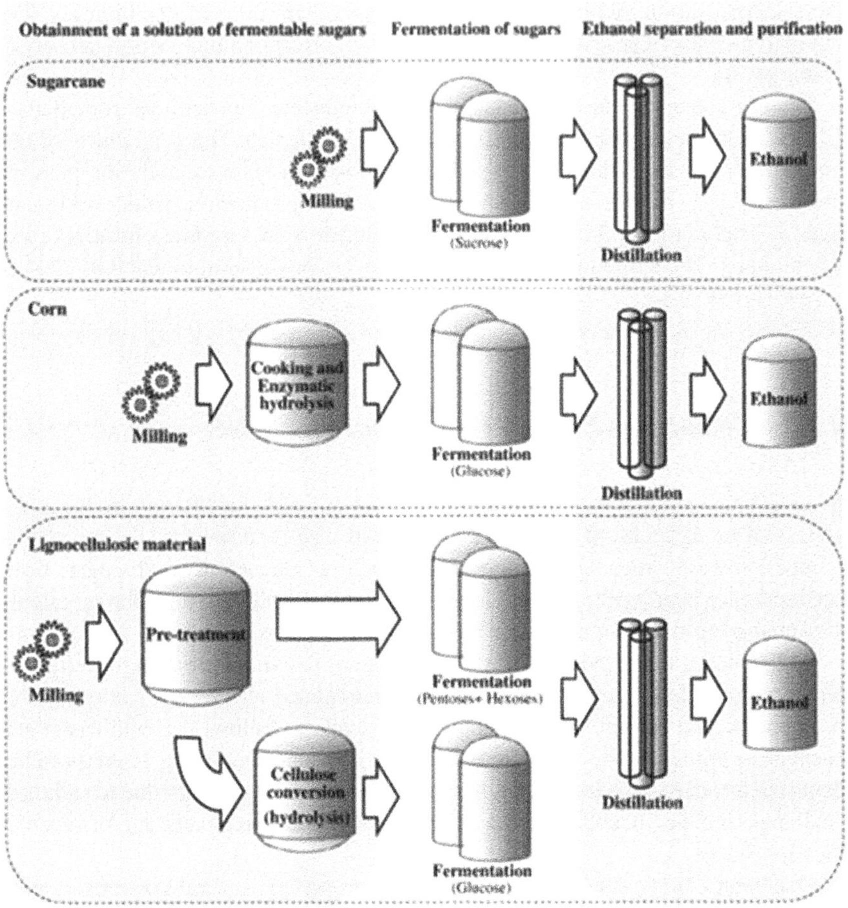

Fig. 4.2 Summarized flowchart of bioethanol production from the main feedstocks [9]

4.2.2.1 From Biomass to Simple Sugars

Case of Sugar-Containing Biomass

One of the main advantages of using sugar crops as feedstock for bioethanol production is the easy obtainment of fermentable sugars. Indeed, only a milling process is needed to extract those sugars in the generated juice, which could be directly used for fermentation [115].

As well, the molasses, obtained as a by-product after the extraction of sugars, could also be fermented directly [116]. Such straightforward process, in addition to the large cultivated area and high biomass production yields of those crops, explains their major contribution in the global bioethanol production (~ 60%).

Simple Sugars from Sugarcane

In practice, and in order to avoid the deterioration in the sugar content, sugarcane needs to be processed within 24–72 h after harvesting. The first step is *milling* where the cane stalks are shredded into fine chips. The shredded biomass is fed into the crushing mill to rupture the juice cells. Then, rollers are used to separate the sugar-loaded juice from the residual fibrous material, bagasse, which could be used to generate heat and electricity or as feedstock for more bioethanol production, based on its residual polysaccharidic content. Under optimum milling conditions, more than 95% of the sugar content in the canes is recuperated in the juice [117].

The turbid dark green juice from the milling process, *aka.* raw or mixed juice, needs to be clarified in order to remove both soluble and insoluble impurities and to adjust its acidity (pH from 4.5–5 to 7–7.5). This *clarification* process is mainly based on thermochemical treatments using heat and lime (calcium hydroxide) [118]. Other studies managed to improve the clarification of sugarcane juice using phosphoric acid (post-liming) [119] and bentonite and powder activated carbons (pre-liming) [120]. The latter technique for instance was able to reduce the juice turbidity by 80%.

The clarified juice contains around 15 wt% solids. Thus, it needs to be concentrated by *evaporation* before fermentation. The aim is to produce a syrup with an adequate sugar content. In this regard, it was recommended to adjust this sugar content between 14 and 18% in order to achieve optimum fermentation efficiency from the most commonly used fermenting microorganism, *Saccharomyces cerevisiae* yeast, at a temperature of around 33–35 °C [121]. Noting that a second clarifying step could also be applied on the concentrated juice [122].

Afterwards, the syrup or concentrated juice is subjected to a *sterilization* treatment in order to avoid the risk of contamination and then cooled down to an optimum temperature for subsequent microbial fermentation.

Simple Sugars from Sugar Beet

Like sugar cane, sugar beet can also be processed to extract the sugar content and recuperate it in the generated juice or molasses. In most cases, those beet extracts do not require pretreatment procedures as the sugar content easily recuperated in the juice. As well, almost all in the form of sucrose could be spontaneously split by enzyme invertase into glucose and fructose during the initial stage of fermentation [123]. It was reported that raw beet juice contains between 15 and 20% of dry solids which comprises 85 to 90% sugars and 10 to 15% nonsugars (coloring compounds, proteins, and minerals) [124]. Therefore, the raw juice could be used in fermentation after pH adjustment. For molasses, it has to be diluted to a suitable original gravity and pH buffering [125].

In practice, the extraction of sugars from the beet roots starts by the *slicing* process, in which the bulb-shaped roots (~ 2 kg) are cut into thin pieces. Then, for sucrose *extraction*, the sliced biomass could be mixed with fresh water or recycled water from a previous process and bring the mixture temperature up to 70–80 °C [126]. Another technique to extract the sugar content is based on washing the finely sliced roots with a water stream in a countercurrent apparatus (diffusion process) [127].

After extraction, two main compounds are collected: juice and pulp. This residual wet pulp is pressed, thus generating a liquid fraction which is filtered to remove fine pulp particles, heated, and pumped back into the diffuser, in order to reduce the water consumption. As for the solid fraction, it is further dried into pellets and used for animal nutrition. Sugar beet pulp was also use as feedstock to produce other kind of biofuels such as biomethane [128].

The "raw juice" collected after the extraction procedure contains water (65.62%) and sucrose (16.50%), but also solid particles (17.30%), non-usable sugars (0.24%), and some impurities (0.34%), which could have some detrimental effect on the subsequent ethanol conversion process [129]. Henceforth, the raw juice has to be treated to remove the impurities by mixing it with milk of lime (suspension of calcium hydroxide in water). This *clarification* step allows the impurities such as sulfate, phosphate, citrate, and oxalate to precipitate, and large organic molecules such as proteins, saponins, and pectins to aggregate in the presence of multivalent cations [130]. The raw juice is then fed into a *carbonation* tank where carbon dioxide is added to convert the lime into calcium carbonate, enclosing the nonsugar compounds [131]. The generated slurry is then subjected to a *filtration* step to remove the precipitated and agglomerated impurities. Noting that this clarification–carbonation–filtration process may be repeated multiple times, though most commonly it is repeated twice [132]. The filtrate coming from those previous steps is termed "thin juice." It has a sugar content of 12 to 14% and could be and must be thickened by evaporation to produce more concentrated syrup with a sugar content between 65 and 70%, knows as "thick juice."

Overall, the production of bioethanol from sugar beets could be based on one or a combination of the descried sugar beet-processing intermediates (i.e., raw, thin, and thick juices) as well as from its by-products (i.e., molasses) [133].

Case of Starch-Containing Biomass

As we have seen in Sect. 4.2.1.2, the starch content in a biomass could be stored whether in *grains* such is the case for corn, wheat, barley, and sorghum or in *tubers* in plants like cassava, yam, and sweet potato. In the following section, we will present the sugar extraction from corn and cassava, the respective representative of grains and tubers, with the highest starch content (Cf. Table 4.2).

Simple Sugars from Corn

Basically, the sugar extraction process from corn grains is based on the following steps:

1. *Milling*: The grains are grinded or milled using dry or wet processes to produce starch-loaded flour or slurry, respectively. The former is then slurried with water to form a mash. For the dry-grind process, the whole corn kernel is milled and then mixed with water and enzymes. In this case, the components of the kernel

not fermented including the germ, fiber, and protein could be separated and produced as coproducts [134]. As for the wet milling process, the steeping of corn is a necessary prerequisite to the fractionation of corn components. Corn kernels are thus soaked in water containing 0.1 to 0.2% sulfur dioxide (SO_2) at a temperature between 50 and 55 °C and for 24–40 h. This will result in the breakdown of starch granules and the release of starch [135].

2. *Cooking and liquefaction*: Since starch cannot be metabolized directly by yeast, it has to be hydrolyzed into glucose before the microbial fermentation. This process starts by adjusting the pH of the mash to around 6.0, then adding the thermostable alpha-amylase enzymes, and able to break down the starch polymer by quickly and randomly hydrolyzing alpha 1–4 bonds [136]. The mash is then heated above 100 °C using a jet cooker. The high temperature and mechanical shear enhance the rupture starch molecules, especially those of a high molecular weight to lower molecular-weight carbohydrates, called starch sugars or dextrins [137].
3. *Saccharification*: Those complex sugars (polysaccharides) are further processed enzymatically into simple fermentable sugars (monosaccharides) via glucoamylase. To ensure an optimized saccharification process, sulfuric acid is used to lower the pH to 4.5. After 5 h, glucoamylase is added at 0.11% (db), thus hydrolyzing the dextrins to simple glucose at a temperature around 60 °C [138].

Simple Sugars from Cassava

Cassava is a starch-containing root crop widely used as a raw material for many industrial foodstuff and feed applications. More recently, cassava roots were used as a feedstock for bioethanol production. The related process aims at extracting the maximum amount for starch from the roots, to be later liquefied and further hydrolyzed in to fermentable sugars.

In practice, and considering their high moisture content, cassava roots are prone to spoilage. Therefore, the harvested roots are in general cut into pieces and sun-dried in nearby facilities. The dried chips (moisture content $< 14\%$) are less bulky, less costly for transportation, and can be stored for a year in the warehouse [139]. The resulting dried cassava chips possess comparable composition and handling characteristics as corn grains and thus could be processed likewise (i.e., dry milling) in order to extract the starch content.

Once the cassava-derived starch is extracted, the conversion process into simple sugars is like any starch derived from other biomass (corn as we have seen earlier): liquefaction, hydrolysis,or saccharification and purification [140].

When starch is extracted from cassava roots, a fibrous residual biomass is generated: cassava pulp. It accounts for 10–30% of the raw roots (wet weight) [141] and contains around 50–70% starch and 20–30% fibers, on a dry matter basis [142]. This cassava-processing by-product could be further valorized as feedstock for animal feed or the production of other biofuels.

Case of Lignocellulosic Biomass

Lignocellulosic biomass are among the most promising feedstock for bioethanol production considering their renewability, polysaccahridic content (Cf. Table 3.1), and worldwide availability whether as raw materials such as woody and herbaceous bioresources or residual by-products from agro-industrial activities [143]. Currently, the most promising and abundant lignocellulosic feedstocks in the United States, South America, Asia, and Europe are corn stover, sugarcane bagasse, and rice and wheat straws [144].

Nonetheless, it has to be noted that, unlike sugar and starch-containing feedstocks, the use of lignocellulosic biomass from bioethanol production has two main handicaps delaying the full scale industrial application of those bioresources to produce ethanol at competitive prices. Indeed, the presence of a mixed sugar content (hexose and pentose) and many inhibiting compounds in the hydrolysate (prior to fermentation) tend to slow and/or lower the conversion rates and yields.

Regarding the sugar content, the saccharification of lignocellulosic biomass, (i.e., hydrolyzing its cellulose and hemicellulose polymers) results in the release of glucose (C6) and some pentose (C5) sugars from hemicellulose. It was reported that lignocellulosic resources, in particular hardwood and agricultural raw materials, can contain up to 20% of pentose sugars including xylose and arabinose. Those pentose sugars could not be fermented by *Saccharomyces cerevisiae* yeast, the most used fermentation microorganism in the bioethanol industry [145]. Thus, several R&D investigations were performed to use xylose-fermenting microorganisms since it is the most abundant pentose sugar [146].

The second problem is the release of low molecular weight organic acids, furan derivatives, phenolics, and inorganic compounds during pretreatment and/or hydrolysis of the raw lignocellulosic biomass, which inhibit the microbial activity [147].

To deal with those challenges, the conversion of a lignocellulosic biomass into simple fermentable sugars was developed based on two major procedures: Pretreatment and saccharification,

Pretreatment of Lignocellulose

Considering the complex structure of lignocellulosic materials, pretreatment become a necessary step in the bioethanol production process. The main aim is to separate (or solubilize) the major components of biomass (cellulose, hemicellulose, and lignin).

Basically, raw biomass is pretreated to remove lignin (a phenolic macromolecule), hydrolyze hemicelluloses (hetero-polysaccharides from xylose, mannose, glucose, and galactose), and reduce the crystallinity of cellulose (polysaccharide from glucose), all in order to improve recovery of fermentable sugars [148]. Indeed, the extensive hydrogen linkages among cellulose molecules lead to a crystalline and strong matrix structure. As a comparison, starch only requires a temperature between 60 and 70 °C to be converted from crystalline to amorphous texture. Cellulose, on the other hand, requires a temperature of 320 °C in addition to a high pressure (up to 25 MPa) to transform the rigid crystalline structure into an amorphous structure in water [149].

In general, an efficient and cost-effective pretreatment procedure has to reach the following objectives:

1. Preserve the hemicellulose fraction, thus enhancing the overall yields of fermentable sugars.
2. Reduce the formation of inhibiting compounds, mainly related to the lignin degradation.
3. De-crystallize the cellulose fraction and make it more accessible for hydrolysis which leads to high saccharification yields and low energy input [150].

In practice, there are many pretreatment technologies for a variety of raw bioresources. In this section, we will present a selection of procedures for each of the four pretreatment categories: physical, physiochemical, chemical, and biological procedures.

- *Physical pretreatment:*
  - *Milling* is a pollution-free mechanical pretreatment applied to break down the structure of lignocellulosic material, without separating its various fractions (cellulose, hemicellulose, and lignin). The main advantage of this technique, commonly based on ball milling and also sonication, is to decrease the cellulose crystallinity. As well it reduces the particles size of the feedstock (0.2–2 mm), thus improving the efficiency of the downstream physical and/or chemical pretreatments by increasing the contact surface area with the reactants [151]. The high energy costs related to the operation and maintenance of the milling machines constitutes the main disadvantage of milling.
  - *Microwave* is a pretreatment process substituting conventional heating. Its main advantages are decreasing reactions times, thus saving time and energy, as well as homogenizing the heating of the reaction mixture and minimizing the generation of inhibitors [152]. The application of microwave to pretreat lignocellulosic biomass was reported to be one of the most promising pretreatment techniques allowing to change the native structure of cellulose and induce the degradation of hemicellulose and, to a lesser extent, lignin [153].

- *Physicochemical pretreatment:*
  - *Steam explosion* is a hydrothermal process based on a combined high heat and pressure pretreatment with little or no chemical addition. In practice, the lignocellulosic biomass and steam mixture are held for a period of time to promote hemicellulose hydrolysis and terminated by an explosive decompression [154]. Commonly, temperatures between 160 and 240 °C and pressure between 0.7 and 4.8 MPa are applied. As a result, stream explosion pretreatment was proven to induce high solubility of hemicellulose but low lignin solubility [155].

- *Ammonia Fiber Explosion* (AFEX) is a physicochemical alkaline procedure in which lignocellulosic materials are exposed to liquid ammonia at high temperature and pressure, followed by a quick release of pressure. This process was used to pretreat various lignocellulosic biomass including alfalfa, corn stover, wheat, barley, rice straws, and bagasse [42]. First the lignocellulosic materials are pre-wetted at a moisture content of 15–30% and then loaded in a pressure vessel with liquid ammonia at a ratio of 1–2 kg of ammonia per kg of dry biomass. At ambient temperature, pressures exceeding 12 atm are required [113]. The AFEX technique, in addition of requiring short amounts of time, was proven to be efficient in lignin removal (with low generation of inhibitors), retaining appreciable amount of carbohydrates in the substrates [156]. In order to make this process cost-effective, ammonia needs to be recovered using evaporation for instance [157].

- *Chemical pretreatment:*
  This is the most studied pretreatment technique among the other pretreatment categories. It has been extensively used for the delignification of lignocellulosic biomass. The most commonly used procedures are acid and alkali pretreatments.

  - *Acid pretreatment* is based on exposing the lignocellulosic material to various kinds of acids (either diluted or concentrated) including sulfuric, phosphoric, nitric, and oxalic acids. The main objective of acid pretreatment is to solubilize the hemicellulose fraction into xylose and other sugars and improve the subsequent cellulose hydrolysis. Among all the acid-based pretreatments, dilute acid pretreatment using sulfuric acid at moderate temperatures (~120 °C) is the most studied procedure [158, 159]. Its worldwide application relies on its efficiency to solubilize around 80% of the hemicellulosic content [160] and it cost-effectiveness [47].

    Nonetheless, the use of acids to pretreat lignocellulosic biomass is facing some issues that need to be dealt with. First, the high costs related to the maintenance of corroded equipment (vessels and pipes) [161] and the chemical input to neutralize the acidic hydrolysate before fermentations [162]. Second, the acid treatment generates other by-products susceptible of inhibiting the microbial activity during fermentation. Such by-products, including furans, furfural, carboxylic acids, and some phenolic compounds, need to be removed from the lignocellulosic hydrolysate.
  - *Alkali pretreatment* is a delignification process similar to the Kraft paper pulping technology. The main impact of the alkaline pretreatment is therefore the removal of lignin from the biomass, thus improving the reactivity of the remaining polysaccharides. A fraction of hemicellulose could also be solubilized. Several bases could be used including sodium hydroxide, calcium hydroxide (lime), potassium hydroxide, ammonia hydroxide, and sodium hydroxide in combination with hydrogen peroxide.

    Alkaline hydrolysis causes several structural modifications inside the lignocellulosic biomass, mainly the separation of structural linkages between lignin and carbohydrates, cellulose swelling and partial de-crystallization, and solvation of cellulose and hemicelluloses [163], which make the holocellulosic available for the saccharification phase.

Although operating at lower temperatures and pressures, compared with other pretreatment methods, alkaline pretreatment has some limitations including the conversion of the alkali into irrecoverable salts or its incorporation into the biomass as salts [148] and the additional costs related to the complex chemical recovery system or large amounts of catalysts and the maintenance of corroded equipment [164].

- *Organosolv* process is believed to be the most promising method for the pretreatment of lignocellulosic biomass [165]. It is based on the use of organic or aqueous-organic solvent mixture along with inorganic or organic acids as catalysts in order to break down the internal lignin and hemicellulose bonds [166] and the lignin–lignin and lignin–carbohydrates linkages [167]. The commonly used solvents include ethanol, methanol, acetone, and ethylene glycol, and the process is generally operated at temperatures between 180 and 220 °C [168]. Lower temperatures could also be suitable depending on the biochemical composition of the biomass and the used catalyst.

  In addition to delignification, hemicellulose hydrolysis also occurs during the organosolv pulping process, thus improving the subsequent enzymatic digestibility of the cellulose fraction. This pretreatment process has also other advantages such as the possible recovery and recycling of solvents and especially the isolation of lignin as a solid material on the one hand and carbohydrates as syrup on the other, thus constituting valuable chemical feedstocks for various industrial processing schemes.
- *Ionic liquids* are a new class of eco-friendly solvents often fluid at room temperature, with melting points of <100 °C, and consisting entirely of ionic species (organic cations and organic or inorganic anions). They are also termed green solvents because of their negligible vapor pressure, non-flammability, and high thermal and chemical stability [169]. Quite recently, ionic liquids have been confirmed to be efficient as a pretreatment process for dissolution of lignocellulosic materials including wood [170], rice straw [171], cotton stalk [172], and oil palm biomass [173]. Under optimum operating conditions, most of hemicellulose and lignin are solubilized, and the cellulose should remain as solid while reducing its crystallinity. After the treatment, the solvents need to be covered and recycled for two main reasons: (1) avoid any inhibitory effect on the growth of the microbes responsible for enzymatic hydrolysis and fermentation and (2) reduce the overall cost.

• *Biological pretreatment:*
It is a biological delignification process using microorganisms directly or their extracted enzyme. The microbial treatment includes fungi such as white-rot fungi, brown-rot fungi, and soft-rot fungi, as well as bacteria [174]. The main advantages of using biological pretreatment are low energy requirement, low

formation of toxic compounds such as furfural and hydroxylmethyl furfural, and eco-friendly working conditions [175].

One of the most effective microorganisms for biomass delignification are the white rot fungi, through the action of their lignin-degrading enzymes such as peroxidases and laccase [156]. As for the impact of the biological pretreatment of the carbohydrates, the brown rot fungi were reported to attack cellulose and white and soft rot fungi attack both cellulose and lignin [176].

This leads us to talk about the disadvantages of applying the biological process to pretreat lignocellulosic biomass. As we have seen, most lignolytic microorganisms do not only solubilize lignin but also carbohydrates. In addition, it is a very slow process and requires continuous control of growth conditions leading to high operational costs. For those reasons, the biological pretreatment still faces some techno-economic challenges limiting its large-scale application [177], which opens the door wide open for scientists, engineers, and researchers from various backgrounds to find efficient and cost-effective solutions.

Cellulose Hydrolysis

After the pretreatment step, where lignin is removed and hemicellulose hydrolyzed, the cellulose fraction, already de-crystallized (i.e., amorphous), needs to be further processed and hydrolyzed in order to release its fermentable glucose content. But, before getting to this point, some of the pretreatment techniques, mainly chemical-based ones, applied to lignocellulosic biomass generate inhibitory compounds that need to be removed or reduced to tolerable concentrations [178]. The objective is to ensure optimum conditions for the enzymatic hydrolysis of cellulose and the subsequent microbial fermentation, whether in a separate or simultaneous hydrolysis and fermentation strategy.

The commonly released inhibitors are furan derivatives, furfural, and 5-hydroxymethylfurfural, resulting from the degradation of pentoses and hexoses, phenolic compounds, and other aromatic compounds derived from the partial degradation of lignin, along with metals leached from the equipment such as copper, iron, chromium, and nickel [179].

Several detoxification methods were reported for the removal of those inhibitors including adsorption [180], biodegradation [181], electrochemical degradation [182], and nanofiltration and reverse osmosis [183]. Once detoxified, the cellulose fraction is now ready to be hydrolyzed into simple sugars [$(C_6H_{10}O_5)_n + nH_2O \rightarrow n\,(C_6H_{12}O_6)$. This saccharification step could be performed via two main processes: chemical (using diluted or concentrated acids) and enzymatic ones.

- *Chemical hydrolysis*

  This process uses acids to break down cellulose chains. Although having a high sugar recovery efficiency from cellulose (e.g., 76% dry weight from corn stalks [184]), cellulose hydrolysis using concentrated acids was not widely implemented by the industrial sector due to the technical challenges working with concentrated sulfuric or hydrochloric acids, along with the additional facilities and energy input needed to recover and reconcentrate those chemicals.

To overcome those shortcomings, chemical hydrolysis of cellulosic biomass was performed to a greater extent and for many years using diluted acids. Via this process, the remaining hemicellulose fraction (after pretreatment) and the amorphous cellulose are reported to be easily and nearly completely hydrolyzed to fermentable sugars [185]. Commonly, dilute acid hydrolysis occurs in a two-stage process. The first stage should be conducted under mild operating conditions (e.g., 0.7% sulfuric acid at 190 °C) to maximize the sugar recovery from hemicellulose while avoiding the formation of furfural and other inhibitors. In the second stage, the temperature is further increased (~ 215 °C), but at milder acid concentration (0.4%) in order to optimize the recovery of glucose from the more resistant cellulose fraction [186].

Nonetheless, despite some high performing cases, most dilute acid processes are limited to a sugar recovery efficiency about 50% [187], leading to a residual solid fraction mainly composed of lignin and still crystalline cellulose (more difficult to hydrolyze), hence to need for another hydrolysis process.

- *Enzymatic hydrolysis*

Enzymatic hydrolysis of raw lignocellulosic biomass is a very slow process due to cellulose crystallinity and the structural and chemical properties of lignin and hemicellulose. After the pretreatment phase, lignin should be removed and a substantial fraction of hemicellulose hydrolyzed. Thus, the enzymatic hydrolysis will mainly target the nearly intact cellulose fraction and convert it to glucose and also the remaining hemicellulose to be converted into pentoses (xylose and arabionose) and hexoses (glucose, galactose, and mannose).

The conversion of cellulose and hemicellulose is catalyzed by highly specific cellulase and hemicellulose enzymes, typically at mild operating conditions (pH 4.8 and temperature between 45 and 50 °C) [188]. The following types of cellulases (cellulolytic enzymes) could be used in the process of degrading cellulose to glucose [176]:

- Endoglucanases: hydrolyzing the β-(1,4) glycosidic bonds in the amorphous regions of cellulose to produce cello-oligosaccharides with free-chain ends.
- Exoglucanases or cellobiohydrolases: hydrolyzing the β-(1,4) glycosidic bonds from the nonreducing ends of the cello-oligosaccharides to generate cellobiose.
- β-glucosidases: hydrolyzing cellobiose into glucose

Reaction wise, the enzymatic hydrolysis or degradation of cellulose is characterized by a rapid initial phase followed by a slow secondary phase lasting until all substrate is consumed. Several substrate and enzyme-related factors were reported to restrict the continuous catalytic activity of enzymes. This includes the accessible surface area, the strong product inhibition, and slow inactivation of absorbed enzyme molecules [113].

4.2.2.2 From Simple Sugars to Bioethanol

After detailing the required steps to convert biomass into simple sugars, we shall now describe the next and final steps to convert those sugars into bioethanol. First, the sugar content has to be fermented by specific microorganisms under specific conditions. Then, the produced bioethanol has to be separated from the mixture and purified (if necessary).

Fermentation Processes

Fermentation occurs when microorganisms, including yeasts, bacteria, and filamentous fungi, feed on simple sugars and in the process produce ethyl alcohol and other by-products [189]. Since the cellulose and hemicellulose hydrolysis during the pretreatment and saccharification stages resulted in the production of both hexose (6-carbons) and pentose sugars (5-carbons), thus, for optimum conversion yields, both sugars need to be targeted. Most of the fermenting microorganisms can typically use hexose sugars, especially glucose but also mannose and galactose. Others, limited in number, use pentose sugars such as xylose and to a lesser extent arabinose and ribose. Therefore, biomass with high cellulosic content (i.e., generating glucose) are easier to convert to bioethanol, the biomass with high hemicellulosic content (i.e., generating pentose sugars), leading to a higher conversion yield for the former.

Theoretically, 100 g of glucose will produce 51.4 g of bioethanol and 48.8 g of carbon dioxide. But, in practice, the fermenting microorganisms will consume some of the glucose for their growth, thus the actual yield will always be less than 100%.

As a rule, microorganisms prefer glucose over galactose followed by xylose and arabinose [190]. Those microorganisms, *aka.* ethanologens, include many yeasts, fungi, and bacteria. Nonetheless, *Saccharomyces cerevisiae*, commonly known as Bakers' yeast, and bacteria *Zymomonas* spp. are the two commonly used microorganisms to ferment glucose to bioethanol, but unable to ferment pentose sugars [8].

Thus, in order to optimize the production of bioethanol from lignocellulosic biomass, and make the process cost-effective, both pentose and hexose sugars (mainly glucose and xylose) have to be fermented. As we have seen, the first released sugars are the hemicellulose-derived pentose during the pretreatment step. Thus, after recovering the cellulose fraction (to be hydrolyzed), the detoxified hydrolyzate could be used as a direct substrate to produce bioethanol using various pentose-fermenting microorganisms [191]. The use of hydrolyzate without detoxification was also reported in a study using corn stover as lignocellulosic feedstock [192].

At the industrial level, four major integrated fermentation configurations could be adopted: separate hydrolysis and fermentation (SHF), simultaneous saccharification and fermentation (SSF), simultaneous saccharification and co-fermentation (SSCF), and consolidated bioprocessing (CBP).

- *Separate hydrolysis and fermentation* is the conventional approach where the biomass hydrolysis and the sugar fermentation are sequentially carried in different reactors. In this configuration, the biomass is first hydrolyzed using enzymes in a separate reactor, then the hydrolyzate flows from the hydrolysis reactor and enters the fermentation reactor. The mixture is then distilled to recover the produced bioethanol. The residual solid, still containing xylose sugars, is subjected to another round of fermentation [193]. The main advantage of this approach is that each process will be conducted under optimum conditions: 50 °C and pH 5 for the hydrolyzing enzyme and 32 °C and pH 5 for the fermenting yeast *Saccharomyces cereviceae* [194]. Nonetheless, SHF has some shortcomings such as generating inhibitory by-products susceptible of reducing the bioethanol yield. As well, the high risk of contamination could seriously damage the process.
- *Simultaneous saccharification and fermentation* is an alternative process to SHF in which hydrolysis and fermentation are performed in the same reactor (enzyme and yeast are put together) under constant operating conditions (37 °C and pH 5). As a result, glucose is rapidly converted into ethanol [195]. As well, it was reported that SSF process gives higher ethanol production yield when compared with SHF due to the fact that low residual sugar relieves inhibition on the cellulase enzyme [146].
- *Simultaneous saccharification and co-fermentation* is another integration alternative based on the co-fermentation of pentose and hexose sugars at the same time. Thus, after the pretreatment, the hydrolyzed hemicellulose and the solid fraction of cellulose are not separated. Under this configuration, the hemicellulose-derived sugars will be converted into bioethanol while the cellulose is subjected to the previously mentioned SSF process [196]. Generally, the fermentation of pentose and hexose sugars is conducted using two different microorganisms such as the combination of *Candida shehatae* and *S. cerevisiae* which was reported to be as suitable option for SSCF process [197], as well as two mutants of *E. coli* [198]. Nonetheless, since SSCF is performed in a single reactor, adjusting the operating conditions (mainly pH and temperature) for optimum growth of two different microorganisms is challenging. Thus, the use of a single microorganism able of fermenting both C5 and C6 sugars is a much better option. In this context, advances in genetic engineering led to the construction of several co-fermenting microorganisms, such as *S. cerevisiae* IPE003 [199] and recombinant *S. cerevisiae* 1400(pLNH32) and 1400(pLNH33) yeasts [200].
- *Consolidated bioprocessing* is biological integrated process combining on-site enzymes production, cellulose hydrolysis, and fermentation in a single step. This promising configuration is still in its early stages with several studies reporting that bioprocessing technologies for bioethanol production show a trend towards consolidation over time [201] and that this highly integrated process could be

implemented in the near future [202]. Although some studies focused on native cellulolytic microorganisms, but as for SSCF, genetic engineering is playing a major role in improving CBP potentialities. Thus, several microbes were investigated for high CBP efficiency including *Clostridium thermocellum* bacteria [203] and thermo-tolerant *Kluyveromyces marxianus* and *Saccharomyces cerevisiae yeasts* [204].

So far, the industrial implementation of CBP is still limited because finding the optimal conditions within a single bioreactor for all steps in the process (hydrolytic enzyme secretion, saccharification, and fermentation) is a challenging task. But, once resolved, we will have an efficient and cost-effective integrated process to convert biomass into bioethanol.

Bioethanol Recovery and Purification (*Distillation and Dehydrating*)

In general, bioethanol is recovered from the fermented broth mainly via distillation of the final medium composed by water and ethanol (5–12 wt%). The conventional method separates the compounds based on their relative volatilities. Thus, any ideal binary mixture could be heated so that low boiling point components are concentrated in the vapor phase. Then, by condensing this vapor, the less volatile compounds are further concentrated in the produced liquid phase. In a fractional distillation, the evaporation/condensation step is repeated many times so that at each time the resulting vapor is more enriched of the more volatile component until complete purification.

But for bioethanol, there is an issue. The generated ethanol/water in the fermentation solution forms a constant-boiling azeotrope composed of 89.4 mol.% ethanol and 10.6 mol.% water at 78.2 °C and standard atmospheric pressure [205]. Therefore, this nonideal ethanol–water mixture is impossible to generate highly pure ethanol via conventional distillation alone [206]. Indeed, right from the start this azeotropic mixture is partially boiled, the resulting vapor has the same ratio of components as the original (un-boiled) mixture; that's why further separation by conventional distillation is no longer possible.

At this point, some would ask why purifying bioethanol and add an extra cost to produce anhydrous ethanol. The question is indeed legitimate especially knowing that in Brazil, many Flex-Fuel vehicles are running on 100% hydrous ethanol (with 4.0–4.9% V/V of water) [207]. Well, when blended with gasoline, anhydrous ethanol improves the fuel's octane index and reduces CO, HC, and NO_x emissions [208]. Besides, anhydrous ethanol is widely used by many chemical industries as solvent or as a major component in the synthesis of esters, organic, and cyclic compound chains and the production of detergents, paints, cosmetics, aerosols, perfumes, medicine, and food, to name a few [209].

Back to bioethanol purification, it has to be known that the best approach is to start recovering bioethanol from the fermentation broth as it is produced. Thus, as the fermentation process goes on, the ethanol concentration in the mixture decreases which enhances the conversion of the remaining sugars. Indeed, it is well proven that the accumulation of alcohol during fermentation is accompanied by a progressive decrease in the rate of sugar conversion to ethanol [210].

At the industrial level, several processes are applied to produce anhydrous ethanol including extractive distillation [211], azeotropic distillation [212], pressure-swing distillation [213], pervaporation [214], and adsorption [215]. A combination of two or more of these methods is also a promising approach.

Now, after studying the bioethanol production sector from feedstock to end product, let us move to another valuable liquid biofuel: *Biodiesel.*

4.3 Biodiesel

4.3.1 Biodiesel Characteristics

Biodiesel is every renewable fuel processed from renewable, biodegradable, and nontoxic oil source, whether raw such as vegetable oils and animal fats or residual such as recycled waste cooking oils.

Based on the characteristics of the oily feedstock and the operating conditions during its conversion, the resulting biodiesels would have different properties. Still, biodiesel in general has many advantageous assets when compared with fossil petro-diesel. It has a better combustion quality due to the higher cetane number (45–65 vs. 40–55) [216]. As well, biodiesel was proven to be better for the environment since it is produced from renewable biomaterials and its combustion emits less greenhouse gas than petro-diesel, thus reducing health risks associated with air pollution [217]. Nonetheless, because of its high viscosity (3.5–5.5 centistokes at 40 °C), it difficult to directly use biodiesel in a conventional diesel engine; that is why it is frequently blended with fossil diesel with has a lower viscosity (2–3.5 cSt at 40 °C) [216]. One effective solution to lower the oil viscosity is the use of an alcoholic compound, mainly methanol, in the transesterification reaction [218]. Of course, in order to produce a renewable biodiesel, a renewable bio-alcohol has to be applied. Noting that biodiesel–diesel blends are referred to as *Bxx*. The "xx" specifies the amount of biodiesel by volume in the blend. B20, for instance, refers to a blend of 20% biodiesel and 80% (v/v) petro-diesel.

Since the environmental factor is supposed to be one of the main driving forces behind the increasing recourse to biodiesel, several research studies analyzed the exhaust emissions from engines running on fossil diesel on the one hand and a combination of both diesel and biodiesel at various ratios on the other. The related results revealed some interesting facts that need to be seriously taken into account. Most studies proved that the exhaust emissions from diesel engines running on biodiesel blended with petro-diesel contain lower amounts of CO, CO_2, SO_2, unburned hydrocarbons, and particle matter, when compared with engines fed only with fossil diesel [219], and the higher oxygen content in biodiesel is believed to be one the main causes as it promotes a complete combustion process [220].

On the other hand, however, NO_x emissions were reported to increase proportionally with the increase of the biodiesel share in the blend due to the shorter ignition delay, i.e., the more we add biodiesel to the blend, the more the injection timing will be advanced [221]. Such issue could be remediated by adjusting the engine characteristics so that the injection time is retarded. As well, this tendency was also related to the esters in biodiesel having more double bonds which leads to higher NO_x emissions. The addition of ethanol to the biodiesel–diesel blend helped reducing the emissions of nitrogen oxides [222]. Furthermore, some studies even reported insignificant changes in CO emissions and an increase in CO_2 emissions [223].

Currently, engines running on petroleum-derived diesel are the mostly used combustion engines in the transportation sector due to their effective power generation and especially their fuel economy [224]. This tendency is generating an excessive demand for diesel which poses the biggest challenge facing biodiesel as an alternative fuel. Indeed, it is the expected high demand for biodiesel to replace fossil diesel that is pushing researchers and investors all over the world to explore new feedstocks, develop new conversion procedures, and optimize old ones, in order to meet this increasing demand.

Before proceeding, one important historical background has to be debated. Rudolf Diesel, the father of diesel engine, used in his first experiment peanut oil to run his compression ignition engine. Hence, in reality, using biodiesel constitutes a return to the origins and not an alternative to fossil diesel, which for purely economic reasons took the lead as the main fuel for many decades based on the advanced techniques for the refinement of cheap and "back then" abundant crude oil. It was only during the Second World War that biodiesel was used for emergency needs. Other than that, petro-diesel reigned supreme for many decades.

What's the point of this short historical account? Well, we have to have a pragmatic assessment of the biodiesel future based on its past. Thus, the beneficial environmental impact, although important, is not the key issue. The focal point should be how to make the biodiesel production process *cost-effective*. This is why petro-diesel took the lead and this is how biodiesel could reclaim the leading position with time and through an extensive R&D effort.

Let us now start by specifying and describing the potential feedstocks and then detail the various conversion procedures to transform oil and fat into biodiesel.

4.3.2 Biodiesel Feedstocks

4.3.2.1 From Where to Choose

The high demand for biodiesel as an alternative fuel to replace fossil diesel has driven many researchers and industrials around the world to look for suitable local feedstock that would generate high production yields at low cost. This is why selecting the "right" feedstock is the first and most important decision in this promising industry because an unwise decision about the raw material would cause serious profitability losses for the whole production process.

For instance, it was reported that the feedstock price for most production units should not exceed 50% of the production cost. If it increases (50–70%), then the net profit of this industry decreases, and more so if a pretreatment or refining process is required [225]. In this context, a research group conducted an extensive literature review and reported that the cost for just procuring the oil feedstock represented 75% of the overall biodiesel production cost [226]. Therefore, acquiring the raw material with the highest lipid content at the lowest cost is crucial to promote the competiveness of this industry and that of any biofuel.

One of those unwise decisions is to systematically discard edible oils as potential feedstock to produce biodiesel just because of the prefix "edible." As we will see, many edible oils have interesting biochemical composition for potential conversion into fatty acid methyl ester (FAME). Besides, it is both common sense and wise for a country with an abundant oil production to convert it into added-value biodiesel (or any valuable product for that matter) instead of exporting it raw.

So, in the following section, we will present both *edible* and *nonedible* feedstocks for oil extraction, along with other "*unconventional*" raw materials such as waste cooking oil, animal fats, and oils from microorganisms (mainly algae and fungi). Table 4.6 illustrates a selection of the main feedstocks for biodiesel production.

4.3.2.2 Oil Content and Production Yields

Having a substantial number of bioresources as potential feedstocks is good but not sufficient. Indeed, the oil quantity and quality are key factors to promote a feedstock to the rank of biodiesel feedstock. Indeed, the extraction of the oil, to be later transesterified, needs equipment, chemicals, and others inputs. Thus, the more lipids are in the feedstock, the more oil will be obtained and the lower the related extraction cost will be. In addition, the quality of the oil fed to the production process was proven to have a direct impact on the characteristics of the produced biodiesel [227].

Vegetable oils and animal fats mainly consist of triglycerides synthesized through the esterification of one molecule of glycerol with three fatty acids, but also diglycerides and to a lesser extent monoglyceride [228]. At room temperature, these chemical compounds could be either liquid (oil) or solid (fat or butter). In both Tables 3.2 and 3.3, the main fatty acids and their content are given for various vegetable oils and animal fats.

Table 4.7, on the other hand, gives the oil content in several biomasses, along with their annual oil yields.

Based on this data collection, the most interesting feature is the impressive oil production potential of oleaginous microalgae. Indeed, the highest yielding microalgae produce oil 23 times more than the biomass with the highest edible

Table 4.6 Main oily feedstocks of biodiesel

Edible oils	Nonedible oils	Unconventional oil and fats
Soybeans (*Glycine max*)	Jatropha (*Jatropha curcas*)	Waste cooking oil
Rapeseed (*Brassica napus*)	Mahua (*Madhuca indica*)	Beef and mutton tallow
Oil palm (*Elaeis* spp.)	Karanja (*Pongamia pinnata*)	Pork lard
Sunflower (*Helianthus annuus*)	Moringa (*Moringa oleifera*)	Poultry fat
Coconut (*Cocos nucifera*)	Jojoba (*Simmondsia chinensis*)	Duck tallow
Peanut (*Arachis hypogaea*)	Camelina (*Camelina Sativa*)	Fish oil
Safflower (*Carthamus tinctorius*)	Cottonseed (*Gossypium hirsutum*)	*Botryococcus braunii*[a]
Rice bran oil (*Oryza sativum*)	Cumaru (*Dipteryx odorata*)	*Chlorella vulgaris*[a]
Wheat (*Triticum* spp.)	Cardoon (*Cynara cardunculus*)	*Nannochloropsis gaditana*[a]
Barley (*Hordeum vulgare*)	Neem (*Azadirachta indica*)	*Aspergillus oryzae*[b]
Sesame (*Sesamum indicum*)	Passion seed (*Passiflora edulis*)	*Mucor circinelloides*[b]
Linseed (*Linum usitatissimum*)	Tobacco seed	*Lipomyces starkeyi*[b]
Corn (*Zea mays*)	Coffee ground (*Coffea arabica*)	Cyanobacteria[c]
Groundnut	*Croton megalocarpus*	*Rhodococcus opacus*[c]
Canola		Engineered *Escherichia coli*[c]

[a]Oleaginous microalgae
[b]Oleaginous fungi or yeasts
[c]Oleaginous bacteria

oil yielding (palm oil) and 60 times more than the highest nonedible oil yielding biomass.

Unlike other oil-bearing bioresources, microalgae can be grown in open ponds or photobioreactors. Besides, they have a remarkable biomass production yields as most microalgae can double their biomass within 24 h which is a very important asset, especially for certain algal species with oil content exceeding 80%, based on the dry matter [231].

Nonetheless, several studies reported that algal oils are prone to oxidation during storage because of the higher degree of (poly)unsaturation of their fatty acids (with respect to those of vegetable oils) and the fact that oxidation tends to increase proportionally with the degree of unsaturation [232]. Therefore, the addition of antioxidants becomes a necessary step to stabilize those lipids and overcome this serious problem [5].

Now, after showcasing various potential feedstocks along with their oil content and productivity, are there other criteria to help selecting a suitable raw oleaginous biomass for biodiesel production? Yes, many. Indeed, to deal with this multidimensional issue (economy, energy, environment, and society), we need a multidisciplinary effort taking into consideration the *local* economic situation,

Table 4.7 Oil content and annual yields of selected biodiesel feedstocks [226, 229, 230]

Category	Feedstock	Oil content (%)	Oil yield (L/ha/year)
Edible oils	Oil palm	30–60	5950
	Coconut	63–65	2689
	Rapeseed	38–46	1190
	Peanut	45–55	1059
	Sunflower	25–35	952
	Rice bran	15–23	828
	Soybeans	15–20	446
Nonedible oils	Karanja	27–39	225–2250[a]
	Jatropha	35–40 (seeds) 50–60 (kernel)	1892
	Castor	53	1413
	Jojoba	45–50	818
	Cottonseed	18–25	325
	Rubber seed	40–50	80–120[a]
	Mahua	35–42	–
	Moringa	40	–
Algal oils	*Schizochytrium* sp.	50–77	–
	Botryococcus braunii	25–75	–
	Nannochloropsis sp.	31–68	–
	Nitzschia sp.	45–47	–
	Microalgae[b]	70	136,900
	Microalgae[b]	30	58,700

[a]Kg of oil per hectare
[b]Theoretical laboratory yields of unspecified microalgae

industrial know-how, investment opportunities, environmental regulations, and sociopolitical engagement.

In order to give additional tools to help making the right decisions, various processes to extract raw oil and convert it to added-value biodiesel are presented and discussed in the following section.

4.3.3 Biomass-to-Biodiesel Conversion Processes

In practice, the process of production biodiesel from biomass follows two major steps: oil or fat are extracted from the feedstock and subsequently converted into biodiesel. Nonetheless, considering the different properties of the feedstocks (oils and fats) on the one hand and the importance of producing biodiesel with excellent combustion properties on the other, a refining process could be implemented prior to the transesterification stage. Several deacidification, degumming, and filtration techniques are therefore used in order to remove the non-triglyceride fraction including free fatty acid, phospholipids (gums), carbohydrates, water, and impurities such as color pigments, metal complexes, and dirt [233].

4.3.3.1 Oil Extraction

The aim of any extraction technique is to separate the lipid content from the rest including carbohydrates and proteins, and the better the separation is, the higher the production yield and quality of the oil will be. Several factors could affect the efficiency of the separation process such as the structural and chemical characteristics of the raw feedstock [234].

Three main oil extraction techniques could be applied, separately or combined, to obtain oil from oleaginous biomass: mechanical, solvent-based chemical, and enzymatic procedures.

1. *Mechanical extraction*: is based on applying pressure on the oleaginous biomass, whether seeds or kernel. The main advantage of this process is that it preserves the oil characteristics. In practice, engine driven screw presses are commonly used with extraction efficiencies between 68 and 80% of the oil content. In continuous press method, the operating conditions are pressure of about 400 bar with temperatures between 120 and 155 °C, but for the discontinuous press, a higher pressure (4-35 MPa) and a lower temperature (95 °C) [235]. Afterwards, the oil is collected in a trough under the screw, and the de-oiled residual cake is discharged at the end of the barrel.

   For many bioresources, mechanical processing is considered a pretreatment to optimize the extraction yields of other techniques. Nonetheless, it was revealed that oleaginous seeds with lipid content exceeding 20% wt could be partially, and in some case fully, processed via mechanical means [236].
2. *Chemical extraction:* is based on the use of solvents. In general, the chemical extraction technique yields more oil than the mechanical one, but the latter produces oil of better quality [237]. Basically, it consists of placing oleaginous biomass in contact with a solvent which will induce the dissolution of the oil. But, since the diffusion process allowing the oil to be extracted from the cell is very slow, therefore pretreating of the raw material (mainly mechanical) is necessary to facilitate the diffusion process and thus optimizing the efficiency of the solvent-based extraction. This efficiency could be affected by many factors including the feedstock storage, operating temperature, equipment design, as well as the solvent/biomass ratio and the chemical of properties of the solvent itself [238]. In this context, choosing the suitable solvent relies on its aptitude to solubilize the targeted oil, as well as the related safety measures and its cost. It was estimated that applying the solvent extraction in a large-scale process becomes cost-effective if the daily biodiesel production exceeds 50 tons [239].

   Several solvents were used to chemically extract oil from various biomass including n-hexane, trichloroethylene, carbon sulfide, and recently some promising ionic liquids, with ultrasound and microwave assistance in some cases [240]. Currently, hexane is the mostly used solvent for the extraction of vegetable oil. In general, the chemical oil extraction is conducted via three main techniques: immersion, percolation, or their combination [241].

3. *Enzymatic extraction*: Although solvent-based chemical extraction is the widely used procedure for industrial-scale production of biodiesel, some drawbacks are still associated with this process especially pollution emission (not for ionic liquids) and the complex and costly downstream purification and recovery process. The use of enzymes instead was promoted by scientists as a suitable option to overcome those limitations in an eco-friendly manner [242], which is of special interest to "green" industries. The only issue with this method is that it takes more time to complete the reaction than the other procedures.

   In practice, this process is based on contacting mechanically treated oleaginous biomass with specific enzymes. Therefore, the reaction time, temperature, and pH substantially influence the overall oil yield. To illustrate the efficiency of enzymatic oil extraction, a research team worked on the seeds of *Jatropha curcas*. In the first attempt, they use ultrasonication for 10 min at pH 9, followed by aqueous oil extraction, which gave an oil yield of 67%. Then, at the same pH, the exposure time to ultrasounds was reduced to 5 min, followed by aqueous enzymatic extraction using an alkaline protease. As a result, the oil yield was increased to 74% and the reaction time decreased from 18 to 6 h [243].

Although other advanced techniques such as supercritical fluid [244] and microwave-assisted [245] extraction methods showed promising results, mechanical pressing and solvent extraction remain the most commonly used technique for commercial oil extraction. For instance, to extract oil from algae, the marine biomass was first subjected to mechanical pressing using an expeller, and then the produced pulp was mixed with hexane. With this combination, more than 95% of the total algal oil was extracted [246].

4.3.3.2 Transesterification

Vegetable oil cannot be directly used in conventional engine due to too high viscosity (nearly 10 times that of fossil diesel). That is why transesterification of triglycerides to produce fatty alkyl esters, or biodiesel, is widely used since the produced biodiesel has a reduced viscosity and can be mixed with any conventional fuel, without engine modification.

Other technologies could be used for biodiesel production such as micro-emulsion, dilution, and pyrolysis (or catalytic cracking). But, although simple processes, some serious disadvantages limited their large-scale application, including high viscosity and bad volatility for the first two and low purity and high (and costly) temperature requirement for pyrolysis [247]. Ultimately, transesterification is widely considered to be the mostly used procedure to convert oil into biodiesel, but not without some drawbacks related to technical difficulties adjusting the

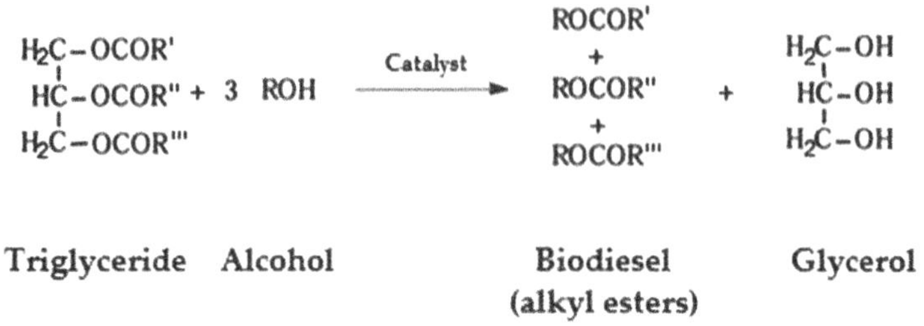

Fig. 4.3 Catalytic transesterification reaction

reaction operating conditions and the generated pollutants (requiring chemical intervention).

In practice, transesterification is carried out using methanol in order to produce biodiesel, based on its availability and lower reaction time. The end products of this reaction, conducted in the presence of a chemical or enzymatic catalyst, are biodiesel (FAME) and glycerol (Cf. Fig. 4.3). Ethanol was also used for the production of biodiesel, producing, therefore, ethyl esters of the corresponding oil, as well as other alcohols such as propanol, isopropanol, tert-butanol, octanol, and butanol.

The oil or fat transesterification could be carried out via two main approaches: catalytic (the widely used) and non-catalytic (the eco-friendly alternative).

1. *Catalytic transesterification*: Considering the poor solubility of alcohols in oil or fat, a catalyst is necessary to enhance their solubility and therefore optimize the reaction rate and yield. Several catalysts were used including *alkali catalysts* (NaOH, KOH, NaMeO, KMeO), acid catalysts (H_2SO_4, H_3PO_4, HCl, and $CaCO_3$), as well as *lipase enzymes* [248]. Currently, the transesterification is widely performed using the homogeneous alkali catalysts such as potassium hydroxide, potassium methoxide, sodium hydroxide, or sodium methoxide [249]. Other researchers pointed out that, in order to optimize the conversion yields and avoid additional purification steps, feedstocks containing significant amounts of free fatty acids (FFA) could be esterified into biodiesel. In this context, the best approach to convert oily feedstocks with high FFA content was based on combining acid-catalyzed esterification (for FFA) and alkali-catalyzed transesterification (for triglycerides) [250].

   In practice, the catalytic transesterification of oil into biodiesel starts by dissolving catalyst in the alcohol and then adding the oil to the mixture in a closed reaction vessel. The reaction is initiated when the temperature reaches 70 °C. After 1–8 h, the reaction is terminated, and the conversion of oils into alkyl esters should be completed. In general, alcohol is used in excess in the reaction to ensure the complete conversion. After removing the catalyst, the end products are crude biodiesel and glycerin, which are separated using the gravimetric method based on their respective densities (glycerol denser). The final step is to purify the crude biodiesel. It starts by a neutralization phase, then applying one of the methods applied for crude biodiesel purification such as the conventional wet and dry washing or the more recent membrane separation technology, and the use of ionic liquids [251].

Besides, biodiesel can be enzymatically produced by lipase-catalyzed transesterification. The main advantages of those biocatalysts are their high efficiency, selectivity, easy glycerol recovery and biodiesel purification, and conducting a continuous process with low energy consumption, as the reaction requires mild conditions (<70 °C) [252]. As well, it was proven that the use of lipase enzymes in oil transesterification is well adapted for low refined oils containing large amounts of water and FFA also under mild conditions and with high conversion yields [253]. However, the relative high cost of lipases and slower reaction rate remain the main challenges facing enzymatic production of biodiesel [254].

2. *Non-catalytic transesterification*: This technique uses alcohols under supercritical conditions to transesterify oil or fat. In this case, supercritical alcohols play the double role of reactant and acid catalyst [255]. As a result, the transesterification reaction is completed in a very short time, 7–15 min for non-catalytic methods and many hours for the catalytic ones (alkali being much faster acid catalysts). In addition, the use of supercritical alcohols (methanol and, to a lesser extent, ethanol) is a more eco-friendly alternative as it reduces the generation of by-products and therefore for further separation and purification processing. It was also experimentally proven that the supercritical method is more tolerant to the presence of water and free fatty acids than the conventional alkali-catalyzed procedure [256].

   Nonetheless, the non-catalytic technique is conducted under higher temperature, pressure, and alcohol amount when compared with the catalytic methods. This leads to a higher operating cost related to a substantial energy and chemicals demand. To address this issue, several methods were recommended including adding co-solvents and the simultaneous removal of glycerol to avoid the glycerolysis reaction (i.e., the produced biodiesel reacts with glycerol to form back monoglyceride instead of methyl esters) [257].

4.4 Gas from Renewable Biomass

After showcasing the numerous bioresources apt to be used in the biodiesel industry, and detailing and comparing the various procedures to extract oil and convert it to added-value biofuel, let us now move to present and discuss the production procedures to convert biomass into gases that could be (1) injected in natural gas grids, (2) used as biofuel or (3) feedstock for the production of other chemicals or (4) for electricity generation.

4.4.1 Biogas

4.4.1.1 General Considerations

Basically, all bioresources reported as suitable feedstocks for bioethanol and biodiesel production are also suitable for biogas production, along with many others. Indeed, there are no strict requirements for biomass with specific biochemical composition for biogas production, as it is the case for bioethanol (polysaccharides) and biodiesel (lipids). This interesting feature laid the ground for extensive biomass, and especially biowastes, valorization schemes to produce biogas from feedstocks as versatile as food waste and sewage sludge [258], municipal solid wastes [259], animal wastes [260], algal biomass [261], and several other bioresources.

The biogas is a carbon-neutral source of renewable energy valued for its energy efficiency and environmental impact. In this context, it was stated that transportation fuels produced from manure and wastes-derived biogas, in addition to those from energy crops, fulfill the European Union sustainability requirement from 2017 onwards, with an expected reduction in greenhouse gas emissions of 60%, when compared with fossil fuels [262]. Furthermore, it was reported that the major part of the EU renewable energy supply by 2020 will originate from bioenergy, from which 25% (at least) could be secured by the production of biogas from wet organic materials [263].

Biogas is mainly produced via the anaerobic digestion of raw biomass, agro-industrial residues, and municipal wastes. [264, 265] and it is predominantly made of biomethane CH_4 (55–70% vol.) and carbon dioxide CO_2 (30–45% vol.), along with trace compounds of hydrogen sulfide, ammonia, hydrogen, oxygen, nitrogen, and water vapor [266].

A multitude of bioresources and wastes were found to be suitable feedstocks for the production of biogas including agricultural crops and by-products, municipal wastes, as well as some marine biomasses (microalgae), as illustrated in Table 4.8.

Although methane and carbon dioxide are both potent greenhouse gasses, extensive R&D efforts turned them into valuable source of energy, as well as a very promising building block in organic synthesis for the case of CO_2 [269, 270], especially considering its tremendous availability on the one hand and the resulting mitigation of global warming and climate change on the other.

4.4.1.2 Biogas Production via Anaerobic Digestion

Anaerobic digestion is a naturally occurring process based on the ability of various bacteria to convert organic materials into biogas through complex biochemical reactions in anaerobic (little to no oxygen) environment. This conversion is carried out through the biological activities of acid and methane-forming bacteria able to

Table 4.8 Biogas yields and biomethane content derived from selected feedstocks [267, 268]

Feedstock		Biogas yield (m^3/ton)	Biomethane content (%)
Agricultural crops and by-products	Barley	169–291	60–70
	Wheat	48–146	–
	Corn	653	54
	Sugar beet leaves	40–50	49–57
	Corn stover	182–436	–
	Grass silage	75–126	–
	Beef cattle manure	16–49	53
	Dairy manure	25–32	54
	Poultry manure	69–96	60
	Animal fat	801–837	-
Municipal wastes	Municipal wastewater sludge	17–146	65
	Household wastes	143–214	–
Microalgae	*Arthrospira platensis*	481	–
	Chlamydomonas reinhardtii	587	–
	Chlorella kessleri	335	–
	Dunaliella salina	505	–
	Euglena gracilis	485	–
	Scenedesmus obliquus	287	-

break complex organic compounds and produce biogas, composed mainly of biomethane and carbon dioxide [271].

Figure 4.4 schematizes the anaerobic digestion of biomass and via (1) hydrolysis, (2) acidogenesis or fermentation, (3) acetogenesis, and (4) methanogenesis.

1. *Hydrolysis*, or liquefaction, is the first stage in the process of converting biomass to biogas. It uses fermentative bacteria to break down heavy insoluble organic compounds into lighter soluble molecules via hydrolytic enzymes such as cellulases (for cellulose), amylases (for starch), lipases (for fat), and proteases (for proteins). Thus, at this stage, carbohydrates, proteins, and fat are transformed into sugars or ethanol, peptides or amino acids, and fatty acids. And it goes without saying how crucial this fractionation step is to optimize the overall biogas production yields.
2. *Acidogenesis* is the next step of anaerobic digestion in which the hydrolyzed organic materials are further broken down by acidogenic bacteria. This fermentative process produces shorter and lighter carbonic acids, alcohols, volatile fatty acids, as well as ammonia, H_2, CO_2, and H_2S. Nonetheless, the organic matter still needs further fractionation.
3. *Acetogenesis* is the last fractionation step before the ultimate conversion into biogas. Acetogenic microorganisms break down the organic acids and alcohols generated during acidogenesis into acetic acid, CO_2, and H_2.

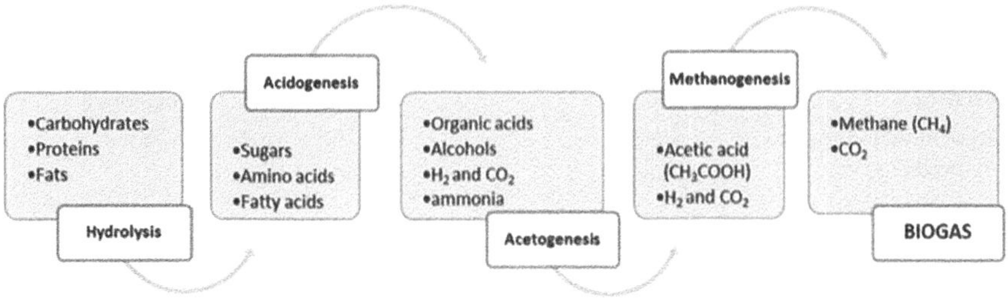

Fig. 4.4 Biomass processing into biogas via anaerobic digestion

4. *Methanogenesis* is based on the aptitude of methanogenic bacteria to cleave acetic acid molecules and generate methane and carbon dioxide ($CH_3COOH \rightarrow CH_4 + CO_2$) or to reduce carbon dioxide with hydrogen ($CO_2 + 4H_2 \rightarrow CH4 + 2H_2O$) [272]. In general, there is a limited amount of hydrogen in digesters, which makes the acetic acid cleavage the main process for biogas production.

At this stage, the produced raw biogas is composed of biomethane (the sought biofuel), as well as CO_2 and many other impurities, hence the necessity to "clean" it.

4.4.1.3 Biogas Purification and Upgrading

The raw biogas needs to be *purified* from the impurities, such as hydrogen sulfide, ammonia, and siloxane, susceptible of causing damages to the equipment (mainly corrosion). It also needs to be *upgraded* by removing the CO_2 fraction in order to increase the calorific value and reduce the emissions of components harmful to human health and the environment [273].

Ultimately, biomethane with higher fuel standard will be produced and subsequently injected in the natural gas grid, used as a transportation biofuel or to fuel engines and gas turbines for producing electricity [274]. Several purification and upgrading techniques were reported, including water scrubbing, pressure swing adsorption (PSA), amine absorption, and membrane separation.

1. *Water scrubbing* is the most widely used technique to remove carbon dioxide from biogas and landfill gas. It is based on the differences of solubility in water of the biogas components. CH_4 solubility in aqueous media is indeed much lower than that of CO_2. Water scrubbing was reported to upgrade biogas to 98% CH_4 [275], but it has some drawbacks. H_2S (more soluble in water than CO_2) ends up being dissolved in water and causes corrosion issues, hence the need to remove H_2S prior to water scrubbing. Besides, during the water regeneration process, the CO_2 fraction has to be collected.
2. *Pressure swing adsorption* (PSA) is based on the ability to separate gaseous compounds (adsorbates) with respect to their molecular sizes by selective adsorption onto the surface of a solid (absorbent) such as activated carbons

and zeolites [276]. Thus, PSA could be applied to separate CH_4 (the larger molecule) from the rest of the gas molecules (CO_2, N_2 and O_2). In general, the adsorption process is carried out under relatively higher pressure (around 800 kPa) while desorption (regeneration) is performed at lower pressure.

Like in water scrubbing, the H_2S fraction, which is irreversibly adsorbed on the adsorbent surface, is detrimental to PSA. Thus, before feeding the biogas to the adsorbent-packed column, the H_2S content has to be removed. This upgrading process results in the production of 96–98% of CH_4, with possible losses estimated around 2 and 4% [277].

3. *Amine absorption* is based on the aptitude of aqueous alkanolamine solutions to absorb CO_2 [278]. One of the main assets of this process is that amine solutions, as chemical solvents, react selectively with CO_2, which effectively limits methane losses (<0.1%) and achieves high CH_4 purities (>95%) [279]. In practice, absorption and desorption columns are combined to ensure the continuous generation of the amine solution through steam-heating the liquid, which require more energy supply.
4. *Membrane separation* is a technology separating gaseous compounds of various molecular dimensions. First, the biogas is pressurized (5 to 20 bar) and fed to the membrane unit. CO_2 and H_2S, as well as other gas components, permeates through the membrane, while larger CH_4 molecules are retained. Polyimide and cellulose acetate-based membranes were reported to be the most suitable membranes on the market for biogas upgrading [280]. For instance, a study investigated the efficiency of a polyvinylamine/polyvinylalcohol blend membrane to selectively remove CO_2. Under optimum operating conditions, the process allowed a methane recovery of 99% with purity of 98% [281].

   But, although technically effective, studies showed that applying membrane modules for biogas purification and upgrading becomes economically competitive only when flow rates are lower than 3500 m^3/h, at standard temperature and pressure conditions (i.e., 0 °C and 1 bar) [282].

4.4.2 *Biological Synthetic Gas (Bio-Syngas)*

4.4.2.1 General Considerations

The production of bio-syngas is achieved by converting renewable bioresources to a gas mixture (bio-syngas) via a thermochemical process, called gasification, at high temperatures (800–1200 °C). This mixture is primarily made of CO and H_2, as well as CH_4, CO_2, light hydrocarbons (ethane and propane), and heavier hydrocarbons (tars). Other kinds of gases including undesirable H_2S and inert N_2 could also be present in the syngas composition. The ratios within bio-syngas mixture depend on the biochemical composition of the feedstock and the operating conditions during the gasification process [283]. Noting that for the carbonaceous residues, constituting a challenging issue to this technology, several valorization schemes were proposed including their use as catalysts for biodiesel production [284].

Basically, any carbonaceous material could be used as feedstock. Conventional syngas is currently being generated from fossil fuels such as coal, natural gas, and naphtha, and bio-syngas is, therefore, the renewable "green" alternative, exhibiting promising prospective on the environmental front, and also contributing to the diversification of the energy supply portfolio, hence reducing the reliance on fossils fuels [285].

4.4.2.2 Gasification Process

In practice, the thermo-chemical conversion of biomass to bio-syngas requires several conversion steps: drying, pyrolysis, oxidation, and reduction [286]. Each stage takes place in a specific compartment in the reactor, as shown in Fig. 4.5.

The main reactions occurring during gasification are endothermic (drying, pyrolysis, and reduction), hence the need for energy source which is generally supplied by the oxidation of part of the feedstock. In this context, the gasification process could be operated in an allo-thermal or auto-thermal way. In the first case, the energy necessary for gasification is provided within the gasifier by partial combustion. As for the second way, gasifiers are externally heated [288]. Several catalysts could be added during the gasification process in order to improve reaction rates and yields [289, 290].

1. *Partial oxidation phase*: The exothermic oxidization of part of the biomass will provide the rest of the endothermic processes with the needed thermal energy. Basically, any carbonaceous compounds in the bio-syngas could be used in the partial oxidation reaction, especially so for biochar and biohydrogen. To ensure a partial oxidation, the flow rate of air supplied to the process is controlled in order to have less oxygen amount than required for complete oxidation by stoichiometry. In this case, the reactions will generate significant amount of CO and intermediates that could be proceeded into other gaseous fuels including H_2 and CH_4 [291].
2. *Drying*: In order to facilitate biomass gasification, a grounded feedstock with a moisture content lower that 10% is recommended for the rest of the processing [292]. In general, when the biomass temperature reaches 150 °C, the drying could be considered complete [293].
3. *Pyrolysis*: During this stage, the dried biomaterial is thermochemically decomposed, under oxygen-free conditions, into compounds with lower molecular weights in solid, liquid (or condensed), and gaseous states. In general, the pyrolysis is carried out using temperatures ranging from 250 to 700 °C, depending on the biochemical composition of the biomass and the overall operating conditions. For instance, pyrolysis of a biomass predominately cellulosic needs to be subjected to a temperature between 600 and 700 °C. As an

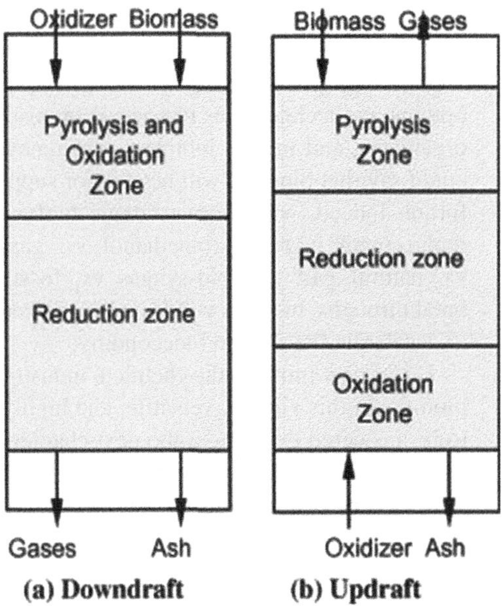

Fig. 4.5 Zonal arrangement of the gasification stages in downdraft and updraft gasifiers [287]

endothermic process, the heat requirement for pyrolysis, as it is the case for drying, comes from the partial oxidation stage.

The gaseous fraction (pyrolysis gas) represents around 70–90 wt% of the feedstock and is mainly constituted of carbon monoxide, carbon dioxide, hydrogen, methane, as well as some or inert gases [294]. As for the condensed and solid fractions (tars and char, respectively), they have a high carbon content and are therefore of high heating value [295].

4. *Reduction*: In this gasification step, all the compounds generated during the previous pyrolysis and oxidation stages react with each other. Specifically, the gas mixture and the biochar will interact to produce the final bio-syngas. The main chemical reactions occurring during this reduction phase are:

   (a) Water gas shift reaction: $CO + H_2O \leftrightarrow CO_2 + H_2$
   (b) Steam reforming reaction: $CH_4 + H_2O \leftrightarrow CO + 3H_2$
   (c) Boudouard reaction: $C + CO_2 \leftrightarrow 2CO$
   (d) Water gas (primary) reaction: $C + H_2O \leftrightarrow CO + H_2$
   (e) Methanation reaction: $C + 2H_2 \leftrightarrow CH_4$

The final step in the conversion of biomass to bio-syngas is to upgrade it to be used in various industrial applications. At this stage, the produced bio-syngas needs to undergo a number of downstream processing steps in order to remove the residual impurities such as barrier filtering, hot gas cleaning, and wet scrubbing [296].

Currently, this renewable bio-syngas is a recognized raw material for combined heat and power processes, and more attention is being paid by industrials to use it as feedstock in biorefineries producing added-value fuels, fine chemicals, and materials, including ethanol, synthetic fuel, jet fuel, Fischer-Tropsch diesel, hydrogen, as well as methanol and synthetic lubricants.

At the end of this chapter, and after presenting and discussing the various options and technologies to convert bioresources (from plants, animals, microorganisms, and marine biomass) into renewable and eco-friendly biofuels, we could say that biomass will be a major supplier of clean fuel and energy in near future. Indeed, with those strategic biofuels, we proved that the eco-friendly replacement is ready (bioethanol vs. gasoline, biodiesel vs. diesel, biogas vs. natural gas, and bio-syngas vs. fossil-based syngas). With more R&D breakthroughs, biofuels will become more competitive, which is the real catalyst to accelerate the shift to bioeconomy.

Let us now move to the chemical industry and see what would be the place of biomass in this vibrant, versatile, and highly profitable sector. The big question to be answered throughout the next chapter is: How can we convert biomass in order to produce various added-value biochemicals able to compete and then replace chemical from fossil feedstocks.

References

1. Kang Y, Khan S, Ma X. Climate change impacts on crop yield, crop water productivity and food security – a review. Prog Nat Sci. 2009;19:1665–74.
2. Eliasson J. The rising pressure of global water shortages. Nature. 2015;517:6.
3. Milà-Villarroel R, Homs C, Ngo J, Martín J, Vidal M. Famine, hunger, and undernourishment. Reference module in food science. In: Caballero B, Finglas PM, Toldrá F, editors. Encyclopedia of food and health. Amsterdam: Academic Press; 2016. p. 581–8.
4. Alonso DM, Bonda JQ, Dumesic JA. Catalytic conversion of biomass to biofuels. Green Chem. 2010;12:1493–513.
5. Brennan L, Owende P. Biofuels from microalgae – a review of technologies for production, processing, and extractions of biofuels and co-products. Renew Sust Energ Rev. 2010;14:557–77.
6. Sticklen MB. Plant genetic engineering for biofuel production: towards affordable cellulosic ethanol. Nat Rev Genet. 2008;9:433–43.
7. Youngs H, Somerville C. Best practices for biofuels. Science. 2014;344:1095–6.
8. Cheng JJ, Timilsina GR. Status and barriers of advanced biofuel technologies: a review. Renew Energy. 2011;36:3541–9.
9. Mussatto SI, Dragone G, Guimaraes PMR, et al. Technological trends, global market, and challenges of bio-ethanol production. Biotechnol Adv. 2010;28:817–30.
10. Khalil SRA, Abdelhafez AA, Amer EAM. Evaluation of bioethanol production from juice and bagasse of some sweet sorghum varieties. Ann Agric Sci. 2015;60:317–24.
11. Bull TA, Glasziou KT. Sugarcane. In: Evans LT, editor. Sugarcane in crop physiology. Some case histories. London: Cambridge University Press; 1975. p. 51–72.
12. Rabelo SC, Andrade RR, Filho RM, Costa AC. Alkaline hydrogen peroxide pretreatment, enzymatic hydrolysis and fermentation of sugarcane bagasse to ethanol. Fuel. 2014;136:349–57.
13. Inyang M, Gao B, Pullammanappallil P, Ding W, Zimmerman AR. Biochar from anaerobically digested sugarcane bagasse. Bioresour Technol. 2010;101:8868–72.

14. Prado RM, Caione G, Campos CNS. Filter cake and vinasse as fertilizers contributing to conservation agriculture. Appl Environ Soil Sci. 2013;2013:1–8.
15. Asadi M. Beet-sugar handbook. Hoboken, NJ: Wiley; 2006.
16. Elliott MC, Weston GD. Biology and physiology of the sugar-beet plant. In: Cooke DA, Scott RK, editors. The sugar beet crop. Berlin: Springer; 1993.
17. Bichsel SE An overview of the U.S. sugar beet industry. Proceedings of the symposium on the chemistry and processing of sugar beet. Denver, Colorado; 1987.
18. Nersesian RL. Energy for the 21st century: a comprehensive guide to conventional and alternative sources. 2nd ed. Armonk, NY: M.E. Sharpe Inc.; 2015.
19. Billa E, Koullas D, Monties B. Structure and composition of sweet sorghum stalk components. Ind Crop Prod. 1997;6:297–302.
20. Bridgers EN, Chinn MS, Veal MW, Stikeleather LF. Influence of juice preparations on the fermentability of sweet sorghum. Biol Eng. 2011;4:57–67.
21. Serna-Saldívar SO, Chuck-Hernández C, Pérez-Carrillo E, Heredia-Olea E. Sorghum as a multifunctional crop for the production of fuel ethanol: current status and future trends. In: Lima MAP, editor. Bioethanol. London: In Tech; 2012. p. 51–74.
22. Cinelli BA, Castilho LR, Freire DMG, Castro AM. A brief review on the emerging technology of ethanol production by cold hydrolysis of raw starch. Fuel. 2015;150:721–9.
23. Buleon A, Colonna P, Planchot V, Ball S. Starch granules: structure and biosynthesis. Int J Biol Macromol. 1998;23:85–112.
24. Baldwin TL, Bower BS, Chotani GK Expression of granular starch hydrolyzing enzymes in trichoderma and process for producing glucose from granular starch substrates. WO 2005052148 A2 (2005).
25. Thatoi H, Dash PK, Mohapatra S, Swain MR. Bioethanol production from tuber crops using fermentation technology: a review. Int J Sustainable Energy. 2016;35:443–68.
26. Robertson MJ. Relationships between internode elongation, plant height and leaf appearance in maize. Field Crop Res. 1994;38:135–45.
27. Shaw RH. Climate requirement. In: Sprague GF, Dudly JW, editors. Corn and corn improvement. 3rd ed. Madison, WI: American Society of Agronomy; 1988.
28. Dowswell CR, Paliwal RL, Cantrell RP. Maize in the third world. Boulder, CO: Westview Press; 1996.
29. Renewable Fuel Association. Industry Statistics – World Fuel Ethanol Production. http://www.ethanolrfa.org/resources/industry/statistics/#1454098996479-8715d404-e546
30. Shewry PR, Hawkesford MJ, Piironen V, et al. Natural variation in grain composition of wheat and related cereals. J Agric Food Chem. 2013;61:8295–303.
31. IDRC communications – facts and figures on food and biodiversity. http://www.idrc.ca/EN/Resources/Publications/Pages/ArticleDetails.aspx?PublicationID=565
32. Koehler P, Wieser H. Chemistry of cereal grains. In: Gobbetti M, Gänzle M, editors. Handbook on sourdough biotechnology. New York: Springer; 2013.
33. Šramková Z, Gregová E, Šturdíka E. Chemical composition and nutritional quality of wheat grain. Acta Chim Slov. 2009;2:115–38.
34. European Biofuels Technology Platform. Bioethanol use in Europe and globally. http://biofuelstp.eu/overview.html
35. Wertz JL, Mercier JP, Bédué O. Cellulose science and technology. Boca Raton, FL: CRC Press; 2010.
36. El-Sharkawy MA. Physiological characteristics of cassava tolerance to prolonged drought in the tropics: implications for breeding cultivars adapted to seasonally dry and semiarid environments. Braz J Plant Physiol. 2007;19:257–86.
37. Osunsami AT, Akingbala JO, Oguntimein GB. Effect of storage on starch content and modification of cassava starch. Strach. 1989;41:54–7.
38. Zhou A, Thomson E. The development of biofuels in Asia. Appl Energy. 2009;86:11–20.
39. Chaisinboon O, Chontanawat J. Factors determining the competing use of Thailand's cassava for food and fuel. Energy Procedia. 2011;9:216–29.

40. Anyanwu CN, Ibeto CN, Ezeoha SL, Ogbuagu NJ. Sustainability of cassava (*Manihot esculenta Crantz*) as industrial feedstock, energy and food crop in Nigeria Renew. Energy. 2015;81:745–52.
41. Kristensen SBP, Birch-Thomsen T, Rasmussen K, Rasmussen LV, Traoré O. Cassava as an energy crop: a case study of the potential for an expansion of cassava cultivation for bioethanol production in Southern Mali. Renew Energy. 2014;66:381–90.
42. Srinivas T. Industrial demand for cassava starch in India. Starch. 2007;59:477–81.
43. Abera S, Rakshit SK. Comparison of physicochemical and functional properties of cassava starch extracted from fresh root and dry chips. Starch. 2003;55:287–96.
44. Kuiper L, Ekmekci B, Hamelinck C et al. Bio-ethanol from cassava. Ecofys Netherlands BV. 2007. http://www.mexicatel.com/EthanolExtraction4rmCassava.pdf
45. Balat M, Balat H. Recent trends in global production and utilization of bio-ethanol fuel. Appl Energy. 2009;86:2273–82.
46. Yan Z, Li J, Li S, et al. Impact of lignin removal on the enzymatic hydrolysis of fermented sweet sorghum bagasse. Appl Energy. 2015;160:641–7.
47. Lynd LR, Elander RT, Wyman CE. Likely features and costs of mature biomass ethanol technology. Appl Biochem Biotechnol. 1996;58:741–61.
48. Cheng JJ, Timilsina GR Advanced biofuel technologies – status and barriers. The World Bank Policy Research Working Paper 5411 (2010).
49. Suess HU. Pulp bleaching today. Berlin: Walter de Gruyter GmbH; 2010.
50. Räisänen T, Athanassiadis D. Basic chemical composition of the biomass components of pine spruce and birch. Forest Refine; 2013. http://www.biofuelregion.se/UserFiles/file/Forest%20Refine/1_2_IS_2013-01-31_Basic_chemical_composition.pdf
51. Frederick Jr WJ, Lien SJ, Courchene CE, DeMartini NA, Ragauskas AJ, Iisa K. Production of ethanol from carbohydrates from loblolly pine: a technical and economic assessment. Bioresour Technol. 2008;99:5051–7.
52. Lan TQ, Gleisner R, Zhu JY, Dien BS, Hector RE. High titer ethanol production from SPORL-pretreated lodgepole pine by simultaneous enzymatic saccharification and combined fermentation. Bioresour Technol. 2013;127:291–7.
53. Shi Z, Yang Q, Ono Y, Funahashi R, Saito T, Isogai A. Creation of a new material stream from Japanese cedar resources to cellulose nanofibrils. React Funct Polym. 2015;95:19–24.
54. Yamashita Y, Sasaki C, Nakamura Y. Effective enzyme saccharification and ethanol production from Japanese cedar using various pretreatment methods. J Biosci Bioeng. 2010;110:79–86.
55. Cagelli L, Lefèvre F. The conservation of P*opulus nigra* L. and gene flow with cultivated poplars in Europe. For Genet. 1995;2:135–44.
56. Zalesny Jr RS, Hall RB, Zalesny JA, McMahon BG, Berguson WE, Stanosz GR. Biomass and genotype × environment interactions of *Populus* energy crops in the Midwestern United States. Bioenerg Res. 2009;2:106–22.
57. Kennedy JH. Cottonwood, an American wood. Washington, DC: U.S. Department of Agriculture, Forest Service; 1985.
58. Wang ZJ, Zhu JY, Zalesny RS, Chen KF. Ethanol production from poplar wood through enzymatic saccharification and fermentation by dilute acid and SPORL pretreatments. Fuel. 2012;95:606–14.
59. Sjostrom E. Wood chemistry: fundamentals and applications. 2nd ed. San Diego: Academic Press; 1993.
60. Matsushita Y, Yamauchi K, Takabe K, et al. Enzymatic saccharification of Eucalyptus bark using hydrothermal pre-treatment with carbon dioxide. Bioresour Technol. 2010;10:4936–9.
61. McIntosh S, Vancov T, Palmer J, Spain M. Ethanol production from Eucalyptus plantation thinnings. Bioresour Technol. 2012;110:264–72.
62. Romani A, Garrote G, Parajo JC. Bioethanol production from autohydrolyzed *Eucalyptus globulus* by simultaneous saccharification and fermentation operating at high solids loading. Fuel. 2012;94:305–12.

63. Lewandowski I, Clifton-Brown JC, Andersson B, et al. Biofuels: environment and harvest time affects the combustion qualities of Miscanthus genotypes. Agron J. 2003;95:1274–80.
64. Brosse N, Dufour A, Meng X, Sun Q, Ragauskas A. Miscanthus: a fast growing crop for biofuels and chemicals production. Biofuels Bioprod Biorefin. 2012;6:580–98.
65. Kim SJ, Kim MY, Jeong SJ, Jang MS, Chung IM. Analysis of the biomass content of various Miscanthus genotypes for biofuel production in Korea. Ind Crop Prod. 2012;38:46–9.
66. Heaton E, Voigt T, Long SP. A quantitative review comparing the yields of two candidate C4 perennial biomass crops in relation to nitrogen, temperature and water. Biomass Bioenergy. 2004;27:21–2730.
67. Heaton EA, Dohlemann FG, Long SP. Meeting US biofuel goals with less land: the potential of Miscanthus. Glob Chang Biol. 2008;14:2000–14.
68. Vogel KP, Brejda JJ, Walters DT, Buxton DR. Switchgrass biomass production in the Midwest USA: harvest and nitrogen management. Agron J. 2002;94:413–20.
69. Somerville C. The billion-ton biofuels vision. Science. 2006;312:1277.
70. Morrow WR, Griffin WM, Matthews HS. Modeling switchgrass derived cellulosic ethanol distribution in the United States. Environ Sci Technol. 2006;40:2877–86.
71. Schmer MR, Vogel KP, Mitchell RB, Perrin RK. Net energy of cellulosic ethanol from switchgrass. Proc Natl Acad Sci USA. 2008;105:464–9.
72. Dale BE. Biomass refining: protein and ethanol from Alfalfa. Ind Eng Chem Prod Res Dev. 1983;22:466–72.
73. Lamb JFS, Jung HJG, Riday H. Growth environment, harvest management and germplasm impacts on potential ethanol and crude protein yield in alfalfa. Biomass Bioenergy. 2014;63:114–25.
74. Dien BS, Miller DJ, Hector RE, et al. Enhancing alfalfa conversion efficiencies for sugar recovery and ethanol production by altering lignin composition. Bioresour Technol. 2011;102:6479–86.
75. Zhou S, Weimer PJ, Hatfield RD, Runge TM, Digman M. Improving ethanol production from alfalfa stems via ambient-temperature acid pretreatment and washing. Bioresour Technol. 2014;170:286–92.
76. Rabelo SC, Carrere H, Filho RM, Costa AC. Production of bioethanol, methane and heat from sugarcane bagasse in a biorefinery concept. Bioresour Technol. 2011;102:7887–95.
77. Aguiar MM, Ferreira LFR, Monteiro RTR. Use of vinasse and sugarcane bagasse for the production of enzymes by lignocellulolytic fungi. Braz Arch Biol Technol. 2010;53:1245–54.
78. Amorim HV, Lopes ML, de Castro OJV, Buckeridge MS, Goldman GH. Scientific challenges of bioethanol production in Brazil. Appl Microbiol Biotechnol. 2011;91:1267–75.
79. Abril D, Medina M, Abril A. Sugar cane bagasse prehydrolysis using hot water. Braz J Chem Eng. 2012;29:31–8.
80. Statista. World sugar cane production from 1965 to 2014. http://www.statista.com/statistics/249604/sugar-cane-production-worldwide/
81. Dussan KJ, Silva DDV, Perez VH, da Silva SS. Evaluation of oxygen availability on ethanol production from sugarcane bagasse hydrolysate in a batch bioreactor using two strains of xylose-fermenting yeast. Renew Energy. 2016;87:703–10.
82. Kumar S, Dheeran P, Singh SP, Mishra IM, Adhikari DK. Continuous ethanol production from sugarcane bagasse hydrolysate at high temperature with cell recycle and in-situ recovery of ethanol. Chem Eng Sci. 2015;138:524–30.
83. Kim S, Dale BE. Global potential bioethanol production from wasted crops and crop residues. Biomass Bioenergy. 2004;26:361–75.
84. Kadam KL, McMillan JD. Availability of corn stover as a sustainable feedstock for bioethanol production. Bioresour Technol. 2003;88:17–25.
85. Yu H, Ren J, Liu LA, et al. New magnesium bisulfite pretreatment (MBSP) development for bio-ethanol production from corn stover. Bioresour Technol. 2016;199:188–93.
86. Kaltschmitt M, Reingardt GA, Stelzer T. Life cycle analysis of biofuels under different environmental aspects. Biomass Bioenergy. 1997;12:121–34.

87. Ballesteros I, Negro MJ, Oliva JM, Cabañas A, Manzanares P, Ballesteros M. Ethanol production from steam-explosion pretreated wheat straw. Appl Biochem Biotechnol. 2006;129:496–508.
88. Saha BC, Iten LB, Cotta MA, Wu YV. Dilute acid pretreatment, enzymatic saccharification and fermentation of wheat straw to ethanol. Process Biochem. 2005;40:3693–700.
89. Paschos T, Xiros C, Christakopoulos P. Simultaneous saccharification and fermentation by co-cultures of *Fusarium oxysporum* and *Saccharomyces* cerevisiae enhances ethanol production from liquefied wheat straw at high solid content. Ind Crop Prod. 2015;76:793–802.
90. Juliano BO. Rice hall and rice straw. In: Juliano BO, editor. Rice: chemistry and technology. 2nd ed. St. Paul, MN: AACC International; 1985.
91. Statista. Statistics and facts about rice. http://www.statista.com/topics/1443/rice/
92. Poornejad N, Karimi K, Behzad T. Improvement of saccharification and ethanol production from rice straw by NMMO and [BMIM][OAc] pretreatments. Ind Crop Prod. 2013;41:408–13.
93. Singh R, Srivastava M, Shukla A. Environmental sustainability of bioethanol production from rice straw in India: a review. Renew Sust Energ Rev. 2016;54:202–16.
94. Swain MR, Krishnan C. Improved conversion of rice straw to ethanol and xylitol by combination of moderate temperature ammonia pretreatment and sequential fermentation using *Candida tropicalis*. Ind Crop Prod. 2015;77:1039–46.
95. Suganya T, Varman M, Masjuki HH, Renganathan S. Macroalgae and microalgae as a potential source for commercial applications along with biofuel sproduction: a biorefinery approach. Renew Sust Energ Rev. 2016;55:909–41.
96. Sheehan J, Dunahay T, Benemann J, Roessler P. A look back at the US Department of Energy's aquatic species program – biodiesel from algae. NREL/TP-580–24190. NREL, Golden, CO; 1998.
97. Millar A. Macroalgae. NSW Department of Primary Industries; 2011. http://www.dpi.nsw.gov.au/__data/assets/pdf_file/0009/378774/Macroalgae-Primefact-947.pdf
98. Roesijadi G, Jones SB, Snowden-Swan LJ, Zhu Y. Macroalgae as a biomass feedstock: a preliminary analysis. Washington, DC: U.S. Department of Energy; 2010.
99. Kim NJ, Li H, Jung K, Chang HN, Lee PC. Ethanol production from marine algal hydrolysates using Escherichia coli KO11. Bioresour Technol. 2011;102:7466–9.
100. Lee JY, Li P, Lee J, Ryu HJ, Oh KK. Ethanol production from *Saccharina japonic*a using an optimized extremely low acid pretreatment followed by simultaneous saccharification and fermentation. Bioresour Technol. 2013;127:119–25.
101. Jiang R, Ingle KN, Golberg A. Macroalgae (seaweed) for liquid transportation biofuel production: what is next? Algal Res. 2016;14:48–57.
102. Lee OK, Seong DH, Lee CG, Lee EY. Sustainable production of liquid biofuels from renewable microalgae biomass. J Ind Eng Chem. 2015;29:24–31.
103. Jang SS. Production of mono sugar from acid hydrolysis of seaweed. Afr J Biotechnol. 2012;11:1953–61.
104. Kim GS, Shin MK, Kim YJ et al. Method of producing biofuel using sea algae. WO 2008105618 A1 (2008).
105. Wei N, Quarterman J, Jin YS. Marine macroalgae: an untapped resource for producing fuels and chemicals. Trends Biotechnol. 2013;31:70–7.
106. Short F, Carruthers T, Dennison W, Waycott M. Global seagrass distribution and diversity: a bioregional model. J Exp Mar Biol Ecol. 2007;350:3–20.
107. Mustafa S, Shapawi R. Aquaculture ecosystems: adaptability and sustainability. Hoboken, NJ: Wiley; 2015.
108. Bettaieb F, Khiari R, Hassan ML, et al. Preparation and characterization of new cellulose nanocrystals from marine biomass *Posidonia oceanica*. Ind Crop Prod. 2015;72:175–82.
109. Ncibi MC, Ranguin R, Pintor MJ, Jeanne-Rose V, Sillanpää M, Gaspard S. Preparation and characterization of chemically activated carbons derived from Mediterranean *Posidonia oceanica* (L.) fibres. J Anal Appl Pyrolysis. 2014;109:205–14.

110. Pilavtepe M, Celiktas MS, Sargin S, Yesil-Celiktas O. Transformation of *Posidonia oceanica* residues to bioethanol. Ind Crop Prod. 2013;51:348–54.
111. Pilavtepe M, Sargin S, Celiktas MS, Yesil-Celiktas O. An integrated process for conversion of *Zostera marina* residues to bioethanol. J Supercrit Fluids. 2012;68:117–22.
112. Rossillo-Calle F, Walter A. Global market for bio-ethanol: historical trends and future prospects. Energy Sustain Dev. 2006;10:20–32.
113. Balat M, Balat H, Oz C. Progress in bioethanol processing. Prog Energy Combust Sci. 2008;34:551–73.
114. Gupta VK, Tuohy MG. Biofuel technologies: recent developments. Berlin: Springer; 2013.
115. Tan L, Sun ZY, Okamoto S, et al. Production of ethanol from raw juice and thick juice of sugar beet by continuous ethanol fermentation with flocculating yeast strain KF-7. Biomass Bioenergy. 2015;81:265–72.
116. Içoz E, Tugrul MK, Saral A, Içoz E. Research on ethanol production and use from sugar beet in Turkey. Biomass Bioenergy. 2009;33:1–7.
117. Chen JCP. Outline of raw sugar process and extraction of juice. In: Chen JCP, Chou CC, editors. Cane sugar handbook: a manual for cane sugar manufacturers and their chemists. Hoboken, NJ: Wiley; 1993.
118. Prati P, Moretti RH. Study of clarification process of sugar cane juice for consumption. Food Sci Technol (Campinas). 2010;30:776–83.
119. Thai CC, Bakir H, Doherty WO. Insights to the clarification of sugar cane juice expressed from sugar cane stalk and trash. J Agric Food Chem. 2012;21:2916–23.
120. Laksameethanasan P, Somla N, Janprem S, Phochuen N. Clarification of sugarcane juice for syrup production. Proc Eng. 2012;32:141–7.
121. Wyman CE. Ethanol fuel. In: Cleveland CJ, Ayres RU, Costanza R, et al., editors. Encyclopedia of energy, vol. 2. Philadelphia, PA: Elsevier Science; 2004. p. 541–55.
122. Gonzales JE Method for producing sugar cane juice. US 6245153 B1 (2001).
123. Leiper KA, Schlee C, Tebble I, Stewart GG. The fermentation of beet sugar syrup to produce bioethanol. J Inst Brew. 2006;112:122–33.
124. Hinková A, Bubník Z. Sugar beet as a raw material for bioethanol production. Czech J Food Sci. 2001;19:224–34.
125. Dodic S, Popov S, Dodic J, Rankovic J, Zavargo Z, Mucibabic RJ. Bioethanol production from thick juice as intermediate of sugar beet processing. Biomass Bioenergy. 2009;33:822–7.
126. Di Nicola G, Santecchia E, Santori G, Polonara F. Advances in the development of bioethanol: a review. In: Bernardes MAD, editor. Biofuel's engineering process technology. Rijeka, Croatia: InTech Publisher; 2011.
127. Cheesman OD. Environmental impacts of sugar production: the cultivation and processing of sugarcane and sugar beet: background. In: Cheesman OD, editor. Environmental impacts of sugar production: the cultivation and processing of sugarcane and sugar beet. Surrey: CABI Bioscience; 2004. p. 1–10.
128. Zieminski K, Romanowska I, Kowalska-Wentel M, Cyran M. Effects of hydrothermal pretreatment of sugar beet pulp for methane production. Bioresour Technol. 2014;166:187–93.
129. Ogbonna JC, Mashima H, Tanaka H. Scale up of fuel ethanol production from sugar beet juice using loofa sponge immobilized bioreactor. Bioresour Technol. 2001;76:1–8.
130. Krajnc D, Glavi P. Assessment of different strategies for the co-production of bioethanol and beet sugar. Chem Eng Res Des. 2009;87:1217–31.
131. Asadi M. Beet-sugar handbook. Hoboken, NJ: Wiley; 2007.
132. Loginova K, Loginova M, Vorobiev E, Lebovka NI. Better lime purification of sugar beet juice obtained by low temperature aqueous extraction assisted by pulsed electric field. LWT – Food Sci Technol. 2012;46:371–4.

133. Dziugan P, Balcerek M, Pielech-Przybylska K, Patelski P. Evaluation of the fermentation of high gravity thick sugar beet juice worts for efficient bioethanol production. Biotechnol Biofuels. 2013;6:158–68.
134. Rajagopalan S, Ponnampalam E, McCalla D, Stowers M. Enhancing profitability of dry mill ethanol plants. Appl Biochem Biotechnol. 2005;120:37–50.
135. Yasri NG, Yaghmour A, Gunasekaran S. Effective removal of organics from corn wet milling steepwater effluent by electrochemical oxidation and adsorption on 3-D granulated graphite electrode. J Environ Chem Eng. 2015;3:930–7.
136. Naguleswaran S, Li J, Vasanthan T, Bressler D, Hoove R. Amylolysis of large and small granules of native triticale, wheat and corn starches using a mixture of alpha-amylase and glucoamylase. Carbohydr Polym. 2012;88:864–74.
137. Bothast RJ, Schlicher MA. Biotechnological processes for conversion of corn into ethanol. Appl Microbiol Biotechnol. 2005;67:19–25.
138. Kwiatkowski JR, McAloon AJ, Taylor F, Johnston DB. Modeling the process and costs of fuel ethanol production by the corn dry-grind process. Ind Crop Prod. 2006;23:288–96.
139. Sriroth K, Wanlapatit S, Piyachomkwan K. Cassava bioethanol. In: Lima MAP, Natalense APP, editors. Bioethanol. Rijeka, Croatia: In Tech Publisher; 2012.
140. Cates ES, Dinwiddie JA, Aux G, Batie C, Crabb G Process for starch liquefaction and fermentation. US 7915020 B2 (2011).
141. Kosugi A, Kondo A, Ueda M, et al. Production of ethanol from cassava pulp via fermentation with a surface-engineered yeast strain displaying glucoamylase. Renew Energy. 2009;34:1354–8.
142. Rattanachomsri U, Tanapongpipat S, Eurwilaichitr L, Champreda V. Simultaneous non-thermal saccharification of cassava pulp by multi-enzyme activity and ethanol fermentation by Candida tropicalis. J Biosci Bioeng. 2009;107:488–93.
143. Lynd LR. Overview and evaluation of fuel ethanol from cellulosic biomass: technology, economics, the environment and policy. Energy Environ. 1996;21:403–65.
144. Limayem A, Ricke SC. Lignocellulosic biomass for bioethanol production: current perspectives, potential issues and future prospects. Prog Energy Combust Sci. 2012;38:449–67.
145. Tesfaw A, Assefa F. Current trends in bioethanol production by *Saccharomyces cerevisiae*: substrate, inhibitor reduction, growth variables, coculture, and immobilization. Int Scholar Res Not. 2014;2014:1–11.
146. Dien BS, Cotta MA, Jeffries TW. Bacteria engineered for fuel ethanol production: current status. Appl Microbiol Biotechnol. 2003;63:258–66.
147. Larsson S, Sainz AQ, Reimann A, Nilvebrant NO, Jonsson LJ. Influence of lignocellulose-derived aromatic compounds on oxygen-limited growth and ethanolic fermentation by *Saccharomyces cerevisiae*. Appl Biochem Biotechnol. 2000;84:617–32.
148. Ncibi MC. Bioconversion of renewable bioresources and agricultural by-Products into bioethanol. Recent Pat Chem Eng. 2010;3:165–79.
149. Kang Q, Appels L, Tan T, Dewil R. Bioethanol from lignocellulosic biomass: current findings determine research priorities. Sci World J. 2014;2014:1–13.
150. Banerjee S, Mudaliar S, Sen R, et al. Commercializing lignocellulosic bioethanol: technology bottlenecks and possible remedies. Biofuels Bioprod Biorefin. 2010;4:77–93.
151. Talebnia F, Karakashev D, Angelidaki I. Production of bioethanol from wheat straw: an overview on pre-treatment, hydrolysis and fermentation. Bioresour Technol. 2010;101:4744–53.
152. Balcu I, Macarie CA, Segneanu AE, Oana R. Combined microwave-acid pretreatment of the biomass. In: Shaukai SS, editor. Progress in biomass and bioenergy production. Rijeka, Croatia: In Tech Publisher; 2011. p. 223–2238.
153. Lu X, Xi B, Zhang Y, Angelidaki I. Microwave pretreatment of rape straw for bioethanol production: focus on energy efficiency. Bioresour Technol. 2011;102:7937–40.
154. Avellar BK, Glasser WG. Steam-assisted biomass fractionation I: process considerations and economic evaluation. Biomass Bioenergy. 1998;14:205–18.

155. Agbor VB, Cicek N, Sparling R, Berlin A, Levin DB. Biomass pretreatment: fundamentals toward application. Biotechnol Adv. 2011;29:675–85.
156. Sarkar N, Ghosh SK, Bannerjee S, Aikat K. Bioethanol production from agricultural wastes: an overview. Renew Energy. 2012;37:19–27.
157. Mabee WE, Gregg DJ, Arato C, et al. Updates on softwood-to-ethanol process development. Appl Biochem Biotechnol. 2006;129:55–70.
158. Li P, Cai D, Luo Z, Qin P, et al. Effect of acid pretreatment on different parts of corn stalk for second generation ethanol production. Bioresour Technol. 2016;206:86–92.
159. Bouza RJ, Gub Z, Evans JH. Screening conditions for acid pretreatment and enzymatic hydrolysis of empty fruit bunches. Ind Crop Prod. 2016;84:67–71.
160. Gaur R, Soam S, Sharma S, et al. Bench scale dilute acid pretreatment optimization for producing fermentable sugars from cotton stalk and physicochemical characterization. Ind Crop Prod. 2016;83:104–12.
161. Alvira P, Tomas-Pejo E, Ballesteros M, Negro MJ. Pretreatment technologies for an efficient bioethanol production process based on enzymatic hydrolysis: a review. Bioresour Technol. 2010;101:4851–61.
162. Mosier N, Wyman C, Dale B, et al. Features of promising technologies for pretreatment of lignocellulosic biomass. Bioresour Technol. 2005;96:673–86.
163. Zheng Y, Pan Z, Zhang R. Overview of biomass pretreatment for cellulosic production. Int J Agric Biol Eng. 2009;2:51–68.
164. Liu C, van der Heide E, Wang H, Li B, Yu G, Mu X. Alkaline twin-screw extrusion pretreatment for fermentable sugar production. Biotechnol Biofuels. 2013;6:97–108.
165. Hage RE, Brosse N, Chrusciel L, Sanchez C, Sannigrahi P, Ragauskas A. Characterization of milled wood lignin and ethanol organosolv lignin from Miscanthus. Polym Degrad Stab. 2009;94:1632–8.
166. Pan XJ, Arato C, Gilkes N, et al. Biorefining of softwoods using ethanol organosolv pulping: preliminary evaluation of process streams for manufacture of fuel grade ethanol and co-products. Biotechnol Bioeng. 2005;90:473–81.
167. Holtzapple MT, Humphrey AE. The effect of organosolv pretreatment on the enzymatic hydrolysis of poplar. Biotechnol Bioeng. 1984;26:670–6.
168. Chen H, Zhao J, Hua T, Zhao X, Liu D. A comparison of several organosolv pretreatments for improving the enzymatic hydrolysis of wheat straw: Substrate digestibility, fermentability and structural features. Appl Energy. 2015;150:224–32.
169. Hallett JP, Welton T. Room-temperature ionic liquids: solvents for synthesis and catalysis. 2. Chem Rev. 2011;111:3508–76.
170. Diego AF, Richard CR, Richard PS, Patrick M, Guillermo M, Robin DR. Can ionic liquids dissolve wood? Processing and analysis of lignocellulosic materials with 1-n-butyl-3-methylimidazolium chloride. Green Chem. 2007;9:63–9.
171. Hou XD, Li N, Zong MH. Significantly enhancing enzymatic hydrolysis of rice straw after pretreatment using renewable ionic liquid–water mixtures. Bioresour Technol. 2013;136:469–74.
172. Haykir NI, Bahcegul E, Bicak N, Bakir U. Pretreatment of cotton stalk with ionic liquids including 2-hydroxy ethyl ammonium formate to enhance biomass digestibility. Ind Crop Prod. 2013;41:430–6.
173. Financie R, Moniruzzaman M, Uemura Y. Enhanced enzymatic delignification of oil palm biomass with ionic liquid pretreatment. Biochem Eng J. 2016;110:1–7.
174. Canilha L, Chandel AK, Milessi TSS, et al. Bioconversion of sugarcane biomass into ethanol: an overview about composition, pretreatment methods, detoxification of hydrolysates, enzymatic saccharification, and ethanol fermentation. J Biomed Biotechnol. 2012;2012:1–15.
175. Salvachua D, Prieto A, Lopez-Abelairas M, Lu-Chau T, Martinez AT, Martinez MJ. Fungal pretreatment: an alternative in second-generation ethanol from wheat straw. Bioresour Technol. 2011;102:7500–6.

176. Sun Y, Cheng J. Hydrolysis of lignocellulosic materials for ethanol production: a review. Bioresour Technol. 2002;83:1–11.
177. Eggeman T, Elander RT. Process economic analysis of pretreatment technologies. Bioresour Technol. 2005;96:2019–25.
178. Soudham VP, Brandberg T, Mikkola JP, Larsson C. Detoxification of acid pretreated spruce hydrolysates with ferrous sulfate and hydrogen peroxide improves enzymatic hydrolysis and fermentation. Bioresour Technol. 2014;166:559–65.
179. Cavka A, Jönsson LJ. Detoxification of lignocellulosic hydrolysates using sodium borohydride. Bioresour Technol. 2013;136:368–76.
180. Monlau F, Sambusiti C, Antoniou N, Zabaniotou A, Solhy A, Barakat A. Pyrochars from bioenergy residue as novel bio-adsorbents for lignocellulosic hydrolysate detoxification. Bioresour Technol. 2015;187:379–86.
181. Zhang D, Ong YL, Li Z, Wu JC. Biological detoxification of furfural and 5-hydroxyl methyl furfural in hydrolysate of oil palm empty fruit bunch by *Enterobacter* sp. FDS8. Biochem Eng J. 2013;72:77–82.
182. Lee KM, Min K, Choi O, et al. Electrochemical detoxification of phenolic compounds in lignocellulosic hydrolysate for Clostridium fermentation. Bioresour Technol. 2015;187:228–34.
183. Nguyen N, Fargues C, Guiga W, Lameloise ML. Assessing nanofiltration and reverse osmosis for the detoxification of lignocellulosic hydrolysates. J Membr Sci. 2015;487:40–50.
184. Demirbas A. Bioethanol from cellulosic materials: a renewable motor fuel from biomass. Energy Sources. 2005;27:327–37.
185. Jiang LQ, Fang Z, Li XK, Luo J, Fan SP. Combination of dilute acid and ionic liquid pretreatments of sugarcane bagasse for glucose by enzymatic hydrolysis. Process Biochem. 2013;48:1942–6.
186. Chandel AK, Es C, Rudravaram R, Narasu ML, Rao LV, Ravindra P. Economics and environmental impact of bioethanol production technologies: an appraisal. Biotechnol Mol Biol Rev. 2007;2:14–32.
187. Badger PC. Ethanol from cellulose: a general review. In: Janick J, Whipkey A, editors. Trends in new crops and new uses. Alexandria, VA: ASHS Press; 2002. p. 17–21.
188. Beguin P, Aubert JP. The biological degradation of cellulose. FEMS Microbiol Rev. 1994;13:25–58.
189. Jeffries TW, Jin YS. Ethanol and thermotolerance in the bioconversion of xylose by yeasts. Adv Appl Microbiol. 2000;47:221–68.
190. Gong CS, Cao NJ, Du J, Tsao GT. Ethanol production from renewable resources. Adv Biochem Eng Biotechnol. 1999;65:207–41.
191. Kundu C, Lee JW. Bioethanol production from detoxified hydrolysate and the characterization of oxalic acid pretreated Eucalyptus (*Eucalyptus globulus*) biomass. Ind Crop Prod. 2016;83:322–8.
192. Agbogbo FK, Haagensen FD, Milam D, Wenger KS. Fermentation of acid-pretreated corn stover to ethanol without detoxification using *Pichia stipitis*. Appl Biochem Biotechnol. 2008;145:53–8.
193. Hamelinck CN, Hooijdonk GV, Faaij AP. Ethanol from lignocellulosic biomass: techno-economic performance in short-, middle- and long-term. Biomass Bioenergy. 2005;28:384–410.
194. Menon V, Rao M. Trends in bioconversion of lignocellulose: biofuels, platform chemicals and biorefinery concept. Prog Energy Combust Sci. 2012;38:522–50.
195. Cantarella M, Cantarella L, Gallifuoco A, Spera A, Alfani F. Comparison of different detoxification methods for steam-exploded poplar wood as a substrate for the bioproduction of ethanol in SHF and SSF. Process Biochem. 2004;39:1533–42.
196. Teixeira LC, Linden JC, Schroeder HA. Simultaneous saccharification and cofermentation of peracetic acid-pretreated biomass. Appl Biochem Biotechnol. 2000;84:111–27.

197. Cardona CA, Sanchez OJ. Fuel ethanol production: process design trends and integration opportunities. Bioresour Technol. 2007;98:2415–57.
198. Eiteman MA, Lee SA, Altman E. A co-fermentation strategy to consume sugar mixtures effectively. J Biol Eng. 2008;2:3–11.
199. Liu ZH, Chen HZ. Simultaneous saccharification and co-fermentation for improving the xylose utilization of steam exploded corn stover at high solid loading. Bioresour Technol. 2016;201:15–26.
200. Ho NWY, Chen ZD Stable recombinant yeasts for fermenting xylose to ethanol. US 8652772 B2 (2014).
201. Waldron K. Bioalcohol production biochemical conversion of lignocellulosic biomass. Cambridge: Woodhead Publishing; 2010.
202. Parisutham V, Kim TH, Lee SK. Feasibilities of consolidated bioprocessing microbes: from pretreatment to biofuel production. Bioresour Technol. 2014;161:431–40.
203. Akinosho H, Yee K, Close D, Ragauskas A. The emergence of *Clostridium thermocellum* as a high utility candidate for consolidate bioprocessing applications. Front Chem. 2014;2:1–17.
204. Hu N, Yuan B, Sun J, Wang SA, Li FL. Thermotolerant *Kluyveromyces marxianus* and *Saccharomyces cerevisiae* strains representing potentials for bioethanol production from Jerusalem artichoke by consolidated bioprocessing. Appl Microbiol Biotechnol. 2012;95:1359–68.
205. Treybal RE. Mass transfer operations. 3rd ed. Singapore: McGraw-Hill Books; 1980.
206. Madson PW. Ethanol distillation: the fundamentals. In: Jacques KA, Lyons TP, Kelsall DR, editors. The alcohol textbook. 4th ed. Nottingham: Nottingham University Press; 1995. p. 319–36.
207. Melo TCC, Machado GB, Belchior CRP. Hydrous ethanol–gasoline blends – combustion and emission investigations on a Flex-Fuel engine. Fuel. 2012;97:796–804.
208. Masum BM, Kalam MA, Masjuki HH, Ashrafur Rahman SM, Daggig EE. Impact of denatured anhydrous ethanol–gasoline fuel blends on a spark-ignition engine. RSC Adv. 2014;4:51220–7.
209. Kumar S, Singh N, Prasad R. Anhydrous ethanol: a renewable source of energy. Renew Sust Energ Rev. 2010;14:1830–44.
210. Osman YA, Ingram LO. Mechanism of ethanol inhibition of fermentation in *Zymomonas mobilis* CP4. J Bacteriol. 1985;164:173–80.
211. Yuan S, Zou C, Yin H, Chen Z, Yang W. Study on the separation of binary azeotropic mixtures by continuous extractive distillation. Chem Eng Res Des. 2015;93:113–9.
212. Gomis V, Pedraza R, Saquete MD, Font A, García-Cano J. Ethanol dehydration via azeotropic distillation with gasoline fractions as entrainers: a pilot-scale study of the manufacture of an ethanol–hydrocarbon fuel blend. Fuel. 2015;139:568–74.
213. Kiran B, Jana AK. A hybrid heat integration scheme for bioethanol separation through pressure-swing distillation route. Sep Purif Technol. 2015;14:307–15.
214. Magalad VT, Gokavi GS, Nadagouda MN, Aminabhavi TM. Pervaporation separation of water–ethanol mixtures using organic–inorganic nanocomposite membranes. J Phys Chem C. 2011;115:14731–44.
215. Kupiec K, Rakoczy J, Komorowicz T, Larwa B. Heat and mass transfer in adsorption–desorption cyclic process for ethanol dehydration. Chem Eng J. 2014;241:485–94.
216. Lapuerta M, Armas O, Rodriguez-Fernandez J. Effect of biodiesel fuels on diesel engine emissions. Prog Energy Combust Sci. 2008;34:198–223.
217. Omidvarborna H, Kumar A, Kim DS Characterization and exhaust emission analysis of biodiesel at different temperatures and pressures: laboratory study. J Hazard Toxic Radioact Waste 2015;19(2):1–6.
218. Van Gerpen J. Biodiesel processing and production. Fuel Process Technol. 2005;86:1097–107.
219. Kumar A, Nerella VKV. Experimental analysis of exhaust emissions from transit buses fuelled with biodiesel. Open Environ Eng J. 2009;2:81–96.

220. Wang W, Clark NN, Lyons DW, et al. Emissions comparisons from alternative fuel uses and diesel buses with a chassis dynamometer testing facility. Environ Sci Technol. 1997;31:3132–7.
221. Kegl B, Hribernik A. Experimental analysis of injection characteristics using biodiesel fuel. Energy Fuel. 2006;20:2239–40.
222. He BQ. Advances in emission characteristics of diesel engines using different biodiesel fuels. Renew Sust Energ Rev. 2016;60:570–86.
223. Can O, Öztürk E, Solmaz H, Aksoy F, Çinar C, Yücesu HS. Combined effects of soybean biodiesel fuel addition and EGR application on the combustion and exhaust emissions in a diesel engine. Appl Therm Eng. 2016;95:115–24.
224. Datta A, Mandal BK. A comprehensive review of biodiesel as an alternative fuel for compression ignition engine. Renew Sustain Energy Rev. 2016;57:799–821.
225. Anuar MR, Abdullah AZ. Challenges in biodiesel industry with regards to feedstock, environmental, social and sustainability issues: a critical review. Renew Sust Energ Rev. 2016;58:208–23.
226. Atabani AE, Silitonga AS, Badruddin IA, Mahlia TMI, Masjuki HH, Mekhilef S. A comprehensive review on biodiesel as an alternative energy resource and its characteristics. Renew Sust Energ Rev. 2012;16:2070–93.
227. Gomes MCS, Arroyo PA, Pereira NC. Influence of oil quality on biodiesel purification by ultrafiltration. J Membr Sci. 2015;496:242–9.
228. Firestone D. Gas chromatographic determination of mono- and diglycerides in fats and oils: summary of collaborative study. J AOAC Int. 1994;77:677–80.
229. Schenk PM, Thomas-Hall SR, Stephens E, et al. Second generation biofuels: high-efficiency microalgae for biodiesel production. Bioenergy Res. 2008;1:20–43.
230. Chisti Y. Biodiesel from microalgae. Biotechnol Adv. 2007;25:294–306.
231. Spolaore P, Joannis-Cassan C, Duran E, Isambert A. Commercial applications of microalgae. J Biosci Bioeng. 2006;101:87–96.
232. Stansell GR, Gray VM, Sym SD. Microalgal fatty acid composition: implications for biodiesel quality. J Appl Phycol. 2012;24:791–801.
233. Pagliero C, Ochoa N, Marchese J, Mattea M. Degumming of crude soybean oil by ultrafiltration using polymeric membranes. JAOCS. 2001;78:793–6.
234. Ncibi MC, Sillanpää M. Recent research and developments in biodiesel production from renewable bioresources. Recent Pat Chem Eng. 2013;6:183–91.
235. Singh J, Bargale PC. Development of a small capacity double stage compression screw press for oil expression. J Food Eng. 2000;43:75–82.
236. Santori G, Di Nicola G, Moglie M, Polonara F. A review analyzing the industrial biodiesel production practice starting from vegetable oil refining. Appl Energy. 2012;92:109–32.
237. Meher LC, Vidya SSD, Naik SN. Optimization of alkali catalyzed transesterification of *Pongamia pinnata* oil for production of biodiesel. Bioresour Technol. 2006;97:1392–7.
238. Martin GJO. Energy requirements for wet solvent extraction of lipids from microalgal biomass. Bioresour Technol. 2016;205:40–7.
239. Achten WMJ, Verchot L, Franken YJ, Mathijs E, Singh VP, Aerts R, Muys B. Jatropha bio-diesel production and use. Biomass Bioenergy. 2008;32:1063–84.
240. Fauzi AHM, Amin NAS. An overview of ionic liquids as solvents in biodiesel synthesis. Renew Sust Energ Rev. 2012;16:5770–86.
241. Pramparo M, Gregory S, Mattea M. Immersion vs. percolation in the extraction of oil from oleaginous seeds. J Am Oil Chem Soc. 2002;79:955–60.
242. Taher H, Al-Zuhair S, Al-Marzouqi AH, Haik Y, Farid MM. A review of enzymatic transesterification of microalgal oil-based biodiesel using supercritical technology. Enzyme Res. 2011;2011:1–25.
243. Shah S, Sharma A, Gupta MN. Extraction of oil from *Jatropha curcas* L. seed kernels by combination of ultrasonication and aqueous enzymatic oil extraction. Bioresour Technol. 2005;96:121–3.

244. Rubio-Rodríguez N, De Diego SM, Beltran S, Jaime I, Sanz MT, Rovira J. Supercritical fluid extraction of fish oil from fish by-products: a comparison with other extraction methods. J Food Eng. 2012;109:238–48.
245. Cardoso-Ugarte GA, Juarez-Becerra GP, Sosa-Morales ME, Lopez-Malo A. Microwave-assisted extraction of essential oils from herbs. J Microw Power Electromagn Energy. 2013;47:63–72.
246. Govindarajan L, Raut N, Alsaeed A. Novel solvent extraction for extraction of oil from algae biomass growth in desalination reject stream. J Algal Biomass Util. 2009;1:18–28.
247. Robles MA, González MPA, Esteban CL, Molina GE. Biocatalysis: towards ever greener biodiesel production. Biotechnol Adv. 2009;27:398–408.
248. Karmakar A, Karmakar S, Mukherjee S. Properties of various plants and animals feedstocks for biodiesel production. Bioresour Technol. 2010;101:7201–10.
249. Helwani Z, Othman MR, Aziz N, Kim J, Fernando WJN. Solid heterogeneous catalysts for transesterification of triglycerides with methanol: a review. Appl Catal A. 2009;363:1–10.
250. Berchmans HJ, Hirata S. Biodiesel production from crude *Jatropha curcas* L. seed oil with a high content of free fatty acids. Bioresour Technol. 2008;99:1716–21.
251. Stojkovic IJ, Stamenković OS, Povrenović DS, Veljković VB. Purification technologies for crude biodiesel obtained by alkali-catalyzed transesterification. Renew Sust Energ Rev. 2014;32:1–15.
252. Akoh CC, Chang SW, Lee GC, Shaw JF. Enzymatic approach to biodiesel production. J Agric Food Chem. 2007;55:8995–9005.
253. Ribeiro BD, de Castro AM, Coelho MA, Freire DM. Production and use of lipases in bioenergy: a review from the feedstocks to biodiesel production. Enzyme Res. 2011;2011:1–16.
254. Narwal SK, Gupta R. Biodiesel production by transesterification using immobilized lipase. Biotechnol Lett. 2013;35:479–90.
255. Kusdiana D, Saka S. Effects of water on biodiesel fuel production by supercritical methanol treatment. Bioresour Technol. 2004;91:289–95.
256. Kafuku G, Lee KT, Mbarawa M. Non-catalytic and catalytic transesterification: A reaction kinetics comparison study. Int J Green Energy. 2015;12:551–8.
257. da Silva C, Oliveira JV. Biodiesel production through non-catalytic supercritical transesterification: current state and perspectives. Braz J Chem Eng. 2014;31:271–85.
258. Eriksson O, Bisaillon M, Haraldsson M, Sundberg J. Enhancement of biogas production from food waste and sewage sludge – environmental and economic life cycle performance. J Environ Manag. 2016;175:33–9.
259. Melikoglu M. Vision 2023: assessing the feasibility of electricity and biogas production from municipal solid waste in Turkey. Renew Sust Energ Rev. 2013;19:52–63.
260. Abdeshahian P, Lim JS, Ho WS, Hashim H, Lee CT. Potential of biogas production from farm animal waste in Malaysia. Renew Sust Energ Rev. 2016;60:714–23.
261. Neves VT, Sales EA, Perelo LW. Influence of lipid extraction methods as pre-treatment of microalgal biomass for biogas production. Renew Sust Energ Rev. 2016;59:160–5.
262. European Union Directive 2009/28/EC – Promotion of the use of energy from renewable sources, 2001/77/EC and 2003/30/EC (2009)
263. Holm-Nielsen JB, Al ST. Oleskowicz-Popiel P. Bioresour Technol. 2009;100:5478–84.
264. Ali G, Nitivattananon V, Abbas S, Sabir M. Green waste to biogas: renewable energy possibilities for Thailand's green markets. Renew Sust Energ Rev. 2012;16:5423–9.
265. Mao C, Feng Y, Wang X, Ren G. Review on research achievements of biogas from anaerobic digestion. Renew Sust Energ Rev. 2015;45:540–55.
266. Ncibi MC, Sillanpää M. Recent patents and research studies on biogas production from bioresources and wastes. Recent Innov Chem Eng. 2014;7:2–9.
267. Navaratnasamy M, Edeogu I, Papworth L. Economic feasibility of anaerobic digesters. http://www.thecropsite.com/articles/1773/economic-feasibility-of-anaerobic-digesters/#sthash.XPf8WY9D.dpuf

268. Mussgnug JH, Klassen V, Schlüter A, Kruse O. Microalgae as substrates for fermentative biogas production in a combined biorefinery concept. J Biotechnol. 2010;150:51–6.
269. Liu Q, Wu L, Jackstell R, Beller M. Using carbon dioxide as a building block in organic synthesis. Nat Commun. 2015;6:1–15.
270. Li H, Opgenorth PH, Wernick DG, et al. Integrated electromicrobial conversion of CO_2 to higher alcohols. Science. 2012;335:1596.
271. Burke D. Dairy waste anaerobic digestion handbook. Olympia, WA: Environmental Energy Company; 2001. p. 1–57. http://www.makingenergy.com/Dairy%20Waste%20Handbook.pdf
272. Molino A, Nanna F, Ding Y, Bikson B, Braccio G. Biomethane production by anaerobic digestion of organic waste. Fuel. 2013;103:1003–9.
273. Tippayawong N, Thanompongchart P. Biogas quality upgrade by simultaneous removal of CO_2 and H_2S in a packed column reactor. Energy. 2010;35:4531–5.
274. Andriani D, Wresta A, Atmaja TD, Saepudin A. A review on optimization production and upgrading biogas through CO_2 removal using various techniques. Appl Biochem Biotechnol. 2014;172:1909–28.
275. Nock WJ, Walker M, Kapoor R, Heaven S. Modeling the water scrubbing process and energy requirements for CO_2 Capture to upgrade biogas to biomethane. Ind Eng Chem Res. 2014;53:12783–92.
276. Kim YJ, Nam YS, Kang YT. Study on a numerical model and PSA (pressure swing adsorption) process experiment for CH_4/CO_2 separation from biogas. Energy. 2015;91:732–41.
277. Allegue LB, Hinge J Biogas and bio-syngas upgrading. Aarhus, Denmark; 2012. http://www.teknologisk.dk/_root/media/52679_Report-Biogas and syngasupgrading.pdf
278. Liu Y, Li H, Wei G, Zhang H, Li X, Jia Y. Mass transfer performance of CO_2 absorption by alkanolamine aqueous solution for biogas purification. Sep Purif Technol. 2014;133:476–83.
279. Scholz M, Melin T, Wessling M. Transforming biogas into biomethane using membrane technology. Renew Sustain Energy Rev. 2013;17:199–212.
280. Sun Q, Li H, Yan J, Liu L, Yu Z, Yu X. Selection of appropriate biogas upgrading technology – a review of biogas cleaning, upgrading and utilization. Renew Sust Energ Rev. 2015;51:521–32.
281. Deng L, Hägg MB. Techno-economic evaluation of biogas upgrading process using CO2 facilitated transport membrane. Int J Greenhouse Gas Control. 2010;4:638–46.
282. Baker RW. Membrane technology and applications. 3rd ed. Hoboken, NJ: Wiley; 2012.
283. Molino A, Chianese S, Musmarra D. Biomass gasification technology: the state of the art overview. J Energy Chem. 2016;25:10–25.
284. Luque R, Pineda A, Colmenares JC. Carbonaceous residues from biomass gasification as catalysts for biodiesel production. J Nat Gas Chem. 2012;21:246–50.
285. Molino A, Braccio G. Synthetic natural gas SNG production from biomass gasification – thermodynamics and processing aspects. Fuel. 2015;139:425–9.
286. Colla L, Zanella D, Cavazzi M, Pelizza ML Apparatus and method for recovering energy from biomass, in particular from vegetable biomass. EP 2589646 A1 (2013).
287. Sharma AK. Equilibrium modeling of global reduction reactions for a downdraft (biomass) gasifier. Energy Convers Manag. 2008;49:832–42.
288. Aylott M. Biomass gasification in the UK – where are we now? Biomass Magazine; 2010. http://biomassmagazine.com/articles/5149/biomass-gasification-in-the-ukundefinedwhere-are-we-now
289. Lv P, Yuan Z, Wu C, Ma L, Chen Y, Tsubaki N. Bio-syngas production from biomass catalytic gasification. Energy Convers Manag. 2007;48:1132–9.
290. Courson C, Makaga E, Petit C, Kiennemann A. Development of Ni catalysts for gas production from biomass gasification reactivity in steam- and dry-reforming. Catal Today. 2000;63:427–37.

291. Guan Q, Wei C, Chai X. Energetic analysis of gasification of biomass by partial oxidation in supercritical water. Chin J Chem Eng. 2015;23:205–12.
292. Puig AM, Bruno JC, Coronas A. Review and analysis of biomass gasification models. Renew Sust Energ Rev. 2010;14:2841–51.
293. Hamelinck CN, Faaij APC, den Uil H, Boerrigter H. Production of FT transportation fuels from biomass; technical options, process analysis and optimisation, and development potential. Energy. 2004;29:1743–71.
294. Schmid JC, Wolfesberger U, Koppatz S, Pfeifer C, Hofbauer H. Variation of feedstock in a dual fluidized bed steam gasifier – influence on product gas, tar content, and composition Environ. Prog Sustain Energy. 2012;31:205–15.
295. Roos CJ. Clean heat and power using biomass gasification for industrial and agricultural projects. U.S. Department of Energy; 2010. http://www.energy.wsu.edu/Documents/BiomassGasification_2010.pdf
296. de Jong W. Biosyngas generation via gasifaction of biomass, gas cleaning, and fuel gas upgrading. In: Hu YH, Ma X, Fox EB, Guo X, editors. Production and purification of ultraclean transportation fuels. Washington, DC: ACS Publications; 2011.

Chapter 5
Biochemicals

Abstract Various chemicals are being produced from fossil resources, primarily petroleum (petrochemicals) as well as coal and natural gas. With the decade-long advances in the petrochemical industry, the extent of utilization of the end products produced from platform chemicals from fossil resources is so versatile that it affects every aspect in our today's life, including plastics, pesticides, dyes, personal care products, and even vitamins and aspirin.

In this chapter, various profitable opportunities to convert biomass and derived wastes into value-added biochemical compounds are showcased and discussed with respect to their production procedures (chemical and/or enzymatic) and yields and applied separation and purification technologies. This includes the production of organic acids (glycolic, 3-hydroxypropionic, and succinic acids), pharmaceuticals and biocosmetics (antibiotics, antibiotics, and antioxidants), fuel and food additives and biopesticides, all from renewable and available plants, aquatic biomass, and microorganisms.

In this context, joint academic and industrial research and development is highlighted as the main key to implement a bio-based economy able to produce a wide range of alternative green chemicals at a large scale, either within the current chemical, pharmaceutical, and cosmetic industries or via more sustainable bioprocessing activities in new integrated biorefineries.

5.1 Introduction

One of the main challenges facing the growing bioeconomy is to develop efficient and cost-effective green technologies for the sustainable production of commodity items from biomass.

As we have seen in the first chapter (*cf.* Figs. 1.5 and 1.6), numerous platform and building block chemicals are produced from fossil resources, primarily petroleum and also coal and natural gas. The extent of utilization of the end products derived from fossil resources is so versatile that it affects every aspect in our today's life. Thus, providing the green alternatives of those petroleum-derived chemicals will have two important consequences:

M. Sillanpää, C. Ncibi, *A Sustainable Bioeconomy*,
DOI 10.1007/978-3-319-55637-6_5

1. Benefit from the fully operational infrastructure and decade-long expertise of commodity production from various chemicals. Therefore, we just need to focus on producing a wide range of platform chemicals from renewable resources at competitive costs and in total respect of the environment.
2. Quicken the shift towards a bio-based economic model of production and consumption, which will speed up the planet healing, ease geopolitical tensions worldwide, create green jobs, and mitigate climate change.

The current situation is that numerous petrochemicals are still being produced from petroleum and used as building blocks or platform chemicals for the production of products as diverse as plastics, pesticides, paints, tires, shampoos, diapers, and even vitamins and aspirin (to name a few). With its limited reserves, heavy environmental legacy (spills, global warming, etc.), and prompted geopolitical tensions, the "days" of petroleum are numbered. The direct consequence is the orientation (or reorientation to be accurate) to the only true renewable resource of fixed carbon, biomass, to produce commodities.

Nonetheless, the competition with fossil resources is not over. Indeed, coal and natural gas are next in line to replace petroleum, which means trading one addiction for another, unless the biomass biorefining industry is ready on time to provide alternative green chemicals at a larger scale. Several industries could benefit from this new orientation mainly the chemical, pharmaceutical, and cosmetic industries. The produced biochemicals could be used as precursors for a variety of bulk chemicals and polymers including coatings, adhesives, and diverse plastic, polymer, and resin products [1–3].

In the present chapter, we will showcase and discuss the various profitable opportunities to convert biomass and valorize wastes into value-added chemical compounds including fine chemicals (organic acids), pharmaceuticals and bio-cosmetics, fuel and food additives, and biopesticides.

5.2 Fine Chemicals: Organic Acids

Organic acids are mainly produced via fermentative processes (chemical synthesis in some cases) and used as precursors for a variety of bulk chemicals and polymers including coatings, adhesives, and diverse plastic, polymer, and resin products. Among the top 12 value-added chemicals from biomass reported by the US Department of Energy, eight are organic acids [4], which illustrate how valuable those chemicals are. Consequently, one of the main challenges facing bioeconomy is to develop a sustainable and efficient (production wise) platform to produce organic acids from complex (even recalcitrant) biomass and derived by-products and wastes to the various industries at competitive cost.

5.2.1 Glycolic Acid (GA)

GA is an α-hydroxy acid (smallest of its group) containing both alcohol and carboxyl groups. Its chemical properties are of special interest for a wide range of industrial applications such as the leather, textile, and personal care products industries. Poly lactic-*co*-glycolic acid (PLGA), a copolymer of polyglycolic acid (PGA) and polylactic acid (PLA), is reported to be the best biomaterial available for drug delivery, design and performance wise [5]. Glycolic acid can also be polymerized to polyglycolic acid (PGA), a valuable polymer for food packaging considering its high gas barrier properties and mechanical strength [6]. Nowadays, most of the production of GA is secured from petrochemical resources, but the worldwide production yields are far from sufficient. Indeed, for the PGA production sector, the annual market requirements are estimated at 500,000 tons, whereas the annual GA production is only around 40,000 tons. Thus, the expansion of the GA production industry (forecasted market size of $200 million by 2020 [7]) is necessary and well justified, not from fossil resources this time but from renewable biomass using formative routes. The objective is to join sustainability and profitability.

In the beginning of the orientation, several microorganisms were reported in the literature for their ability to synthesize GA. But the high cost and the recourse to petrochemical compounds in the process such as ethylene glycol [8] and glycolonitrile [9] seriously constrained this production scheme. Thus, based on the bioeconomy philosophy, the use of available and low-cost biomass is well positioned to ensure a continuous, eco-friendly, and flexible pathway to produce GA, or any other biochemical for that matter.

In general, the transition towards bio-based industrial process to produce organic acids was tightly linked to the innovative sustainable bioprocesses using renewable resources (fractionation to simple sugars and then fermentation to organic acids).

Recent R&D came up with new bio-based pathways to produce GA, mainly based on engineered microorganisms. For instance, the fermentative production of GA from renewable resources can be achieved using an engineered strain of *E. coli* with an overexpressed glyoxylate shunt in the tricarboxylic acid cycle [10]. Other scientists also proved the feasibility of synthesizing glycolic acid from renewable resources using the yeasts *S. cerevisiae* and *Kluyveromyces lactis* [11]. The related results showed that the engineered microorganisms were tolerant to high acid concentration (up to 50 g/L). The *S. cerevisiae* strain, provided with d-xylose and ethanol as carbon source, produced around 1 g/L of GA. Under the same cultivation conditions, the other yeast strain, *K. lactis*, produced a much higher yield approximating 15 g of GA per liter.

5.2.2 3-Hydroxypropionic Acid (3-HPA)

3-HPA is a three-carbon platform chemical identified as one of the top-value chemicals from biomass [4]. The interesting fact about this compound is that, for economic and environmental reasons, there is no viable route to produce 3-HPA from fossil fuel feedstocks. This cleared the way for sustainable biological routes to fill the void. In this regard, the two main feedstocks for the production of 3-HPA are glucose and glycerol. The most cost-effective feedstock is bioglycerol, a valuable by-product of the biodiesel industry. The bio-based routes to convert glycerol to 3-HAP, only requiring two enzymatic activities (glycerol dehydratase and aldehyde dehydrogenase), were based on the use of recombinant strains of *E. coli* [12] and *Klebsiella pneumoniae* [13]. Glucose was also used as feedstock to produce 3-HPA using *E. coli* through the malonyl-CoA route [14], as illustrated in Fig. 5.1, or using *S. cerevisiae* through the β-alanine pathway [15].

The production yields of 3-HPA, and other selected organic acids, are provided in Table 5.1, at the end of this "fine chemicals" section.

At this point, it has to be noted that some challenges are still facing the large-scale production of 3-HPA such as engineering more acid-tolerant microbial stains, enhancing the internal redox balance, and applying more efficient and cost-effective purification techniques to separate 3-HPA from the culture broth [15].

For the industrial sector, the bifunctionality of 3-HPA (acid and alcohol functions) is of special interest as it enables its conversion to many intermediate platform molecules by simple chemical transformations, thus widening its

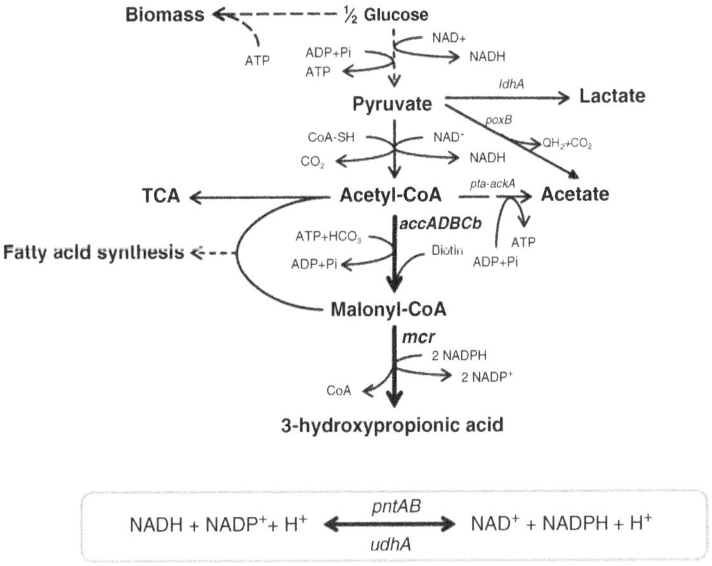

Fig. 5.1 Metabolic pathway of 3-hydroxypropionic acid production from glucose through malonyl-CoA route with recombinant *E. coli* strains [14]

industrial application. One option is to reduce 3-HPA to 1,3-propanediol, a chemical intermediate used for production of resins, engine coolants, and water-based inks [16] and combined with terephthalic acid to produce polytrimethylene terephthalate (PTT), used for the manufacture of fibers and resins [17]. Another option is the dehydration of 3-HPA to acrylic acid, and its ester and amide derivatives, which could be used in the production of polymers and copolymers in the surface coatings, absorbents, textiles, and adhesives industries [18]. The polymerization of 3-HPA into poly-3-hydroxypropionic acid is an interesting procedure for an added-value manufacturing scheme. Indeed, the produced 3-HPA-based bioplastic was reported to have strong mechanical properties and good susceptibility to biodegradation [19].

Overall, scientists are confirming that the prospect of 3-HPA as a novel platform molecule for many industries is more tangible than a decade ago, based on the advances made on two main fronts: biological conversions and catalytic chemical processes [20].

5.2.3 *Succinic Acid (SA)*

SA ($C_4H_6O_4$), or 1,4-butanedioic acid, is a versatile platform chemical used for the production bulk chemicals such as 1,4-butanediol, tetrahydrofuran, adipic acid, and γ-butyrolactone, as well as a monomer of some biodegradable polymers [21]. Several industries profited for the use of SA as precursor for the production of various commodities in the food, pharmaceutical, and polymer industries [22]. Currently, SA is mainly produced from butane through maleic anhydride hydrogenation. This "conventional" chemical route using a fossil resource was proven to be expensive (more so when the petroleum reserves start to decline) and also to contribute in some environmental problems [23]. Based of those setbacks, the orientation towards producing SA from low-cost biomass via biological routes has become a major trend in this field.

The industrial biological production of SA started with the use of fermentable sugars extracted from corn or sugarcane as substrates. Then, with the breakthroughs in the sector of biomass fractionation, an evident shift towards nonfood feedstocks to produce simple sugars (mainly lignocellulosic and algal biomass) was observed. In this context, various natural resources were investigated for the production of SA including sugarcane bagasse [24], bioglycerol [25], duckweed [26], and macroalgae *Laminaria japonica* [27].

Several microorganisms are being used to produce SA including *Actinobacillus succinogenes* and *Mannheimia succiniciproducens* (bacteria) and *Saccharomyces cerevisiae* (yeast). For instance, *A. succinogenes*, a natural succinate-producing strain was able to produce substantial amounts of succinic acid from a broad range of reducing sugars including arabinose, fructose, glucose, lactose, xylose, and sucrose [28]. Another study applied genome shuffling to improve the fermentative production of SA using *A. succinogenes* [29]. The related results showed that after

48 h in fed-batch fermentation, the genome shuffled strain improved the SA production by 73% when compared with the original strain, corresponding to production rate and yield of 1.99 g/L/h and 95.6 g/L, respectively. What is genome shuffling in brief? It is a technique based on deliberate genomic recombination (as opposed the spontaneous mutation in nature) aiming at generating, screening, and isolating new mutant strains with enhanced tolerance and increased potential to produce metabolites [30].

On the other hand, the use of metabolic engineered strains of *E. coli* and *Corynebacterium glutamicum* enabled a high-yielding production of SA, up to 127 g/L for the former and 146 for the latter [31].

Figure 5.2 gives a wider perspective on the various applications and products derived from succinic acid.

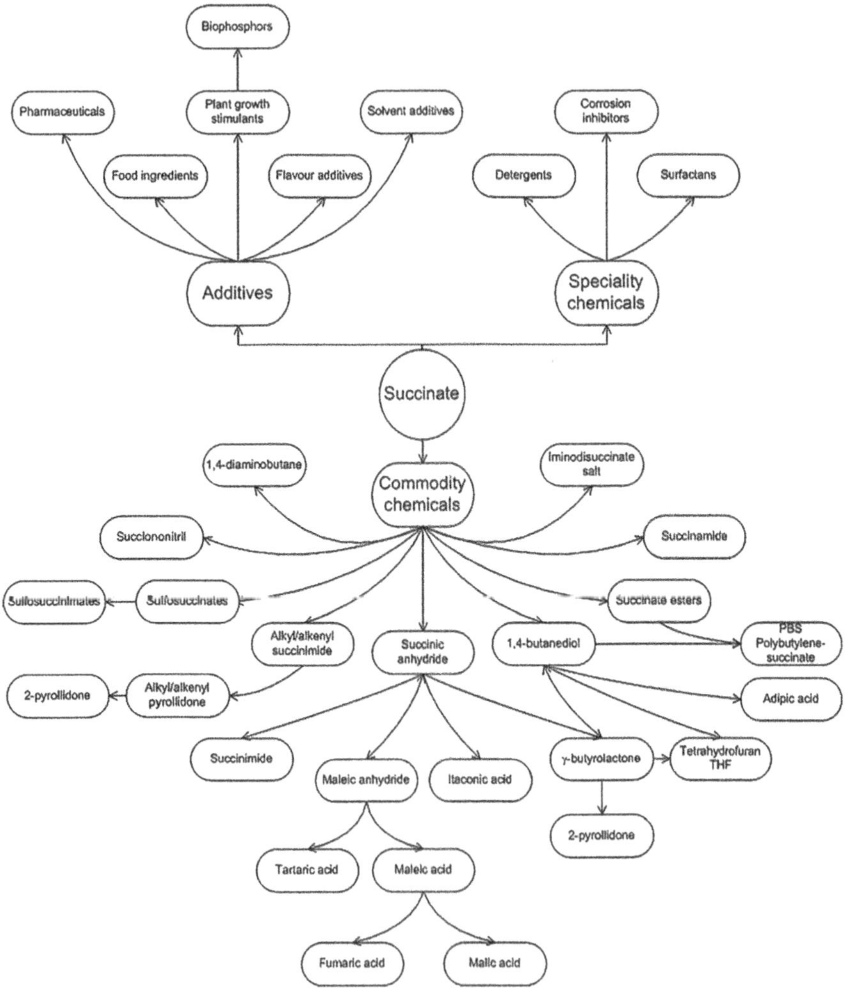

Fig. 5.2 Overview of possible applications and derived products from succinic acid [32]

5.2.4 Production Data for Selected Organic Acids

Table 5.1 illustrates the performances of various microorganisms to produce different organic acids for different renewable feedstocks including sugars from biomass (glucose, D-xylose and sucrose), starch, and sugarcane molasses, as well as from glycerol, a by-product of biodiesel production.

5.3 Pharmaceuticals from Biomass

5.3.1 Aspirin from Wood

Willow tree (*Salix* spp.) was considered as a medicinal plant for its ability to biosynthesize bioactive phytochemicals, including salicin [49]. Willow tree and salicin have been associated with salicylic acid, the key precursor molecule that has contributed to the discovery of acetylsalicylic acid, commonly known as aspirin [50].

These molecules (salicylic acid and related salicylates) were proven to be bioactive chemicals found in the bark and leaves of willow and poplar (*Populus* spp.) trees, as well as in some fruits, grains, and vegetables [51].

Many of those natural resources were included in the human diet. Thus, the ingested salicylates acted as analgesic (painkillers), anti-inflammatory, and antipyretic (reduce fever) compounds, thousands of years before their identification. Recently, the use of willow bark in phytotherapy as an analgesic drug has been gaining interest as an alternative renewable source to the synthetic analog, aspirin (acetylsalicylic acid). Besides, it was proven that willow salicylates did not induce side effects such as irritation and stomach ailments, generally caused by aspirin [52].

5.3.2 Bioactive Compounds from the Sea: Chitin and Chitosan

Chitin, the second most available polysaccharide on earth after cellulose, is a renewable compound found in many terrestrial and marine organisms including insects (exoskeletons), mollusks (endoskeletons), crustaceans (shells), and some fungi and algae [53], with an overall production estimated at 10^{11} tons of chitin per year [54].

Table 5.1 Organic acid production from various feedstocks using various microbial strains

Organic acid	Molecule	Feedstock	Microbial strain	Achieved titer (g/L)	Equiv. yield (g/g)	References
Glycolic acid ($C_2H_4O_3$)	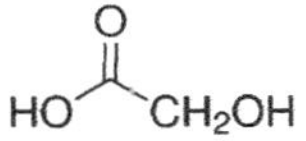	Glucose	*Escherichia coli*	52.2	0.38	[33]
		Xylose	*Kluyveromyces lactis*	15.0	0.28	[11]
		Glucose	*Corynebacterium glutamicum*	5.3	0.18	[34]
Pyruvic acid ($C_3H_4O_3$)	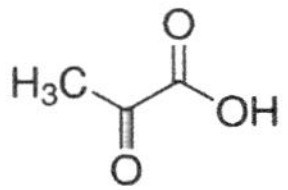	Glucose	*Saccharomyces cerevisiae*	135	0.54	[35]
		Glucose	*E. coli*	90.0	0.68	[36]
		Glucose	*C. glutamicum*	44.0	0.47	[37]
Lactic acid ($C_3H_6O_3$)	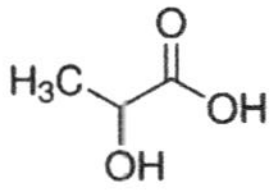	Glucose	*Rhizopus oryzae*	231	0.93	[38]
		Glucose	*C. glutamicum*	195	0.90	[39]
		Molasses	*Bacillus coagulans*	168	0.88	[40]
3-Hydroxypropionic acid ($C_3H_6O_3$)	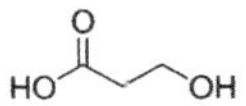	Glucose	*E. coli*	48.4	0.53	[41]
		Glycerol	*E. coli*	57.3	0.88	[12]
		Glycerol	*Klebsiella pneumoniae*	48.9	0.40	[42]
Succinic acid ($C_4H_6O_4$)	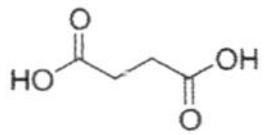	Glucose	*Mannheimia succiniciproducens*	52.4	0.76	[43]
		Glycerol	*Basfia succiniciproducens*	8.4	1.2	[44]
		Cassava starch	*E. coli*	127	0.86	[45]
Itaconic acid ($C_5H_6O_4$)	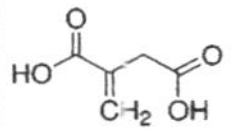	Glucose	*E. coli*	4.3	0.15	[46]
		Glucose	*Aspergillus terreus*	80.0	0.57	[47]
Muconic acid ($C_6H_6O_4$)	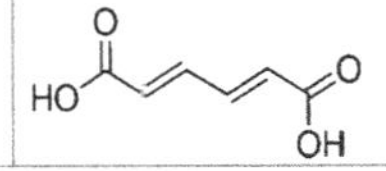	Glucose	*E. coli*	59.2	0.24	[48]
		Glucose	*P. putida*	32.4	1.16	

At a commercial level, chitin is mainly recovered from the by-products of the large seafood industry, especially from the crustacean-processing industries. For instance, 10^3 tons of chitin could be recuperated from shellfish waste annually [55]. The chitin content in various bioresources can be found in Table 3.5.

Several industrial valorization schemes were reported in the literature in the agricultural, food, and textile industries [56, 57]. In this section, the medical and pharmaceutical applications of chitin and chitosan will be discussed.

Many studies reported the promising use and incorporation of chitin and derived compounds in several medical and pharmaceutical chemicals and products such as wound dressing materials [58], matrix for enzymes immobilization [59], drug carriers [60], and medical components to repair wounded skin, nerves, cartilages, and bones [61].

In order to generate new biopolymers from chitin with improved biochemical properties, several chemical modification procedures were investigated. Thus, many research studies worked on dissolving chitin in various solvents and applied different chemical treatments including sulfonation for anticoagulant activity [62], TEMPO-mediated oxidation for antioxidant activity [63], and trimethylsilylation for enhanced solubility in common organic solvents [64]. *O*-acylation with fatty acids is another method reported in the literature for the improvement of chitin solubility. Several catalysts were used including methanesulfonic acid and perchloric acid. The generated compounds are formylated, propionylated, butyrylated, or valerylated chitins [65, 66], as well as mixed esters such as chitin co-(acetate/propionate) and chitin co-(acetate/butyrate) [67].

The main objectives from such modification is to preserve the fundamental skeleton of chitin and therefore its physiochemical and biochemical characteristics while increasing its water solubility and adding new functional properties [68]. The interesting outcome of such scientific endeavor is that both chitin and its chemical derivates are nontoxic, biocompatible, and biodegradable and have a versatile biological activity [69]. In general, chemical deacetylation of chitin to produce the commonly known derivative, chitosan, is the main modification in the industrial sector (Fig. 5.3).

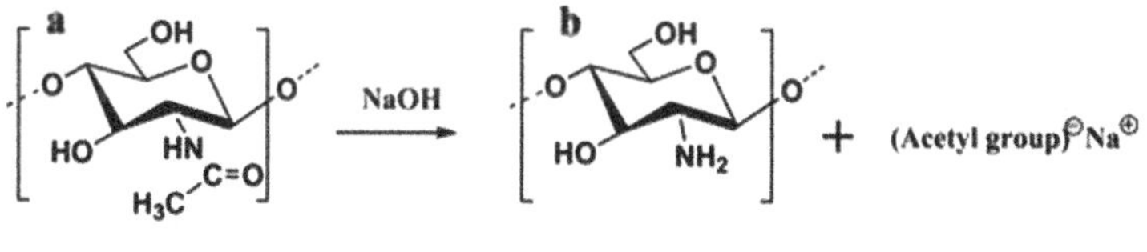

Fig. 5.3 Deacetylation of chitin (**a**) into chitosan (**b**)

5.3.3 Pharmaceutical Enzymes

Enzymes are biomolecules catalyzing and accelerating chemical reactions. Basically, any biochemical reaction in a living cell requires the catalytic intervention of specific enzymes. In the literature, enzymes could be grouped into six categories:

1. *Oxidoreductases*: catalyzing oxidation-reduction reactions via the transfer of electron between biological molecules [70].
2. *Transferases*: catalyzing the transfer of functional groups from one molecule to another. For instance, glutathione S-transferases, extracted from Arabian camels, have an interesting catalytic ability in conjugating the tripeptide glutathione to less toxic and water-soluble compounds easily excluded from the cell through membrane-based glutathione conjugate pumps [71].
3. *Hydrolases*: catalyzing the breaking of one compound into two molecules via the addition of water. Some of those hydrolases include lipases, nucleases, and peptidases, breaking down lipids, nucleic acids, and proteins [72].
4. *Lyases*: catalyzing the cleavage of a single molecule into two compounds. When the product is more important than the substrate, the reverse reaction occurs, and it is catalyzed by *synthases*. For example, cystathionine γ-synthase and cystathionine β-lyase have similar structures and many active-site residues; nonetheless they catalyze distinct side chain rearrangements in the two-step transsulfuration route converting cysteine to homocysteine, the precursor of methionine (an α-amino acid that is used in the biosynthesis of proteins) [73].
5. *Isomerases*: catalyzing the conversion of a molecule from one isomer to another. In general, isomers can be grouped into two main classes: stereoisomers and structural isomers. The former have the same ordering of bonds but different 3-D arrangement of bonded atoms, while the latter have a different ordering of bonds. Isomerases are able to catalyze the conversion reactions for both isomer classes. Glucose isomerase for instance is an important catalyst in the food industry for the conversion of glucose to fructose [74].
6. *Ligases*: catalyzing addition or synthesis reactions between two molecules with the breakdown of a pyrophosphate bond in ATP or a similar triphosphate. One of the main biological activity of those enzymes is DNA ligation, occurring during DNA replication, recombination, and repair events. For instance, in human cells, the DNA ligases encoded by three *LIG* genes are responsible for joining interruptions in the phosphodiester bonding forming the DNA backbone [75].

Table 5.2 Pharmaceutical application of enzymes [80, 81]

Generic name	Brand name	Treatment	Manufacturer
Asparaginase	Elspar	Acute lymphoblastic leukemia	Merck & Co., Inc., USA
Alteplase	Activase	Acute ischemic stroke	Genentech Inc., USA
Anistreplase	Eminase	Myocardial infarction	Wulfing Pharma GmbH, Germany
Imiglucerase	Cerezyme	Gaucher's disease	Genzyme Inc., USA
Rasburicase	Elitek	Hyperuricemia disease	GlaxoSmithKline Inc., Britain
Galsulfase	Naglazyme	Mucopolysaccharidosis VI (MPS VI)	BioMarin Pharmaceuticals, USA
Pegaspargase	Oncaspar	Acute lymphoblastic leukemia	Ben Venue Laboratories Inc., USA
Streptokinase	Streptase	Coronary artery thrombosis	Pfizer Inc., USA
Urokinase	Kinlytic	Pulmonary embolism	Microbix Biosystems Inc., Canada
Pegademase bovine	Adagen	Severe combined immunodeficiency (SCID)	Enzon Inc., USA
Sacrosidase	Sucraid	Congenital sucrase-isomaltase deficiency	Orphan Medical Inc., USA
Agalsidase beta	Fabrazyme	Fabry's disease	Genzyme Inc., USA

The important role of enzymes in many industries is well known, and it is based on several valuable properties such as high-substrate specificity, low toxicity, and rapid and efficient reaction at low concentrations and under mild operating condition (pH and temperature). Around hundred enzymes are being used in various industrial applications, among which over half are extracted from fungi, one-third from bacteria, as well as from animal (8%) and plant (4%) sources [76]. Noting that some microbial strains produce enzymes within the cell (intracellular enzymes), others produce externally (extracellular enzymes) [77].

The concept of enzyme utilization for therapeutic applications is relatively new. In the 1960s, lysosomal enzymes were included in replacement therapies for genetic deficiencies [78]. In 1987, the first recombinant enzyme drug, Activase® (Alteplase), was approved for commercialization. This injectable pharmaceutical drug is used to treat heart attacks, strokes, and blood clots [79].

Table 5.2 presents a selection of pharmaceutical enzymes developed and used for the treatment of various diseases.

5.3.4 Antibiotics and Bacteriocins

In order to adapt to limited nutrient supply and defend themselves against prokaryotic enemies, many bacteria produce antibacterial chemicals. Many of those biochemicals were discovered, but only some of them showed therapeutic properties, later co-opted by humans as antibiotics [82]. Since the well-known discovery of penicillin by Alexander Fleming in 1928, more than 5000 different antibiotics have been isolated from cultures of bacteria, fungi, and plant cells, among which 60% is synthesized by the genus *Streptomyces* alone [83].

Lactic acid bacteria can produce a variety of compounds during fermentation including organic acids, hydrogen peroxide, and bacteriocin [84], with the latter being a proteinaceous toxin exhibiting antimicrobial potency [85]. In this context, it has to be underlined that although bacteriocins might be considered as antibiotics, still major differences do exist. Indeed, bacteriocins restrict their activity to species closely related to the synthesizing strain, whereas antibiotics have a wider bioactivity spectrum.

In addition, bacteriocins are synthesized during the primary phase of growth; however, antibiotics are usually secondary metabolites, and the former are quickly digested by proteases in the digestive tract [86]. Besides, studies revealed that bacteriocins are inactivated by enzymes found in the gastrointestinal tract such as trypsin and pepsin [87].

Application wise, this differentiation is very important to avoid confusion with therapeutic antibiotics, which could lead to allergic reactions and other medical complications [88]. Overall, and based on the specific characteristics of each compound, antibiotics are of interest for the pharmaceutical industry, whereas bacteriocins are mainly used in the food industry (biopreservation) and also for therapeutic applications. Indeed, scientists started recently to explore bacteriocins as viable alternative to antibiotics [89]. The prospects of synergetic activity between antibiotics and bacteriocins to enhance current insufficient infection therapies were also investigated [90].

Tables 5.3 and 5.4 summarize the classification, chemical characterization, mode of action, and synthesizing microbial strain of several antibiotics and bacteriocins, respectively.

5.3.5 Vitamins

Vitamins are organic substances of great interest to human health, although required in small amounts. Since they cannot be synthesized at all or in sufficient quantity in the body, they must be included in the diet or prescribed as pharmaceutical-grade (single/multi) vitamins to cure deficiencies and avoid serious health complications.

Table 5.3 Overview of selected antibiotics for therapeutic applications [91–94]

Class	Antibiotics	Molecular structure	Mode of action	Producing strains
ß- lactams	Penicillins (e.g., amoxicillin, flucloxacillin) and cephalosporins (e.g., cephalexin)	(Penicillin core) 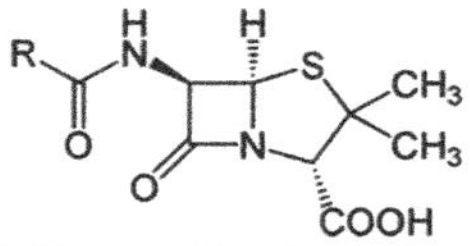	Inhibit bacteria cell wall biosynthesis	*Penicillium chrysogenum*, *Penicillium notatum*, and *Acremonium chrysogenum*
Aminoglycosides	Streptomycin, neomycin, kanamycin, and paromomycin	(Streptomycin) 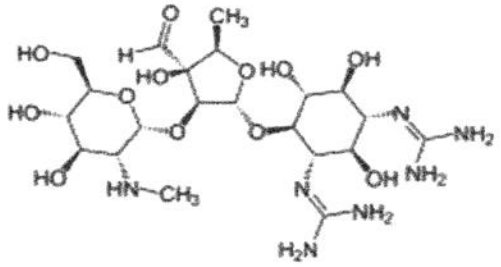	Inhibit the synthesis of proteins by bacteria, leading to cell death	*Streptomyces cattleya*, *S. fradiae*, and *S. griseus*
Ansamycins	Geldanamycin, rifamycin, and naphthomycin	(Geldanamycin) 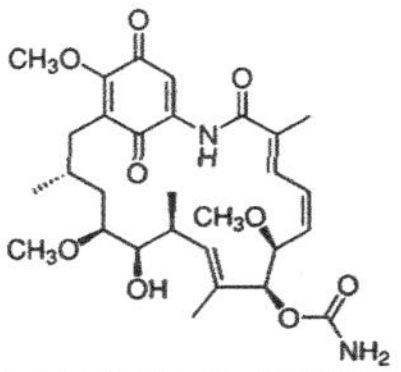	Inhibit the synthesis of RNA by bacteria, leading to cell death	*Amycolatopsis mediterranei* and *Streptomyces hygroscopicus*

(continued)

Table 5.3 (continued)

Class	Antibiotics	Molecular structure	Mode of action	Producing strains
Tetracyclines	Tetracycline, doxycycline, lymecycline, and oxytetracycline	(Oxytetracycline) 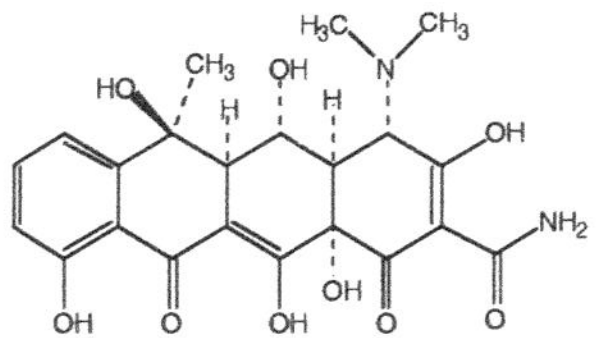	Inhibit synthesis of proteins by bacteria, preventing growth	*Streptomyces aureofaciens* and *S. rimosus*
Quinolones	Ciprofloxacin, levoflɔxacin, trovafloxacin	(Ciprofloxacin) 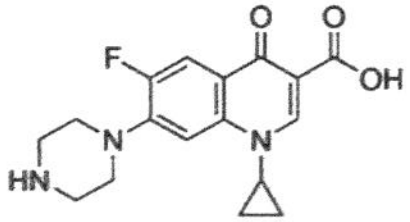	Interfere with bacteria DNA replication and transcription	*S. rimosus* M4018 and *S. rimosus* TM-55

Table 5.4 Overview of selected bacteriocins for therapeutic and food biopreservation applications [95–97]

Bacteriocin	Mode of action	Producing strains
Nisin A	Inhibition of cell wall formation	*Lactococcus lactis*
Nukacin ISK-1		*Staphylococcus warneri*
Lacticin 481		*L. lactis* CNRZ 481
Lactococcin A, B	Membrane disruption	*L. lactis* subsp.
Geobacillin I		*Geobacillus thermodenitrificans* NG80–2
Mesentericin Y105		*Leuconostoc mesenteroides* Y105
Garvicin A	Inhibition of septum formation	*Lactococcus garvieae*
Lactococcin 972		*L. lactis*
Pep5	Pore former	*Staphylococcus epidermidis* 5
Epidermin		*S. epidermidis* Tu3298
Colicins	DNase and RNase	*E. coli*
Microcin B17	Prevents DNA decatenation	

Indeed, vitamin deficiencies can lead to severe diseases and even death in some serious cases. For instance, deficiency of vitamin A, involved in immune functions, vision, reproduction, and cellular communications, constitutes a major health issue in low-income countries, thus putting around 130 million children at increased risk of infectious diseases and mortality [98]. Among the disorders related to vitamin deficiencies, there are also blindness (vitamin A), beriberi (vitamin B_1), pellagra (vitamin B_3), anemia (vitamin B_6), scurvy (vitamin C), and rickets (vitamin D) [99].

Although vitamins originate primarily from plant tissues, most of the currently produced vitamins come from synthetic routes using fossil feedstocks, for practical (large-scale production) and economic reason (chemical synthesis is more cost-effective than extraction and purification). But despite those industrial advantages, the synthetic vs. natural dilemma is still being debated [100] and not only from a consumer perspective, inclined towards natural products. Indeed, several in vivo studies revealed some physiological limitations using synthetic vitamins due to a different bioavailability potential when compared with natural ones [101], that is, the proportion of vitamin absorbed by the intestines and made available for metabolic processes within the body.

Thus, one of the major orientations in bioeconomy is to provide humanity with vitamins from natural resources at a competitive cost. One option is to increase the amount of vitamins from plants by improving the agricultural practices and therefore optimizing the growth conditions of conventional species or through use of transgenic techniques, commonly known as biofortification [102]. The other option is to explore the potentialities of marine biomass, especially microalgae to produce vitamins at a larger scale and with no competition over soil and water [103].

Overall, most scientists agree that within the bioeconomy concept, the biotechnological route for the production of vitamins, especially from microalgae and microbes, is most viable alternative to replace chemically synthesized vitamins. Table 5.5 shows a selection of water and fat-soluble vitamins produced via

Table 5.5 Biotechnological processes for the production of water and fat-soluble vitamins [104, 105]

Vitamin	Enzyme (microorganism)	Procedure
Water-soluble vitamins:		
Vitamin B_2 (riboflavin)	(*Eremothecium ashbyii*, *Ashbya gossypii*, and *Bacillus* sp.)	Fermentative production from glucose
Vitamin B_3 (niacin or nicotininc acid)	Nitrilase (Rhodococcus rhodochrous)	Hydrolysis of 3-cyanopyridine to form corresponding acid (nicotinic acid) and ammonia
Vitamin B_7 (biotin)	(*Serratia marcescens*)	Fermentative production from glucose by genetically engineered bacterium
Vitamin B_{12} (cobalamin)	(*Propionibacterium shermanii* and *Pseudomonas denitrificans*)	Fermentative production from glucose
Vitamin C	2,5-diketo-D-gulonic acid reductase (*Corynebacterium* sp.)	Enzymatic conversion of 2,5-diketo-D-gluconate obtained through fermentative process to 2-keto-L-gulonic, followed by chemical conversion to L-ascorbic acid
Fat-soluble vitamins:		
Vitamin E (α-tocopherol)	(Freshwater microalgae *Euglena gracilis*)	Fermentative production from glucose
Vitamin K_2	(Mutated strain of *Bacillus subtilis*)	Fermentation using soybean extract

biotechnological procedures, i.e., fermentation and microbial or enzymatic transformation.

5.4 Biocosmetics

5.4.1 Cosmetic Ingredients from Biowastes: Antioxidants

Biocosmetic are cosmetic product containing biomass-derived ingredients or natural products having mechanisms of action based on biologic principles. These products, essentially used for cleansing, protecting, and moisturizing the skin, could be formulated and sold in various forms including lotions, creams and powders, and so forth. From the early times, humans used natural resources for cosmetic purposes including natural pigments, honey, milk, clay, and many others. Then, with the R&D in the chemical field during the "petroleum century," chemicals and products from the petrochemical industry were extensively used in cosmetics manufacturing.

Recently, the consumers' awareness about the health risks associates with some chemical ingredients in cosmetic products started to increase. Indeed, it was reported that makeup, shampoo, skin lotion, nail polish, and other cosmetic and personal care products contain chemical ingredients that "lack safety data" [106]. Phthalates, for instance, are a group of chemicals susceptible of having endocrine-disrupting effect, and it has been linked to early puberty in girls and a risk factor for later-life breast cancer [107]. Nonetheless, chemicals like dibutyl phthalate and diethyl phthalate are still found in cosmetic products like nail polish and synthetic fragrance, respectively.

In response to this increasing awareness from consumers, many industrialists in the cosmetics sector started to consider using biochemicals as ingredients in their products. In this section, chemicals from biomass that could be of interest as ingredient in cosmetics formulation and production will be considered. And since bioeconomy is a holistic concept, the so-called organic cosmetics, benefiting for the organic or natural label to gain new clientele and make more profit without following sustainable production procedure will not be included.

The exposure to free radicals whether from endogenous (cellular metabolism) and exogenous sources (UV radiation or pollution) was proven to damage the skin on the cellular and tissue levels. Although the human body can prevent this kind of damage, it could be easily overwhelmed, especially with the current way of living (unhealthy diets, higher risk of exposure to pollutants, etc.), which could induce an oxidative stress or immunosuppression, and can even trigger carcinogenesis [108].

The use of topical cosmetics supplemented with antioxidants helps in providing an additional protection to the skin by neutralizing reactive oxygen species. Antioxidants were also reported to have other valuable bioactive properties for pharmaceutical and cosmetic applications such as their anticarcinogenicity, antimutagenicity, antiallergenicity, bleaching, and antiaging [109].

Several bioactive molecules with antioxidants properties were extracted from diverse bioresources including different classes of polyphenols (flavonoids, anthocyanins, and phenolic acids), which were proven to have high antioxidant capacity and considerable therapeutic properties in protecting skin cells from damages induced by oxidative stress [110].

For instance, a growing interest by researchers and the cosmetic industry in polyphenolic compounds from the flavonoids family was remarked. The green tea extract, Catechin, showed a protective action against ultraviolet B radiation-induced photocarcinogenesis [111]. Note that tea polyphenols are unstable molecules (i.e., short biological activity). Thus, topical formulation of polyphenol has been stabilized by butylated hydroxytoluene in order to reduce its susceptibility to oxidation [112]. As a consequence, and depending on the degree of stabilization, not all cosmetic products containing green tea extracts will exhibit the same level of antioxidation activity.

Other kinds of bioresources were revealed to be promising sources for antioxidants including the invasive species lawn pennywort (*Hydrocotyle sibthorpioides*) [113], various seaweeds [114], and the residues of many fruit- and vegetable-processing industries [115].

5.4.2 *Cosmetic Ingredients from the Sea: Chitin and Collagen*

5.4.2.1 Chitin and Chitosan

Chitin, mainly obtained from crustaceans, and its well-known derivative chitosan have a wide range of industrial applications in the food, agricultural, biomedical, and chemical sectors [56]. Several chemical and biological properties of those chitinous materials were of interest for the cosmetic industry as thickening, gelling, foaming, and antimicrobial and moisturizing agents [116].

Several research studies explored the option of using insoluble chitin as an active carrier for cosmetic products. In one of those studies, chitin-based nanofibrils were embedded with antioxidant ingredients, namely, melatonin, lutein, and Ectoin, for skin protection. The results revealed that, due to their scavenging activity, those nanofibrils did enhance the penetration of the active ingredients through the skin layer, which improved the protection against UV solar radiation, responsible for photoaging and wrinkles [117]. Also, chitin polymers were included in various cosmetic and personal care products for skin care (creams and lotions), hair care (shampoos, dyes, and sprays), as well as oral care (toothpastes and mouthwashes) [118, 119].

Nonetheless, the main restriction for a wider application of chitin remains its very poor solubility in all commonly used solvents [120], hence the need for chemical modification to generate water-soluble chitin derivates such as carboxymethylchitin and chitosan. Note that, even though chitosan can be dissolved in aqueous media at pH < 6.5, its involvement in biological applications requiring neutralization would lead to changes in the shape and size of any chitosan-based material [121].

For instance, it was reported that after neutralization, chitosan-based films become stiff, which is an undesired feature for cosmetic applications. The addition of glycerol, which has a plasticizer property, reduced the stiffness of the films. Besides, since glycerol is polar molecule with hygroscopic properties, it makes the stratum corneum (outermost layer of the skin) softer and more pliable [122], which further justifies the incorporation of glycerol to the chitosan films for the treatment of dry skin conditions.

Further investigations on the application of chitosan in cosmetics revealed that the high molecular weight of this natural polymer tend to limit the transepidermal water loss [123], thus maintaining the skin humidity and therefore its softness. Chitosan was also reported to be a suitable ingredient for allergic skin as high molecular weight compounds were found to reduce skin irritation [124], in addition to its antibacterial and wound healing properties [125].

5.4.2.2 Collagen

Collagen is the most abundant protein in vertebrates, constituting around 25% of the total proteins. The conventional collagen is mainly extracted from the skins of cows, pigs, and other terrestrial animals [126]. Recently, and due the recurrent serious diseases affecting cattle, the use of animal-derived collagen was heavily restricted [127].

As a consequence, collagens from marine resources received increased attention and several studies successfully extracted collagens from the skin, muscle, or bones of various fish species including seabass (*Lates calcarifer*) [128], blacktip shark (*Carcharhinus limbatus*) [129], as well as several carp species [130].

As far as the cosmetic industry is concerned, collagens have an interesting set of properties that include high water retention capacity, antiradical activity, biocompatibility, and weak antigenicity [131]. In addition, collagen hydrolyzate, frequently incorporated in cosmetic formulations, is believed to protect the structure and the function of the skin, hence persevering or enhancing its appearance [132].

5.5 Fuel Additives from Platform Biomolecules

The addition of additives to neat or blended fuels and biofuels is one of the standard procedures to improve the engine performances and reduce emissions. Various kinds of fuel additives were developed including metal organic compound, oxygenates, ignition promoter, wax dispersant, antiknock agents, and lead scavengers.

For instance, one of the main characteristics of good fuel, suitable for spark-ignited internal combustion engines, is its knock resistance [133]. In general, knock-resistant fuels (mainly alcohols) have high octane numbers, and when used as antiknock additives to non-oxygenated gasoline, the octane number of the fuel increases [134]. This octane boosting effect of fuel additives is one illustration, and in the present section, the production of fuel additives from platform biomolecules derived from biomass will be presented.

5.5.1 Additives from Bioglycerol

Most of the biodiesel production is carried out via the transesterification of triglycerides (vegetable oils or animal fats) with low molecular weight alcohol (methanol or ethanol) in the presence of alkali-based catalysts [34]. During this conversion process, bioglycerol is generated as a by-product with an equivalent yield of 10 wt% of the total biodiesel production [135].

The R&D efforts to valorize this by-product and make the biodiesel production process more profitable and competitive [136] faced a big challenge. The direct combustion of crude bioglycerol in engines was not recommended because of its high viscosity, low heating value, and high autoignition [137]. Besides, the incomplete combustion of glycerol generates carcinogenic acrolein [138]. Therefore, new routes to valorize this by-product into value-added chemicals were considered. Among the possible conversion schemes, the chemical transformation of glycerol into value-added oxygenated fuel additives received great attention [139, 140].

In this context, several reaction pathways for catalytic conversion of glycerol were proposed including selective oxidation, hydrogenolysis (to propanediol), dehydration (to acrolein), pyrolysis, gasification, steam reforming, thermal reduction (to syngas), selective transesterification, and etherification (to fuel oxygenates) [141–144].

Two main catalytic conversions were proposed for the production of fuel additives from bioglycerol:

5.5.1.1 Etherification

This procedure converts glycerol into lower viscosity and higher volatility compounds, which could be used as fuel additives or solvents [145]. Some studies showed that high acidity and the use of porous catalysts tend to favor the conversion reaction [146]. The glycerol etherification can be carried out with various alcohols, e.g., tert-butyl alcohol. The regenerated products could be used as oxygenated additives for fuels including diesel and biodiesel, thus establishing eco-friendly and cost-effective production system.

In this regard, there is an increasing interest in the production of alkyl ethers of glycerol by etherification with isobutylene, which reacts with glycerol in the presence of acid catalysts to produce mono-, di-, and tri-tert-butyl glycerol ethers (MTBG, DTBG, and TTBG, respectively) [147, 148]. The higher ethers (DTBG and TTBG) were reported to be suitable fuel additives for diesel and biodiesel formulation. The incorporation of those oxygenated compounds in conventional diesel fuel is on special interest as it helps in reducing the emissions of carbon monoxide, particulate matter, nitrogen oxide, hydrocarbon, and aldehyde resulting from incomplete combustion [149], as well as improving the cold flow properties [150].

5.5.1.2 Acetylation

Bioglycerol acetylation with acetic acid or acetic anhydride is gaining interest among industrialists due to the importance of the generated products (cf. Fig. 5.4), namely, mono-, di-, and triacetyl glycerol (MAG, DAG, and TAG, respectively) [152], which could be used in a wide range of applications ranging from moisturizers to fuel additives [153].

Among the three compounds derived from bioglycerol acetylation, studies revealed that TAG can be added to conventional diesel in order to improve the

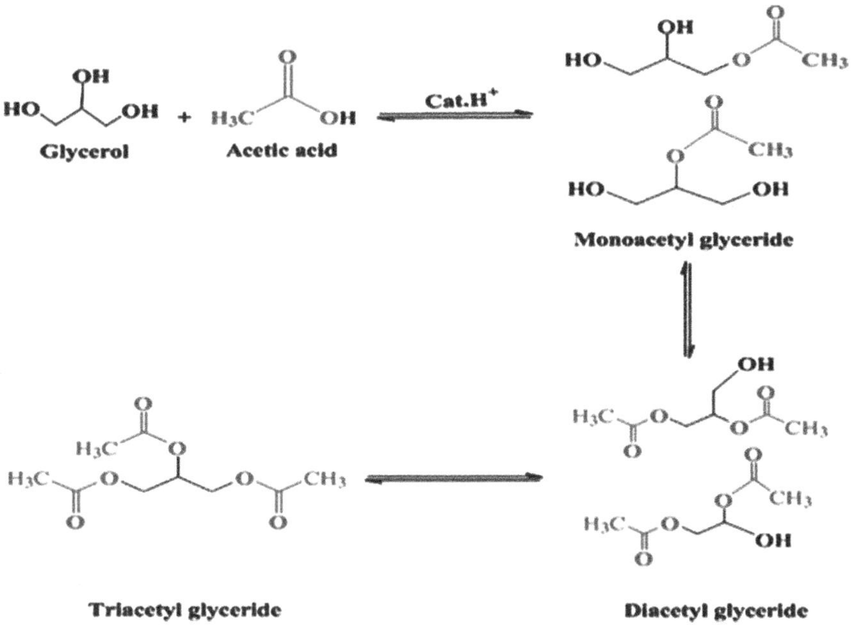

Fig. 5.4 Glycerol acetylation with acetic acid [151]

fuel combustion, reduce NOx emission in exhaust gas, and enhance the fuel cold and viscosity properties [154, 155]. The blend of TAG with fuel helped in ensuring a complete combustion as it serves as an efficient antiknocking agent. Also, adding 10% of TAG to biodiesel fuel was reported to considerably improve the performance of direct-injection diesel engines [156].

5.5.2 Additives from 5-Hydroxymethylfurfural (HMF)

HMF is an interesting platform chemical that can be obtained from sugars extracted from cellulosic biomass such as fructose, glucose, and sucrose [157]. At ambient conditions, HMF is a solid compound with very poor blending properties, which limits its application as fuel additive. Therefore, HMF was used as a platform molecule and subjected to catalytic chemical processes (oxidation and/or reduction) in order to convert carbohydrates into furanic compounds such as 2,5-dimethylfuran (DMF), 2,5-furandicarboxylic acid (FDCA), and 5-ethoxymethylfurfural (EMF) [158, 159].

Among these chemical compounds, EMF, the product of catalytic HMF etherification with ethanol, was reported to be a promising biofuel and additive for diesel [160]. Indeed, this ether has a high cetane number and good oxidation stability and is present in liquid form at room temperature. Also, EHM has a high energy density

(8.7 kWh/L), comparable with that of conventional gasoline (8.8 kWh/L) and diesel fuel (9.7 kWh/L), and considerably higher than that of bioethanol, the reference for liquid biofuel (6.1 kWh/L) [161]. In this context, DMF has also properties that are similar to petroleum-derived gasoline fuels and could be recommended as suitable fuel additive for petrol engines due to its high octane number [162]. The same study also stated that DMF is better fuel than ethanol with respect to production efficiency, energy density, handling, and storage.

Despite the potentialities of HMF as a platform molecule for the production of various interesting chemical compounds (fuel substitutes and fuel additives in the energetic sector), its large-scale production from cellulosic biomass is still limited, mainly due to high production costs. Indeed, although the catalytic production of HMF from hexoses (C6 sugars) was proved to be feasible in water, ionic liquids, and other organic solvents [163, 164], some limitations still exist.

For instance, the dehydration of fructose in pure water generates some by-products such as humins and levulinic acid, thus decreasing the overall production yield of HMF [165]. Also, even though the use of ionic liquids was proved to generate high yields of HMF at bench scale, the shift to an industrial scale production is not cost-effective yet [166].

Note that levulinate esters, product of the esterification of levulinic acid with alcohols, exhibit interesting properties that make them suitable to be used as fuel additives including low toxicity, high lubricity, flash point stability, and moderate flow properties under low temperature conditions [167]. Also, levulinate esters, especially ethyl and methyl levulinates, were also mentioned in the scientific literature as cold flow improvers in biodiesel [168] and oxygenate additives for both gasoline and diesel fuels [169].

5.6 Food Additives

5.6.1 Sweeteners: Xylitol

Scientists are suggesting that children's liking for all that is sweet, starting with mother's milk then fruit jams, chocolate, and ice cream, reflects a basic biological need. Indeed, the obvious preference for sweet-tasting foods and beverages during childhood is universal and evident among infants and children worldwide [170]. Many research studies are suggesting that liking sweet products by children could be related to the pain-reducing properties of sugars [171].

In this context, sweeteners are basically any chemical compounds interacting with the taste buds located on the tongue to enhance the perception of sweet taste. This specific action enables sweeteners to impart the sweet taste to other products, which is a mode of action highly sought by many food, beverage, and pharmaceutical industries.

Xylitol is a pentahydroxy natural polyol (i.e., five-carbon sugar alcohol) used as a commercial sweetener. It possesses several valuable characteristics including a high sweetening power and solubility, low calorie content, lack of carcinogenicity, and cariostatic properties as a non-fermentable sugar [172, 173].

Therefore, its production is attracting great interest from the food industry as a replacement for sucrose (table sugar) and as diabetic sweetener. It was indeed proved that xylitol has similar taste to sucrose, but with less calorific value (around 33% less) [174]. For instance, xylitol ingestion was revealed to cause a small increase of blood glucose; hence, it is frequently used as an alternative to high-energy supplements in diabetic medications [175].

The industrial production of xylitol from lignocellulosic biomass could be carried out via two main procedures:

1. *Chemical hydrogenation of xylose:*

   This process starts by the extraction of the hemicellulosic polysaccharide xylan from bioresources such as hardwoods or corncobs, which is hydrolyzed into xylose and then catalytically hydrogenated into xylitol under high pressure and temperature using catalysts (commonly Raney nickel) [176, 177].

   The main disadvantages related to this chemical procedure are the high energy demand, expensive downstream processing, and the release of toxic by-products (mainly Ni) [178].
2. *Biotechnological production of xylitol:*

   Many scientists are stating that the microbial production of xylitol is the future production process as it shows several advantages when compared with the chemical processes such as a considerably lower energy requirement and less restriction with the substrate (xylose) purity. Nonetheless, some drawbacks still exist and need to be dealt with, including the need for a detoxification procedure to remove inhibitors from the biomass hydrolyzate that would affect the biological activity of the fermenting microorganisms, in addition to the costly separation of the produced xylitol from the fermentation broth [179].

   Also, like most fermentation reactions, the biological conversion of xylose into xylitol is a time-consuming process that takes days up to weeks, depending of the microbial strain being used. Note that a couple of hours is required to achieve the chemical hydrogenation of xylose and thus produce xylitol [180].

   Regarding the bioconversion of xylose to xylitol, two biosynthetic routes were reported in the scientific literature for the fermentative conversion of xylose to xylitol by yeasts, which are considered to be the most efficient xylitol-producing microorganism (compared with bacteria or fungi). The first pathway involves the reduction of xylose to xylitol by NADH or NADPH-preferring enzyme, xylose reductase. In the second pathway, intermediary xylitol is oxidized to xylulose by NAD+ or NADP+–dependent enzyme, xylitol dehydrogenase [181]. Several studies showed that the key operating conditions affecting xylitol production are pH, temperature, stirring rate, oxygen transfer rate, inoculum ratio, and age, as well as the initial xylose concentration [182, 183].

Table 5.6 Production yields and productivity of xylitol by wild species (W) and metabolically engineered strains (E) of selected microorganisms

Microorganism	Substrate	Xylitol Yield (g/g)	Xylitol Productivity (g/L/h)	References
Bacteria:				
Enterobacter liquefaciens (W)	Xylose	33.3	0.35	[184]
Corynebacterium sp. (W)		80	0.4	[185]
Escherichia coli JM109 (E)		0.27	0.67	[186]
Gluconobacter oxydans (W)	Arabitol	51.4	1.9	[187]
Gluconobacter oxydans pSAXDHUP (E)		0.25	1.19	[188]
Escherichia coli ZUC99(pATX210) (E)	Arabinose	0.95	0.95	[189]
Fungi:				
Petromyces albertensis (W)	Xylose	39.8	0.166	[190]
Penicillium crustosum (W)		0.52	0.005	[191]
Penicillium janthinellum (W)		0.29	0.006	
Penicillium citrinum (W)		0.27	0.004	
Aspergillus niger (W)		0.36	0.004	
Fusarium oxysporum (W)		1	0.021	[192]
Yeasts:				
Candida sp. 559–9 (W)	Xylose	173	1.44	[193]
Candida boidinii NRRL Y-17213 (W)		53.1	0.16	[194]
Kluyveromyces marxianus YZJ015 (E)		0.83	1.49	[195]
Saccharomyces cerevisiae S3-TAL-TKL (E)		0.82	0.04	[196]
Pichia pastoris GS225 (E)	Glucose	0.078	0.29	[197]

Table 5.6 illustrates the capabilities of xylitol production by several microorganisms including wild and recombinant bacteria, fungi, and yeasts.

5.6.2 *Flavoring Agents: Vanillin*

5.6.2.1 Background and Current Status

Vanillin is simply the highest volume aroma chemical produced in the world [198]. It is mainly used by the food (yogurt, chocolate, ice cream, etc.) and cosmetic (perfumes, shampoos, soaps, etc.) industries as a flavoring or fragrance ingredient that is highly liked by consumers worldwide.

From a historical perspective, natural vanillin is produced from glucovanillin, after subjecting the beans of the tropical orchid *Vanilla planifolia* to a multistage curing process [199]. Thus, the agronomic technicalities needed during vine cultivation, pollination during flowering, along with the extreme care during beans harvesting and curing, explain why natural vanillin represents less than 1% of world market demand and has a market price almost 300 times higher than synthetic alternatives [200].

In order to meet the increasing demand for this flavoring compound, new synthetic pathways were developed. During the late 1930s, vanillin was synthesized from lignin-containing sulfite liquor, a by-product from the pulp and paper industry [201]. In the 1970s, vanillin was successfully prepared from eugenol (an essential oil extracted from cloves) via its isomerization to isoeugenol and then oxidation [202].

Currently, among the 20,000 tons of vanillin produced per year worldwide [203], 85% comes from guaiacol and 15% are lignin derivates (roughly 3000 tons annually) [204]. Although the guaiacol-based process to produce vanillin is almost by-product-free, which means simpler and less costly separation phase, this route is intrinsically unsustainable as it is depends on a petroleum-derived compound.

Thus, the viable option to produce vanillin is from renewable sources (other than vanilla beans for the practical and economic reasons stated earlier). In this context, researchers from around the world are working on two main routes to produce vanillin *from* biomass (controlled oxidation of lignin) and *via* biomass (microbial synthesis).

5.6.2.2 Vanillin Production from Lignin

Lignin extracted from various lignocellulosic resources as by-product of the pulp and paper industry (spent black liquor) or prior to fermentation procedures (e.g., sugarcane and wood processing) can be oxidized to obtain phenolic compounds. As a feedstock, and based on the extraction procedure, two main lignin types are commercially available for further processing into value-added chemicals including:

- *The sulfur-free lignins*, obtained from the fractionation and conversion of lignocellulosic materials into biofuels [205], organosolv pulping [206], as well as soda pulping of bioresources such as nonwood agricultural residues and nonwood fibers [207].
- *The sulfur-containing lignins*, mainly obtained from kraft [208] and sulfite [209] pulping processes. Kraft pulping remains the main procedure for pulp and paper production, as it allows the removal of around 90% of the lignin content under strong alkaline conditions [210].

Overall, in organosolv pulping, the lignocellulosic biomass is subjected to milder delignification processes, compared to kraft and sulfite pulping. Thus, the lignin fraction resulting from the organosolv treatment is the one undergoing a lesser transformation.

In general, organosolv lignin is characterized by a lower content of hydroxyl groups, higher molecular weight, and lower condensation degree, when compared with lignin extracted through harsher delignification procedures. Additionally, the absence of organic sulfur (either as thiol groups in kraft lignin or as sulfonic groups in lignosulfonates) is an advantage from the point of view of lignin valorization [211].

The chemical oxidation of lignin is a reaction occurring at high pH values (almost 14), high temperatures (up to 150 °C), and high pressure (more than 3 and up to 10 bar) [212, 213]. Several oxidants were reported in the related literature including air, oxygen, nitrobenzene, or metallic oxides [214, 215]. Under those alkaline oxidative operating conditions, lignin is degraded and oxidized, and among the by-products of the involved reactions is vanillin [216].

It is worth mentioning that due to the complex molecular structure of lignin and the drastic oxidative conditions, clearly identifying the reaction mechanism is a complicated endeavor. Nonetheless, some assumptions were reported to describe to the vanillin production via lignin oxidation including the following five stage routes [217]:

1. The formation of phenoxyl radicals through the detachment of one electron from phenoxyl anion
2. The formation of quinone methide via disproportionation of phenoxyl radical
3. The formation of coniferyl alcohol by nucleophilic addition of hydroxide ion
4. The probable formation of γ-carbonyl by oxidation of coniferyl alcohol
5. And, the formation of vanillin by the retro-aldol cleavage of the α- and β-unsaturated aldehydes

5.6.2.3 Microbial Production of Vanillin

The increasing demand from consumers to natural or organic products had a great impact on the industry of food and beverage-flavor additives, especially for the case of vanillin. Also, strict labeling legislations and clearing differentiation between synthetic/artificial products and organic/natural ones encouraged researchers and industrialists to develop new ways to produce natural vanillin (other than from vanilla pods) [218], using renewable feedstocks and sustainable production processes. Marketing wise, and in order to have an idea about the economic repercussion for the label "natural," the current natural vanillin available on the market (extracted from vanilla pods) has price range between 1200 and 4000 US dollars per kg, while the price of the synthetic product is less than 15 dollars per kg [219].

In this context, the production of vanillin from natural raw materials via biotechnological pathways seems to be the most viable option to win the "natural" label by US (FDA) and European legislations [220, 221]. The biotechnological techniques used for "biovanillin" production include:

1. In vitro cell or tissue culture process involving vanillin or its precursor-producing plant systems
2. Biotransformation of different plant-derived materials to generate vanillin by utilizing the enzymatic arsenal of plant cell cultures or microbial cultivation systems
3. Metabolic engineering in microorganisms, as well as plants, through introducing phenylpropanoid pathway genes to generate vanillin [222]

At this point, it has to be noticed that a biotechnological solution to produce vanillin via heterologous (i.e., derived from a different organism or specie) expression of the native vanilla orchid genes in microbes is still a challenge for the scientific community since the pathway is still unknown [223].

Several biomass-derived substrates were successfully used to produce vanillin through microbial bioconversion, including chemical compounds structurally similar to vanillin such as eugenol and ferulic acid (cf. Fig. 5.5), as well as from glucose [225] and curcumin [226].

Nevertheless, numerous research teams believe that ferulic acid, a secondary metabolite highly present in the cell walls of plants such as wheat, maize, and rice [227], is the most suitable natural substrate for bioconversion into vanillin. Indeed, a wide spectrum for microorganisms were reported in the literature for the production of vanillin from ferulic acid, and Table 5.7 presents a selection of those microorganisms along with their bioconversion efficiencies.

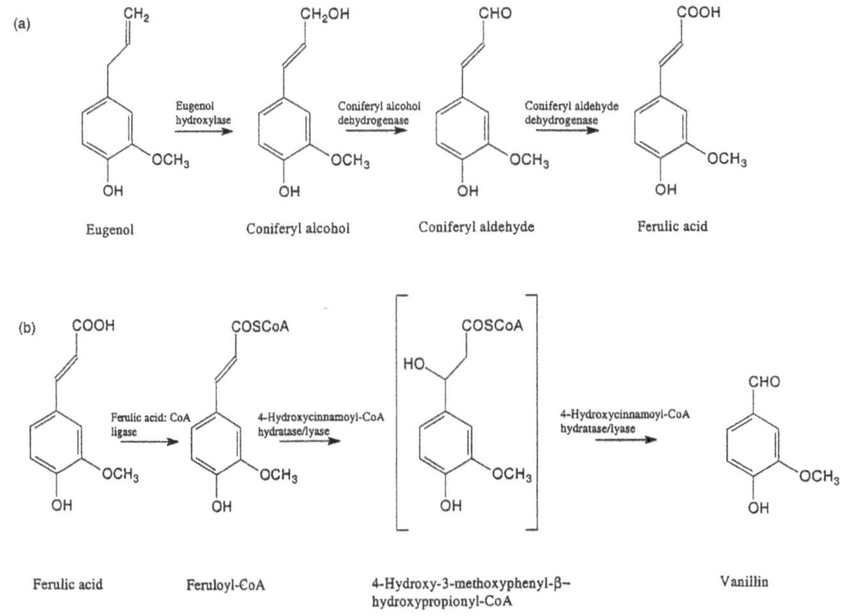

Fig. 5.5 The pathway from eugenol to ferulic acid (**a**) and from ferulic acid to vanillin (**b**) in *Pseudomonas* strains [224]

Table 5.7 Bioconversion of ferulic acid to vanillin using various microorganisms [228]

Microorganism	Bioconversion efficiency (%)
(B) *Bacillus subtilis* B7-S	63.30
(B) *Bacillus subtilis* B7	42.45
(B) *Enterobacter cloacae* Y219	29.25
(B) *Bacillus coagulans*	24.74
(B) *Streptomyces thermophila*	16.72
(B) *Escherichia coli* BL21	16.17
(B) *Bacillus stearothermophilus*	12.61
(B) *Pseudomonas putida*	12.44
(B) *Corynebacterium glutamicum*	10.27
(F) *Artomyces pyxidatus* KX320SH	9.06
(F) *Kluyveromyces marxianus* D10	7.64
(Y) *Rhodotorula rubra*	7.57
(F) *Cantharellus* sp. KX560JY	7.45
(F) *Candida kefyr*	3.99
(F) *Kluyveromyces lactis* B9	3.86

(B) bacteria, (F) fungi, and (Y) yeasts

5.7 Biopesticides

5.7.1 Chemical Pesticide vs. Biopesticides

Since the beginning, humanity's endeavor to domesticate wild plants for their consumption was challenged by the damaging activities of various pests including insects, fungi, weeds, nematodes, bacteria, and viruses, leading to drastic decreases in the quantity and quality of produced goods.

Nowadays, the situation did not change. Indeed, pests are still causing severe damage to cultivated crops around the world. It was estimated that in average 35–40% of all potential food and fiber crops are lost due to pests [229]. The substantial increase in global trade has resulted in numerous cases of accidental introduction of non native pests to new countries [230], which is a more serious threat than in native ecosystems, as natural predators are not necessarily present in the new areas. In other cases, plants and insects that are intentionally introduced in new locations become pests themselves.

Controlling these aggressive species, native or introduced, presents a serious challenge to the agricultural sector around the world, especially that fulfilling the food demand of a growing world population (more than 10 billion by 2100 [231]) will one the primary missions of bioeconomy.

During the last decades, and with the advancements in the agrochemical sector, the pest crisis was resolved to a great extent using chemical pesticides produced by big industries such as Bayer and BASF from Germany, Syngenta from Switzerland, and many US companies like Dow AgroSciences, Monsanto, and DuPont. But the subsequent impact on the environment and on humans, whether throughout the food chain or via direct exposure, was alarming and eventually led to the ban of many pesticides. But due the persistent nature of many of them, scientists are still monitoring the precarious impact of chemical pesticides on soils, water resources, and wildlife [232, 233].

Dichlorodiphenyltrichloroethane (DDT), for example, is a good showcase. Developed for large-scale applications as synthetic insecticide in the 1940s, it had a great impact in the worldwide campaign to combat malaria, typhus, and the other insect-borne human diseases among both military and civilian populations [234]. But the spectacular success of DDT as a pesticide and its extensive use around the world came with a high price. Indeed serious cases of pollution were reported, in addition to the appearance of many DDT-resistant insect species [235].

Overall, the heavy reliance on chemical pesticides during the last decades, mainly in intensive agricultural practices, had a heavy impact on the environment. The highly debated episode on honeybees is the most mediated one. From an agronomic perspective, contaminated soils and groundwater resulted in the loss of many productive lands. Also, residual fraction of pesticides in food products raised many safety concerns among consumers, which led governments to impose trade restrictions on crops imported from countries that are still using banned pesticides or exporting products with high pesticide residues [236].

As a direct consequence, the development of alternative eco-friendly pesticides became a necessity, hence the production and marketing of bio-based pesticides from various natural resources. Biopesticides are naturally occurring biological compounds able to control pests via nontoxic mode of actions. The US Environmental Protection Agency (EPA) categorizes biopesticides into three major groups: microbial, biochemical, and plant-incorporated protectants [237].

Biopesticides could also be grouped into three other groups based on their active substance, namely, microorganisms, biochemicals, and semiochemicals (such as insect pheromones) [238].

In this chapter, the term biopesticide is applied for any living or dead organisms with a pesticidal action, whether as predators or better competitors over available resources or via synthesized substances, from natural or genetically engineered plants or microbes.

Table 5.8 illustrates the advantages in using biopesticides and the disadvantages of chemical pesticides. Some limitations related to bio-based pesticides are also reported in order to have a fair assessment of their efficiency and to highlight the aspects that need to be optimized via worldwide R&D effort.

5.7.2 Pesticides from Plants and Microbes

Microbial biopesticides are pesticides derived from bacteria, fungi, oomycetes, viruses, and protozoa used for the biological control of pests, including insects, weeds, and plant pathogens [241]. In the agricultural sector, bacterial biopesticides

Table 5.8 Biopesticides vs. chemical pesticides: advantages and limitations [239, 240]

Biopesticides		Chemical pesticides	
Advantages	Disadvantages	Advantages	Disadvantages
– Nontoxic and non-pathogenic to nontarget organisms	– Their pest specificity could limit their marketing potentialities and therefore increase their prices	– Cost-effective and less labor input is required	– Reduction in beneficial insects due to the toxicity of these pesticides to nontarget pests, resulting in changes in biodiversity
– Specific to a single group or species of pests, therefore not affecting directly beneficial animals such as predators and parasitoids	– Decompose quickly, hence the need for special formulations and storage procedures	– Highly effective in controlling target pest populations	– Drift of sprays of chemical pesticides can cause severe problems in nearby crops, waterways, and many organisms in the contaminated ecosystem
– Could be used in habitats where chemical pesticides are prohibited including urban areas and near homes, schools, and lakes		– Easily available in large quantities, at high quality, and at reasonable price	– Chemical pesticides do leave residues in food, either by direct application or by bio-magnification
– Residues of microbial pesticides are nonhazardous		– Longer residual activity, providing greater pest control under field conditions	– Overuse of chemical pesticides induces resistance in target pests
– Less likely to have resistance issues			

are the mostly used microbial pesticides as it covers around 74% of the worldwide demand. Second comes fungal biopesticides with 10% market share, followed by viral biopesticides (5%) and other biopesticides [242].

The most widely applied microbial (bacterial) biopesticide is the entomopathogenic (harmful to insects) bacterium *Bacillus thuringiensis*. It was reported that the worldwide use of this entomopathogen is estimated at 13,000 tons [243]. The mode of action of this bacterial pesticide is characterized by the production, during the bacterial sporulation, of crystalline protein inclusions toxic to various insect species. These proteins (*aka*. protoxins) are transformed into toxic peptides in the midgut after ingestion of *B. thuringiensis*, leading to the death killing the susceptible insect via lysis of gut cells [244].

Table 5.9 presents a wide selection of biopesticides derived from various plants and microorganism (bacteria, fungi, viruses, and nematodes), along with their modes of action and commercially available products.

Table 5.9 Selection of plant-derived and microbial pesticides [245, 246]

Plant/microorganism	Targeted pest	Pesticidal activity	Commercial product (manufacturer)
Plants (phytochemical active principles):			
Neem – *Azadirachta indica* (Azadirachtin)	Broad range of insects	Insect growth regulator	Azatrol EC (Gordon's Professional)
Citrus peels (d-Limonene)	Fleas, aphids, fire ants and mosquitoes	Mosquito larvicide and insect repellent	Orange Guard 103 (Orange Guard)
Sabadilla Seed (vetarine)	Thrips and other insects	Neurotoxic causing paralysis and death	Veratran D (MGK)
Capsicum (capsaicin)	Insects	Repellent	Insect repellent (Hot Pepper Wax)
Bacteria:			
Agrobacterium radiobacter	*Agrobacterium tumefaciens* (Crown galls)	Antagonist[a]	Galltrol-A (AgBioChem) Dygall (AgBioResearch)
Bacillus popilliae	Larvae of various beetles	Stomach poison	Doom (Fairfax Biological)
Bacillus sphaericus	Mosquitoes	Larvicide	VectoLex (Valent Biosciences)
Bacillus subtilis	*Rhizoctonia*, *Fusarium*, *Alternaria*, and *Pythium*-causing root rot	Fungicide	Serenade (Agra Quest) Epic (Gustafson)
Bacillus thuringiensis var. *galleriae*	Cotton bollworm (*Helicoverpa armigera*) and tobacco hornworm (*Manduca sexta*)	Stomach poison	Spicturin (ISCB)
B. thuringiensis var. *tenebrionis*	Coleopteran beetles	Stomach poison	DiTera (Valent Biosciences) Trident (Mycogen)
B. thuringiensis var. *kurstaki*	Lepidopteran larvae and some leaf beetles	Stomach poison	Condor (Certis) Cordalene (Agrichem) Batik (Mycogen)
Pseudomonas syringae	Molds and rots attacking fruits during storage	Antagonist	Bio-Save (Jet Harvest Solutions)
Fungi:			
Alternaria destruens	Parasitic plants of the genus *Cuscuta*	Herbicide	Smolder G (Sylvan Bioproducts)

(continued)

Table 5.9 (continued)

Plant/microorganism	Targeted pest	Pesticidal activity	Commercial product (manufacturer)
Gliocladium virens	Soil pathogens causing damping off and root rot	Antagonist	Soil Guardl2G (Certis)
Metarhizium anisopliae	locusts and grasshoppers	Fungal infection	Bio-Blast (EcoScience) Pacer MA (Agri Life)
Myrothecium verrucaria	Various nematodes	Nematicidal	DiTera (Valent Biosciences)
Streptomyces griseoviridis	Wilt, seed rot, and stem rot	Antagonist	Mycostop (Verdera Oy)
Trichoderma harzianum	Wound pathogens	Antagonist	Root Shield (BioWorks)
Trichoderma viride	Rot diseases	Mycoparasitic	Ecosom TV (Agri Life)
Viruses:			
Granulosis virus	Leaf roller and codling moth	Insecticide	Capex (Ander-matt) Cyd-X (Certis)
Nucleopolyhedrovirus (NPV) isolated from *Anagrapha falcifera*	Lepidopterans	Insecticide	AfMNVP (Certis)
NPV from *Anticarsia gemmatalis*	Velvetbean caterpillar and sugarcane borer	Insecticide	Multigen (Embrapa) Polygen (Agrogen)
NPV from *Heliothis zea* and *H. virescens*	Bollworms	Insecticide	Biotrol (Certis) Elcar (Novartis)
NPV from *Syngrapha falcif*	*Helicoverpa* and *Cydia* spp.	Larvicide	NPVSf (Certis)
Nematodes:			
Heterorhabditis bacteriophora	Lepidopteran larvae and Japanese beetles	Entomopathogen	Heteromask (BioLogic) Terranem (Koppert)
Heterorhabditis megidis	Black vine weevils and soil insects	Entomopathogen	Larvanem (Koppert)
Phasmarhabditis hermaphrodita	Slugs	Slug-eating nematode	Nemaslug (Becker Underwood)
Steinernema carpocapsae	Weevils, cutworms, and termites	Entomopathogen	Bio-Safe-N (Certis) Hortscan (BioLogic)

[a]Antagonist: outcompeting species for available space and/or nutrients

After detailing the production of biofuels and biochemicals in the previous two chapters, the next chapter will deal with another important industrial activity within bioeconomy, which is the production of a wide range of biomaterials from renewable biomass and derived wastes in a sustainable yet profitable manner.

References

1. Ghaffar SH, Fan M, McVicar B. Bioengineering for utilisation and bioconversion of straw biomass into bio-products. Ind Crop Prod. 2015;77:262–74.
2. Sukan A, Roy I, Keshavarz T. Dual production of biopolymers from bacteria. Carbohydr Polym. 2015;126:47–51.
3. Lépine E, Riedl B, Wang XM, et al. Synthesis of bio-adhesives from soybean flour and furfural: relationship between furfural level and sodium hydroxide concentration. Int J Adhes Adhes. 2015;63:74–8.
4. U.S. Department of Energy. Top value added chemicals from biomass. 2004. http://www.nrel.gov/docs/fy04osti/35523.pdf
5. Makadia HK, Siegel SJ. Poly Lactic-co-Glycolic Acid (PLGA) as biodegradable controlled drug delivery carrier. Polymers (Basel). 2011;3:1377–97.
6. Robertson GL. Food packaging: principles and practice. 3rd ed. Boca Raton, FL: CRC Press; 2012.
7. Alonso S, Rendueles M, Díaz M. Microbial production of specialty organic acids from renewable and waste materials. Crit Rev Biotechnol. 2015;35:497–513.
8. Kataoka M, Sasaki M, Hidalgo AR, Nakano M, Shimizu S. Glycolic acid production using ethylene glycol-oxidizing microorganisms. Biosci Biotechnol Biochem. 2001;65:2265–70.
9. He YC, Xu JH, Su JH, Zhou L. Bioproduction of glycolic acid from glycolonitrile with a new bacterial isolate of *Alcaligenes* sp. ECU0401. Appl Biochem Biotechnol. 2010;160:1428–40.
10. Soucaille P. Glycolic acid production by fermentation from renewable resources. US Patent 20090155867A1 (2009).
11. Koivistoinen OM, Kuivanen J, Barth D, et al. Glycolic acid production in the engineered yeasts *Saccharomyces cerevisiae* and *Kluyveromyces lactis*. Microb Cell Factories. 2013;12:82–97.
12. Kim K, Kim SK, Park YC, Seo JH. Enhanced production of 3- hydroxypropionic acid from glycerol by modulation of glycerol metabolism in recombinant *Escherichia coli*. Bioresour Technol. 2014;156:170–5.
13. Huang Y, Li Z, Shimizu K, Ye Q. Simultaneous production of 3-hydroxypropionic acid and 1,3-propanediol from glycerol by a recombinant strain of *Klebsiella pneumonia*. Bioresour Technol. 2012;103:351–9.
14. Rathnasingh C, Mohan RS, Lee Y, et al. Production of 3-hydroxypropionic acid via malonyl-CoA pathway using recombinant *Escherichia coli* strains. J Biotechnol. 2012;157:633–40.
15. Kumar V, Ashok S, Park S. Recent advances in biological production of 3-hydroxypropionic acid. Biotechnol Adv. 2013;31:945–61.
16. Kraus GA. Synthetic methods for the preparation of 1,3-Propanediol. Clean. 2008;36:648–51.
17. Kurian JV. Sorona polymer: present status and future perspectives. In: Mohanty AK, Misra M, Drzal LT, editors. Natural fibers, biopolymers, and biocomposites. Boca Raton, FL: CRC Press; 2005.
18. Corma A, Iborra S, Velty A. Chemical routes for the transformation of biomass into chemicals. Chem Rev. 2007;107:2411–502.

19. Zhang D, Hillmyer MA, Tolman WB. A new synthetic route to Poly[3-hydroxypropionic acid] (P[3-HP]): ring-opening of 3-HP macrocyclic esters. Macromolecules. 2004;37: 8198–200.
20. Pina CD, Falletta E, Rossi M. A green approach to chemical building blocks. The case of 3-hydroxypropanoic acid. Green Chem. 2011;13:1624–32.
21. Cheng KK, Zhao XB, Zeng J, Zhang JA. Biotechnological production of succinic acid: current state and perspectives. Biofuels Bioprod Biorefin. 2012;6:302–18.
22. Jang YS, Kim B, Shin JH, et al. Bio-based production of C2–C6 platform chemicals. Biotechnol Bioeng. 2012;109:2437–59.
23. Gunnarsson IB, Karakashev D, Angelidaki I. Succinic acid production by fermentation of Jerusalem artichoke tuber hydrolysate with *Actinobacillus succinogenes* 130Z. Ind Crop Prod. 2014;62:125–9.
24. Liu R, Liang L, Cao W, et al. Succinate production by metabolically engineered *Escherichia coli* using sugarcane bagasse hydrolysate as the carbon source. Bioresour Technol. 2013;135: 574–7.
25. Sadhukhan S, Villa R, Sarkar U. Microbial production of succinic acid using crude and purified glycerol from a *Crotalaria juncea* based biorefinery. Biotechnol Report. 2016;10: 84–93.
26. Shen N, Wang Q, Zhu J, et al. Succinic acid production from duckweed (*Landoltia punctata*) hydrolysate by batch fermentation of *Actinobacillus succinogenes* GXAS137. Bioresour Technol. 2016;211:307–12.
27. Bai B, Zhou JM, Yang MH, et al. Efficient production of succinic acid from macroalgae hydrolysate by metabolically engineered *Escherichia coli*. Bioresour Technol. 2015;185: 56–61.
28. Xi YL, Dai WY, Xu R, et al. Ultrasonic pretreatment and acid hydrolysis of sugarcane bagasse for succinic acid production using *Actinobacillus succinogenes*. Bioprocess Biosyst Eng. 2013;36:1779–85.
29. Zheng P, Zhang K, Yan Q, Xu Y, Sun Z. Enhanced succinic acid production by *Actinobacillus succinogenes* after genome shuffling. J Ind Microbiol Biotechnol. 2013;40: 831–40.
30. Zhang YX, Perry K, Vinci VA. Genome shuffling leads to rapid phenotypic improvement in bacteria. Nature. 2002;415:644–6.
31. Becker J, Wittmann C. Advanced biotechnology: metabolically engineered cells for the bio-based production of chemicals and fuels, materials, and health-care products. Angew Chem. 2015;54:3328–50.
32. Beauprez JJ, Mey MD, Soetaert WK. Microbial succinic acid production: natural versus metabolic engineered producers. Process Biochem. 2010;45:1103–14.
33. Dischert W, Soucaille P. Method for producing high amount of glycolic acid by fermentation, US 13/258 366 (2015).
34. Zahoor A, Otten A, Wendisch VF. Metabolic engineering of *Corynebacterium glutamicum* for glycolate production. J Biotechnol. 2014;192:366–75.
35. van Maris AJ, et al. Directed evolution of pyruvate decarboxylase-negative *Saccharomyces cerevisiae*, yielding a C2-independent, glucose-tolerant, and pyruvate-hyperproducing yeast. Appl Environ Microbiol. 2004;70:159–66.
36. Zhu Y, Eiteman MA, Altman R, Altman E. High glycolytic flux improves pyruvate production by a metabolically engineered *Escherichia coli* strain. Appl Environ Microbiol. 2008;74: 6649–55.
37. Wieschalka S, Blombach B, Bott M, Eikmanns BJ. Bio-based production of organic acids with *Corynebacterium glutamicum*. Microb Biotechnol. 2013;6:87–102.
38. Yamane T, Tanaka R. Highly accumulative production of L(+)-lactate from glucose by crystallization fermentation with immobilized *Rhizopus oryzae*. J Biosci Bioeng. 2013;115: 90–5.

39. Tsuge Y, Yamamoto S, Kato N, et al. Overexpression of the phosphofructokinase encoding gene is crucial for achieving high production of D-lactate in *Corynebacterium glutamicum* under oxygen deprivation. Appl Microbiol Biotechnol. 2015;99:4679–89.
40. Xu K, Xu P. Efficient production of L-lactic acid using co-feeding strategy based on cane molasses/glucose carbon sources. Bioresour Technol. 2014;153:23–9.
41. Lynch MD, Gill RT, Lipscomb TEW. Methods for producing 3- hydroxypropionic acid and other products, US0045231A1 (2014).
42. Huang Y, Li Z, Shimizu K, Ye Q. Co-production of 3- hydroxypropionic acid and 1,3-propanediol by *Klebseilla pneumoniae* expressing *aldH* under microaerobic conditions. Bioresour Technol. 2013;128:505–12.
43. Cheng KK, Wang GY, Zeng J, Zhang JA. Improved succinate production by metabolic engineering. Biomed Res Int. 2013;2013:538790.
44. Scholten E, Dagele D. Succinic acid production by a newly isolated bacterium. Biotechnol Lett. 2008;30:2143–6.
45. Chen C, Ding S, Wang D, Li Z, Ye Q. Simultaneous saccharification and fermentation of cassava to succinic acid by *Escherichia coli* NZN111. Bioresour Technol. 2014;163:100–5.
46. Okamoto S, Chin T, Hiratsuka K, et al. Production of itaconic acid using metabolically engineered *Escherichia coli*. J Gen Appl Microbiol. 2014;60:191–7.
47. Huang X, Lu X, Li Y, Li X, Li JJ. Improving itaconic acid production through genetic engineering of an industrial *Aspergillus terreus* strain. Microb Cell Factories. 2014;13:119.
48. Xie NZ, Liang H, Huang RB, Xu P. Biotechnological production of muconic acid: current status and future prospects. Biotechnol Adv. 2014;32:615–22.
49. Mahdi JG. Biosynthesis and metabolism of β-d-salicin: a novel molecule that exerts biological function in humans and plants. Biotechnol Report. 2014;4:73–9.
50. Mahdi JG, Mahdi AJ, Bowen ID. Historical analysis of aspirin discovery, its relation to the willow tree and antiproliferative potential. Cell Prolif. 2006;39:147–55.
51. Raskin I. Role of salicylic acid in plants. Annu Rev Plant Physiol Plant Mol Biol. 1992;43: 439–63.
52. Chrubasik S, Eisenberg E, Balan E. Treatment of low back pain exacerbations with willow bark extract: a randomized double-blind study. Am J Med. 2000;109:9–14.
53. Zhang P, Zhou W, Wang P, Wang L, Tang M. Enhancement of chitosanase production by cell immobilization of *Gongronella* sp. JG. Braz J Microbiol. 2013;44:189–95.
54. Revathi M, Saravanan R, Shanmugam A. Production and characterization of chitinase from Vibrio species, a head waste of shrimp *Metapenaeus dobsonii* (Miers, 1878) and chitin of *Sepiella inermis* Orbigny, 1848. Adv Biosci Biotechnol. 2012;3:392–7.
55. Merzendorfer H. Chitin. In: Gabius HJ, editor. The sugar code: fundamentals of glycosciences. Weinheim: Wiley; 2011.
56. Hamed I, Ozogul F, Regenstein JM. Industrial applications of crustacean by-products (chitin, chitosan, and chitooligosaccharides): a review. Trends Food Sci Technol. 2016;48:40–50.
57. Hayes M, Carney B, Slater J, Brück W. Mining marine shellfish wastes for bioactive molecules: chitin and chitosan – Part B: applications. Biotechnol J. 2008;3:878–89.
58. Jayakumar R, Prabaharan M, Kumar PTS, Nair SV, Tamura H. Biomaterials based on chitin and chitosan in wound dressing applications. Biotechnol Adv. 2011;29:322–37.
59. Krajewska B. Application of chitin- and chitosan-based materials for enzyme immobilizations: a review. Enzym Microb Technol. 2004;35:126–39.
60. Dev A, Mohan JC, Sreeja V, et al. Novel carboxymethyl chitin nanoparticles for cancer drug delivery applications. Carbohydr Polym. 2010;79:1073–9.
61. Muzzarelli RAA. Chitins and chitosans for the repair of wounded skin, nerve, cartilage and bone. Carbohydr Polym. 2009;76:167–82.
62. Suwan J, Zhang Z, Li B, et al. Sulfonation of papain treated chitosan and its mechanism for anticoagulant activity. Carbohydr Res. 2009;344:1190–6.
63. Huang J, Chen W, Hu S, et al. Biochemical activities of 6-carboxy β-chitin derived from squid pens. Carbohydr Polym. 2013;91:191–7.

64. Kurita K, Sugita K, Kodaira N, Hirakawa M, Yang J. Preparation and evaluation of trimethylsilylated chitin as a versatile precursor for facile chemical modifications. Biomacromolecules. 2005;6:1414–8.
65. Kaifu K, Nishi N, Komai T, Tokura S, Somorin O. Studies on chitin. V. Formylation, propionylation and butyrylation of chitin. Polym J. 1981;13:241–5.
66. Skołucka-Szary K, Ramięga A, Piaskowska W, et al. Chitin dipentanoate as the new technologically usable biomaterial. Mater Sci Eng C. 2015;55:50–60.
67. Jindal DK, Singh SK. Synthesis and characterization of chitin acetate/propionate mixed esters. PHARMANEST Int J Adv Pharm Sci. 2013;4:1177–85.
68. Jayakumar R, Nwe N, Tokura S, Tamura H. Sulfated chitin and chitosan as novel biomaterials. Int J Biol Macromol. 2007;40:175–81.
69. Riva R, Ragelle H, des Rieux A, Duhem N, Jerome C, Preat V. Chitosan and chitosan derivatives in drug delivery and tissue engineering. Adv Polym Sci. 2011;244:19–44.
70. Garavaglia PA, Cannata JJB, Ruiz AM, et al. Identification, cloning and characterization of an aldo-keto reductase from *Trypanosoma cruzi* with quinone oxido-reductase activity. Mol Biochem Parasitol. 2010;173:132–41.
71. Malik A, Fouad D, Labrou NE, et al. Structural and thermodynamic properties of kappa class glutathione transferase from *Camelus dromedaries*. Int J Biol Macromol. 2016;88:313–9.
72. Kilcawley KN, Wilkinson MG, Fox PF. Determination of key enzyme activities in commercial peptidase and lipase preparations from microbial or animal sources. Enzym Microb Technol. 2002;31:310–20.
73. Manders AL, Jaworski AF, Ahmed M, Aitken SM. Exploration of structure–function relationships in Escherichia coli cystathionine γ-synthase and cystathionine β-lyase via chimeric constructs and site-specific substitutions. Biochim Biophys Acta. 2013;1834:1044–53.
74. Zhao H, Cui Q, Shah V, Xu J, Wang T. Enhancement of glucose isomerase activity by immobilizing on silica/chitosan hybrid microspheres. J Mol Catal B Enzym. 2016;126:18–23.
75. Ellenberger T, Tomkinson AE. Eukaryotic DNA ligases: structural and functional insights. Annu Rev Biochem. 2008;77:313–38.
76. Sanchez S, Demain AL. Enzymes and bioconversions of industrial, pharmaceutical and biotechnological significance. Org Process Res Dev. 2011;15:224–30.
77. Junker B, Moore J, Sturr M, et al. Pilot-scale production of intracellular and extracellular enzymes. Bioprocess Biosyst Eng. 2001;24:39–49.
78. de Duve C. The significance of lysosomes in pathology and medicine. Proc Inst Med Chic. 1966;26:73–6.
79. American Heart Association. Activase (Alteplase) in acute ischemic stroke. https://www.heart.org/idc/groups/heart-public/@wcm/@mwa/documents/downloadable/ucm_438896.pdf
80. Vellard M. The enzyme as drug: application of enzymes as pharmaceuticals. Curr Opin Biotechnol. 2003;14:444–50.
81. Das D, Goyal A. Pharmaceutical enzymes. In: Brar SK, et al., editors. Biotransformation of waste biomass into high value biochemicals. New York: Springer; 2014.
82. Katz L, Khosla C. Antibiotic production from the ground up. Nat Biotechnol. 2007;25:428–9.
83. Diez B, Mellado E, Rodriguez M, Fouces R, Barredo JL. Recombinant microorganisms for industrial production of antibiotics. Biotechnol Bioeng. 1997;55:216–26.
84. Daeschel MA. Antimicrobial substances from lactic acid bacteria for use as food preservatives. Food Technol. 1989;43:164–9.
85. Gillor O, Nigro LM, Riley MA. Genetically engineered bacteriocins and their potential as the next generation of antimicrobials. Curr Pharm Des. 2005;11:1067–75.
86. Parada JL, Caron CR, Medeiros ABP, Soccol CR. Bacteriocins from lactic acid bacteria: purification, properties and use as biopreservatives. Braz Arch Biol Technol. 2007;50: 521–42.
87. Cleveland J, Montville TJ, Nes IF, Chikindas ML. Bacteriocins: safe, natural antimicrobials for food preservation. Int J Food Microbiol. 2001;71:1–20.

88. Deraz SF, Karlsson EN, Hedstrom M, Andersson MM, Mattiasson B. Purification and characterisation of acidocin D20079, a bacteriocin produced by *Lactobacillus acidophilus* DSM 20079. J Biotechnol. 2005;117:343–54.
89. Cotter PD, Ross RP, Hill C. Bacteriocins - a viable alternative to antibiotics? Nat Rev Microbiol. 2013;1:95–105.
90. Arthur TD, Cavera VL, Chikindas ML. On bacteriocin delivery systems and potential applications. Future Microbiol. 2014;9:235–48.
91. Different classes of antibiotics – an overview. http://www.compoundchem.com/2014/09/08/antibiotics/
92. Elander RP. Industrial production of beta-lactam antibiotics. Appl Microbiol Biotechnol. 2003;61:385–92.
93. Tang Z, Xiao C, Zhuang Y. Improved oxytetracycline production in *Streptomyces rimosus* M4018 by metabolic engineering of the G6PDH gene in the pentose phosphate pathway. Enzym Microb Technol. 2011;49:17–24.
94. Fierro F, Gutiérrez S, Diez B, Martín JF. Resolution of four large chromosomes in penicillin-producing filamentous fungi: the penicillin gene cluster is located on chromosome II (9.6 Mb) in *Penicillium notatum* and chromosome 1 (10.4 Mb) in *Penicillium chrysogenum*. Mol Gen Genet. 1993;241:573–8.
95. Cavera VL, Arthur TD, Kashtanov D, Chikindas ML. Bacteriocins and their position in the next wave of conventional cntibiotics. Int J Antimicrob Agents. 2015;46:494–501.
96. Yang SC, Lin CH, Sung CT, Fang JY. Antibacterial activities of bacteriocins: application in foods and pharmaceuticals. Front Microbiol. 2014;5:1–10.
97. Oscáriz JC, Pisabarro AG. Classification and mode of action of membrane-active bacteriocins produced by gram-positive bacteria. Int Microbiol. 2001;4:13–9.
98. Kramer K, Waelti M, de Pee S, et al. Are low tolerable upper intake levels for vitamin A undermining effective food fortification efforts? Nutr Rev. 2008;66:517–25.
99. Asensi-Fabado MA, Munne-Bosch S. Vitamins in plants: occurrence, biosynthesis and antioxidant function. Trends Plant Sci. 2010;15:582–92.
100. Thiel RJ. Natural vitamins may be superior to synthetic ones. Med Hypotheses. 2000;55: 461–9.
101. Carr AC, Vissers MCM. Synthetic or food-derived vitamin C - Are they equally bioavailable? Nutrients. 2013;5:4284–304.
102. Bouis HE, Welch RM. Biofortification – a sustainable agricultural strategy for reducing micronutrient malnutrition in the global south. Crop Sci. 2010;50:S20–32.
103. Fabregas J, Herrero C. Vitamin content of four marine microalgae. Potential use as source of vitamins in nutrition. J Ind Microbiol. 1990;5:259–63.
104. Shimizu S. Vitamins and related compounds: microbial production. In: Rehm HJ, Reed G, editors. Biotechnology: special processes. 2nd ed. Weinheim: Wiley-VCH Verlag GmbH; 2001.
105. Survase SA, Bajaj IB, Singhal RS. Biotechnological production of vitamins. Food Technol Biotechnol. 2006;44:381–96.
106. Barrett JR. Chemical exposures: the ugly side of beauty products. Environ Health Perspect. 2005;113:A24.
107. Breast Cancer Fund. Chemicals in cosmetics. http://www.breastcancerfund.org/clear-science/environmental-breast-cancer-links/cosmetics/
108. Chen L, Hu JY, Wang SQ. The role of antioxidants in photoprotection: a critical review. J Am Acad Dermatol. 2012;67:1013–24.
109. Moure A, Cruz JM, Franco D, et al. Natural antioxidant from residual sources. Food Chem. 2001;72:145–71.
110. Dai J, Mumper RJ. Plant phenolics: extraction, analysis and their antioxidant and anticancer properties. Molecules. 2010;15:7313–52.

111. Wang ZY, Agarwal R, Bichers DR, et al. Protection against ultraviolet B radiation-induced photocarcinogenesis in hairless mice by green tea polyphenols. Carcinogen. 1991;12: 1527–30.
112. Dvorakova K, Dorr RT, Valcic S, Timmermann B, Alberts DS. Pharmacokinetics of the green tea derivative, EGCG, by the topical route of administration in mouse and human skin. Cancer Chemother Pharmacol. 1999;43:331–5.
113. Kumari S, Elancheran R, Kotoky J, Rajlakshmi D. Rapid screening and identification of phenolic antioxidants in *Hydrocotyle sibthorpioides* Lam. by UPLC–ESI-MS/MS. Food Chem. 2016;203:521–9.
114. Chew YL, Lim YY, Omar M, Khoo KS. Antioxidant activity of three edible seaweeds from two areas in South East Asia. LWT – Food Sci Technol. 2008;41:1067–72.
115. Peschel W, Sánchez-Rabaneda F, Diekmann W, et al. An industrial approach in the search of natural antioxidants from vegetable and fruit wastes. Food Chem. 2006;97:137–50.
116. Chen RH, Chen WY. Film formation time and skin hydration effects and physico-chemical properties of moisture masks containing different water-soluble chitosans. J Cosmet Sci. 2000;51:1–13.
117. Morganti P, Fabrizi G, Palombo P, et al. Chitin-nanofibrils: a new active cosmetic carrier. J Appl Cosmetol. 2008;26:113–28.
118. Dutta PK, Dutta J, Tripathi VS. Chitin and chitosan: chemistry, properties and applications. J Sci Ind Res. 2004;63:20–31.
119. Gautier S, Xhauflaire-Uhoda E, Gonry P, Piérard GE. Chitin-glucan, a natural cell scaffold for skin moisturization and rejuvenation. Int J Cosmet Sci. 2008;30:459–69.
120. Pillai CKS, Paul W, Sharma CP. Chitin and chitosan polymers: chemistry, solubility and fiber formation. Prog Polym Sci. 2009;34:641–78.
121. Jayakumara R, Prabaharan M, Nair SV, Tokura S, Tamura H, Selvamurugan N. Novel carboxymethyl derivatives of chitin and chitosan materials and their biomedical applications. Prog Mater Sci. 2010;55:675–709.
122. Choi EH, Man MQ, Wang FS, et al. Is endogenous glycerol a determinant of stratum corneum hydration in humans? J Investig Dermatol. 2005;125:288–93.
123. Fujii M, Shimizu T, Nakamura T, Endo F, Kohno S, Nabe T. Inhibitory effect of chitosan-containing lotion on scratching response of hairless mice with atopic dermatitis-like dry skin. Biol Pharm Bull. 2011;34:1890–4.
124. Contri RV, Frank LA, Kaiser M, Pohlmann AR, Guterres SS. The use of nanoencapsulation to decrease human skin irritation caused by capsaicinoids. Int J Nanomedicine. 2014;12: 951–62.
125. Anjum S, Arora A, Alam MS, Gupta B. Development of antimicrobial and scar preventive chitosan hydrogel wound dressings. Int J Pharm. 2016;508:92–101.
126. Hiram M, Montoya U, Luis J, et al. Jumbo squid (*Dosidicus gigas*) mantle collagen: extraction, characterisation, and potential application in the preparation of chitosan-collagen biomaterial. Bioresour Technol. 2010;101:4212–9.
127. Huang YR, Shao CY, Chen HH, Huang BC. Isolation and characterisation of acid and pepsin-solubilised collagens from the skin of balloon fish (*Diodon holocanthus*). Food Hydrocoll. 2011;25:1507–13.
128. Chuaychan S, Benjakul S, Kishimura H. Characteristics of acid- and pepsin-soluble collagens from scale of seabass (*Lates calcarifer*). LWT Food Sci Technol. 2015;63:71–6.
129. Kittiphattanabawon P, Benjakul S, Visessanguan W, Shahidi F. Isolation and properties of acid- and pepsin-soluble collagen from the skin of blacktip shark (*Carcharhinus limbatus*). Eur Food Res Technol. 2010;230:475–83.
130. Liu D, Zhou P, Li T, Regenstein JM. Comparison of acid-soluble collagens from the skins and scales of four carp species. Food Hydrocoll. 2014;41:290–7.
131. Morimura S, Nagata H, Uemura Y, Fahmi A, Shigematsu T, Kida K. Development of an effective process for utilization of collagen from livestock and fish waste. Process Biochem. 2002;37:1403–12.

132. Li GY, Fukunaga S, Takenouchi K, Nakamura F. Comparative study of the physiological properties of collagen, gelatin and collagen hydrolysate as cosmetic materials. Int J Cosmet Sci. 2005;27:101–6.
133. Gautam M, Martin DW. Combustion characteristics of higher-alcohol/gasoline blends. Proc Inst Mech Eng. 2000;214:497–511.
134. Barannik VP, Makarov VV, Petrykin AA, Shamonia A. Aliphatic alcohols – antiknock additives to gasoline. Chem Technol Fuels Oils. 2005;41:452–5.
135. Khayoon MS, Hameed BH. Acetylation of glycerol to biofuel additives over sulfated activated carbon catalyst. Bioresour Technol. 2011;102:9229–35.
136. Gupta M, Kumar N. Scope and opportunities of using glycerol as an energy source. Renew Sust Energ Rev. 2012;16:4551–6.
137. Quispe CAG, Coronado CJR, Carvalho Jr JA. Glycerol: production, consumption, prices, characterization and new trends in combustion. Renew Sust Energ Rev. 2013;27:475–93.
138. Roberts WL, Metzger B, Turner TL. Process for combustion of high viscosity low heating value liquid fuels. US 8496472 B2 (2013).
139. Bradin D, Grune GL, Trivette M. Alternative fuel and fuel additive compositions. US 0013591A1 (2009).
140. Hasheminejad M, Tabatabaei M, Mansourpanah Y, Far MK, Javani A. Upstream and downstream strategies to economize biodiesel production. Bioresour Technol. 2011;102: 461–8.
141. Pagliaro M, Ciriminna R, Kimura H, Rossi M, Della PC. From glycerol to value-added products. Angew Chem Int Ed Eng. 2007;46:4434–40.
142. Zheng Y, Chen X, Shen Y. Commodity chemicals derived from glycerol, an important biorefinery feedstock. Chem Rev. 2008;108:5253–77.
143. Barrault J, Jerome F. Design of new solid catalysts for the selective conversion of glycerol. Eur J Lipid Sci Technol. 2008;110:825–30.
144. Behr A, Gomes JP. The refinement of renewable resources: new important derivatives of fatty acids and glycerol. Eur J Lipid Sci Technol. 2010;112:31–50.
145. Ayoub M, Khayoon MS, Abdullah AZ. Synthesis of oxygenated fuel additives via the solventless etherification of glycerol. Bioresour Technol. 2012;112:308–12.
146. Gonzalez MD, Cesteros Y, Salagre P. Establishing the role of Brønsted acidity and porosity for the catalytic etherification of glycerol with tert-butanol by modifying zeolites. Appl Catal A Gen. 2013;450:178–88.
147. Lee HJ, Seung D, Jung KS, Kim H, Filimonov IN. Etherification of glycerol by isobutylene: tuning the product composition. Appl Catal A Gen. 2010;390:235–44.
148. Liu J, Yang B, Li C. Kinetic study of glycerol etherification with isobutene. Ind Eng Chem Res. 2013;52:3742–51.
149. Melero JA, Vicente G, Morales G, et al. Acid-catalyzed etherification of bio-glycerol and isobutylene over sulfonic mesostructured silicas. Appl Catal A Gen. 2008;346:44–51.
150. Noureddini H. Process for producing biodiesel fuel with reduced viscosity and a cloud point below thirty-two degrees Fahrenheit. US Patent 6015440 (2000).
151. Khayoon MS, Triwahyono S, Hameed BH, Jalil AA. Improved production of fuel oxygenates via glycerol acetylation with acetic acid. Chem Eng J. 2014;243:473–84.
152. Silva LN, Gonçalves VLC, Mota CJA. Catalytic acetylation of glycerol with acetic anhydride. Catal Commun. 2010;11:1036–9.
153. Rahmat N, Abdullah AZ, Mohamed AR. Recent progress on innovative and potential technologies for glycerol transformation into fuel additives: a critical review. Renew Sust Energ Rev. 2010;14:987–1000.
154. Mufrodi Z, Rochmadi R, Sutijan S, Budiman A. Synthesis acetylation of glycerol using batch reactor and continuous reactive distillation column. Engl J. 2014;18:29–40.
155. Liao X, Zhu Y, Wang SG, Li Y. Producing triacetylglycerol with glycerol by two steps: esterification and acetylation. Fuel Process Technol. 2009;90:988–93.

156. Rao PV, Rao BVA. Effect of adding Triacetin additive with Coconut oil methyl ester (COME) in performance and emission characteristics of DI diesel engine. Int J Thermal Technol. 2011;1:100–6.
157. Raveendra G, Srinivas M, Prasad PSS, Lingaiah N. Heteropoly tungstate supported on tantalum oxide: a highly active acid catalyst for the selective conversion of fructose to 5-hydroxy methyl furfural. Int J Adv Eng Sci Appl Math. 2013;5:232–8.
158. Casanova O, Iborra S, Corma A. Biomass into chemicals: one pot-base free oxidative esterification of 5-hydroxymethyl-2-furfural into 2,5-dimethylfuroate with gold on nanoparticulated ceria. J Catal. 2009;265:109–16.
159. Román-Leshkov Y, Barrett CJ, Liu ZY, Dumesic JA. Production of dimethylfuran for liquid fuels from biomass-derived carbohydrates. Nature. 2007;447:982–5.
160. Mascal M, Nikitin EB. Direct, high-yield conversion of cellulose into biofuel. Angew Chem Int Ed Eng. 2008;47:7924–6.
161. Liu B, Zhang Z. One-pot conversion of carbohydrates into 5-ethoxymethylfurfural and ethyl D-glucopyranoside in ethanol catalyzed by a silica supported sulfonic acid catalyst. RSC Adv. 2013;3:12313–9.
162. James OO, Maity S, Usman LA, et al. Towards the conversion of carbohydrate biomass feedstocks to biofuels via hydroxylmethylfurfural. Energy Environ Sci. 2010;3:1833–50.
163. Zakrzewska ME, Bogel-Lukasik E, Bogel-Lukasik R. Ionic liquid-mediated formation of 5-hydroxymethylfurfural – a promising biomass-derived building block. Chem Rev. 2011; 111:397–417.
164. Rosatella AA, Simeonov SP, Frade RFM, Afonso CAM. 5-Hydroxymethylfurfural (HMF) as a building block platform: biological properties, synthesis and synthetic applications. Green Chem. 2011;13:754–93.
165. Deng T, Cui X, Qi Y, Wang Y, Hou X, Zhu Y. Conversion of carbohydrates into 5-hydroxymethylfurfural catalyzed by $ZnCl_2$ in water. Chem Commun. 2012;48:5494–6.
166. Yong G, Zhang Y, Ying JY. Efficient catalytic system for the selective production of 5-hhydroxymethylfurfural from glucose and fructose. Angew Chem Int Ed. 2008;47:9345–8.
167. Hayes DJ. An examination of biorefining processes, catalysts and challenges. Catal Today. 2009;145:138–51.
168. Joshi H, Moser BR, Toler J, Smith WF, Walker T. Ethyl levulinate: a potential bio-based diluent for biodiesel which improves cold flow properties. Biomass Bioenergy. 2011;35: 3262–6.
169. Mao RLV, Zhao Q, Dima G, Petraccone D. New process for the acid-catalyzed conversion of cellulosic biomass (AC_3B) into alkyl levulinates and other esters using a unique one-pot system of reaction and product extraction. Catal Lett. 2011;141:271–6.
170. Ventura AK, Mennella JA. Innate and learned preferences for sweet taste during childhood. Curr Opin Clin Nutr Metab Care. 2011;14:379–84.
171. Pepino MY, Mennella JA. Sucrose-induced analgesia is related to sweet preferences in children but not adults. Pain. 2005;119:210–8.
172. Lynch H, Milgrom P. Xylitol and dental caries: an overview for clinicians. J Calif Dent Assoc. 2003;31:205–9.
173. Ronda F, Gómez M, Blanco CA, Caballero PA. Effects of polyols and nondigestible oligosaccharides on the quality of sugar-free sponge cakes. Food Chem. 2005;9:549–55.
174. Ansari ARM, Mulla SJ, Pramod GJ. Review on artificial sweeteners used in formulation of sugar free syrups. Int J Adv Pharm. 2015;4:5–9.
175. Amo K, Arai H, Uebanso T, et al. Effects of xylitol on metabolic parameters and visceral fat accumulation. J Clin Biochem Nutr. 2011;49:1–7.
176. Tylli M, Eroma OP, Nygren J, Golde M, Heikkila H. Method for producing xylitol. CA 2259003A1 (1997).
177. Mikkola JP, Salmi T, Sjöholm R, Mäki-Arvela P, Vainio H. Hydrogenation of xylose to xylitol: three-phase catalysis by promoted raney nickel, catalyst deactivation and in-situ sonochemical catalyst rejuvenation. Stud Surf Sci Catal. 2000;130:2027–32.

178. Dhara KS, Wendisch VF, Nampoothiri KM. Engineering of *Corynebacterium glutamicum* for xylitol production from lignocellulosic pentose sugars. J Biotechnol. 2016;230:63–71.
179. Mohamad NL, Kamal SMM, Mokhtar MN. Xylitol biological production: a review of recent Studies. Food Rev Intl. 2015;31:74–89.
180. Baudel HM, de Abreu CAM, Zaror CZ. Xylitol production via catalytic hydrogenation of sugarcane bagasse dissolving pulp liquid effluents over Ru/C catalyst. J Chem Technol Biotechnol. 2005;80:230–3.
181. Pal S, Choudhary V, Kumar A, Biswas D, Mondal AK, Sahoo DK. Studies on xylitol production by metabolic pathway engineered Debaryomyceshansenii. Bioresour Technol. 2013;147:449–55.
182. Martinez EA, Silva SS, Felipe MGA. Effect of the oxygen transfer coefficient on xylitol production from sugarcane bagasse hydrolysate by continuous stirred-tank reactor fermentation. Appl Biochem Biotechnol. 2000;84:633–41.
183. Ding XH, Xia LM. Effect of aeration rate on production of xylitol from corncob hemicellulose hydrolysate. Appl Biochem Biotechnol. 2006;133:263–70.
184. Yoshitake J, Ishizaki H, Shimamura M, Imai T. Xylitol production by *Enterobacter* species. Agric Biol Chem. 1973;37:2261–7.
185. Rangaswamy S, Agblevor FA. Screening of facultative anaerobic bacteria utilizing D-xylose for xylitol production. Appl Microbiol Biotechnol. 2002;60:88–93.
186. Suzuki T, Yokoyama SI, Kinoshita Y, et al. Expression of xyrA gene encoding for d-xylose reductase of *Candida tropicalis* and production of xylitol in *Escherichia coli*. J Biosci Bioeng. 1999;87:280–4.
187. Suzuki S, Sugiyama M, Mihara Y, Hashiguchi K, Yokozeki K. Novel enzymatic method for the production of xylitol from d-arabitol by *Gluconobacter oxydans*. Biosci Biotechnol Biochem. 2002;66:2614–20.
188. Sugiyama M, Suzuki S, Tonouchi N, Yokozeki K. Cloning of the xylitol dehydrogenase gene from *Gluconobacter oxydans* and improved production of xylitol from d-arabitol. Biosci Biotechnol Biochem. 2003;67:584–91.
189. Kim SH, Yun JY, Kim SG, Seo JH, Park JB. Production of xylitol from d-xylose and glucose with recombinant *Corynebacterium glutamicum*. Enzym Microb Technol. 2010;46:366–71.
190. Dahiya JS. Xylitol production by *Petromyces albertensis* grown on medium containing d-xylose. Can J Microbiol. 1991;37:14–8.
191. Sampaio FC, da Silveira WB, Chaves-Alves VM, Passos FML, Coelho JLC. Screening of filamentous fungi for production of xylitol from d-xylose. Braz J Microbiol. 2003;34:325–8.
192. Suihko ML, Suomalainen I, Enari TM. d-xylose catabolism in *Fusarium oxysporum*. Biotechnol Lett. 1983;5:525–30.
193. Ikeuchi T, Azuma M, Kato J, Ooshima H. Screening of microorganisms for xylitol production and fermentation behavior in high concentration of xylose. Biomass Bioenergy. 1999;16: 333–9.
194. Vandeska E, Kuzmanova S, Jeffries TW. Xylitol formation and key enzyme activities in *Candida boidinii* under different oxygen-transfer rates. J Ferment Bioeng. 1995;80:513–6.
195. Zhang J, Zhang B, Wang D, Gao X, Hong J. Xylitol production at high temperature by engineered *Kluyveromyces marxianus*. Bioresour Technol. 2014;152:192–201.
196. Walfridsson M, Anderlund M, Bao X, Hahn-Hagerdal B. Expression ofdifferent levels of enzymes from the *Pichia stipitis* XYL1 and XYL2 genes in *Saccharomyces cerevisiae* and its effects on product formation during xylose utilization. Appl Microbiol Biotechnol. 1997;48: 218–24.
197. Cheng H, Lv J, Wang H, Wang B, Li Z, Deng Z. Genetically engineered *Pichia pastoris* yeast for conversion of glucose to xylitol by a single fermentation process. Appl Microbiol Biotechnol. 2014;98:3539–52.
198. Fache M, Boutevin B, Caillol S. Vanillin production from lignin and its use as a renewable chemical. ACS Sustain Chem Eng. 2016;4:35–46.

199. Anuradha K, Shyamala BN, Naidu MM. Vanilla – its science of cultivation, curing, chemistry, and nutraceutical properties. Crit Rev Food Sci Nutr. 2013;53:1250–76.
200. Rana R, Mathur A, Jain CK, Sharma SK, Mathur G. Microbial production of vanillin. Int J Biotechnol Bioeng Res. 2013;4:227–34.
201. Harold H, Tomlinson GH. Manufacture of vanillin from waste sulphite pulp liquor. US 2069185A (1937).
202. Cicognani G, Fiecchi A, Nano GM. Method of preparing vanillin from eugenol. US 3544621 A (1970).
203. Bomgardner MM. Following many routes to naturally derived vanillin. Chem Eng News. 2014;92:14.
204. Borges da Silva E, Zabkova M, Araújo J, et al. An integrated process to produce vanillin and lignin-based polyurethanes from Kraft lignin. Chem Eng Res Des. 2009;87:1276–92.
205. An YX, Zong MH, Wu H, Li N. Pretreatment of lignocellulosic biomass with renewable cholinium ionic liquids: biomass fractionation, enzymatic digestion and ionic liquid reuse. Bioresour Technol. 2015;192:165–71.
206. Schulze P, Seidel-Morgenstern A, Lorenz H, Leschinsky M, Unkelbach G. Advanced process for precipitation of lignin from ethanol organosolv spent liquors. Bioresour Technol. 2016; 199:128–34.
207. Lu H, Hu R, Ward A, Amidon TE, Liang B, Liu S. Hot-water extraction and its effect on soda pulping of aspen woodchips. Biomass Bioenergy. 2012;39:5–13.
208. Neiva D, Fernandes L, Araújo S, et al. Chemical composition and kraft pulping potential of 12 eucalypt species. Ind Crop Prod. 2015;66:89–95.
209. Fernández-Rodríguez J, García A, Coz A, Labidi J. Spent sulphite liquor fractionation into lignosulphonates and fermentable sugars by ultrafiltration. Sep Purif Technol. 2015;152: 172–9.
210. Gierer J. Chemical aspects of kraft pulping. Wood Sci Technol. 1980;14:241–66.
211. Pinto PCR, da Silva EAB, Rodrigues AE. Lignin as source of fine chemicals: vanillin and syringaldehyde. In: Baskar C, et al., editors. Biomass conversion: the interface of biotechnology, chemistry and materials. Berlin, Heidelberg: Springer; 2012.
212. Mathias AL, Lopretti MI, Rodrigues AE. Chemical and biological oxidation of pinus-pinaster lignin for the production of vanillin. J Chem Technol Biotechnol. 1995;64:225–34.
213. Sales FG, Abreu CAM, Pereira JAF. Catalytic wet-air oxidation of lignin in a three-phase reactor with aromatic aldehyde production. Braz J Chem Eng. 2004;21:211–8.
214. Tarabanko VE, Koropatchinskaya NV, Kudryashev AV, Kuznetsov BN. Influence of lignin origin on the efficiency of the catalytic-oxidation of lignin into vanillin and syringaldehyde. Russ Chem Bull. 1995;44:367–71.
215. Villar JC, Caperos A, García-Ochoa F. Oxidation of hardwood Kraft lignin to phenolic derivatives. Nitrobenzene and copper oxide as oxidants. J Wood Chem Technol. 1997;17:259–85.
216. Bjørsvik HR, Liguori L. Organic processes to pharmaceutical chemicals based on fine chemicals from lignosulfonates. Org Process Res Dev. 2002;6:279–90.
217. Tarabanko VE, Petrukhov D. Study of mechanism and improvement of the process of oxidative cleavage of lignins into the aromatic aldehydes. Chem Sustain Dev. 2003;11: 655–67.
218. Li X, Yang J, Gu W, Huang J, Zhang KQ. The metabolism of ferulic acid via 4-vinylguaiacol to vanillin by *Enterobacter* sp. Px6-4 isolated from Vanilla root. Process Biochem. 2008;43: 1132–7.
219. Zamzuri NA, Abd-Aziz S. Biovanillin from agro wastes as an alternative food flavour. J Sci Food Agric. 2013;93:429–38.
220. Priefert H, Rabenhorst J, Steinbüchel A. Biotechnological production of vanillin. Appl Microbiol Biotechnol. 2001;56:296–314.
221. Dubal AS, Tilkari YP, Momin SA, Borkar IV. Biotechnological routes in flavour industries. Adv Biotechnol. 2008;6:20–31.
222. Singh P, Khan S, Pandey SS, et al. Vanillin production in metabolically engineered Beta vulgaris hairy roots through heterologous expression of *Pseudomonas fluorescens* HCHL gene. Indus Crops Prod. 2015;74:839–48.

223. Gallage NJ, Hansen EH, Kannangara R. Vanillin formation from ferulic acid in *Vanilla planifolia* is catalysed by a single enzyme. Nat Commun. 2014;5:1–14.
224. Walton NJ, Mayer MJ, Narbad A. Vanillin. Phytochemistry. 2003;63:505–15.
225. Li K, Frost JW. Synthesis of vanillin from glucose. J Am Chem Soc. 1998;120:10545–6.
226. Esparan V, Krings U, Struch M, Berger RG. A three-enzyme-system to degrade curcumin to natural vanillin. Molecules. 2015;20:6640–53.
227. Boz H. Ferulic acid in cereals – a review. Czech J Food Sci. 2015;33:1–10.
228. Chen P, Yan L, Wu Z, et al. A microbial transformation using *Bacillus subtilis* B7-S to produce natural vanillin from ferulic acid. Sci Rep. 2016;6:1–10.
229. Pimentel D. Encyclopedia of pest management. Boca Raton, FL: CRC Press; 2002.
230. Mazid S, Kalita JC, Rajkhowa RC. A review on the use of biopesticides in insect pest management. Inter J Sci Adv Technol. 2011;1:169–78.
231. UN. World Population Prospects: the 2010 Revision, United Nations, New York (2011). http://www.un.org/en/development/desa/population/publications/pdf/trends/WPP2010/WPP2010_Volume-I_Comprehensive-Tables.pdf
232. Connell DW, Miller G, Anderson S. Chlorohydrocarbon pesticides in the Australian marine environment after banning in the period from the 1970s to 1980s. Mar Pollut Bull. 2002;45:78–83.
233. Ruiz-Suárez N, Boada LD, Henríquez-Hernández LA, et al. Continued implication of the banned pesticides carbofuran and aldicarb in the poisoning of domestic and wild animals of the Canary Islands (Spain). Sci Total Environ. 2015;505:1093–9.
234. EPA. DDT – a brief history and status. https://www.epa.gov/ingredients-used-pesticide-products/ddt-brief-history-and-status
235. Lumjuan N, Rajatileka S, Changsom D, et al. The role of the *Aedes aegypti* Epsilon glutathione transferases in conferring resistance to DDT and pyrethroid insecticides. Insect Biochem Mol Biol. 2011;41:203–9.
236. IFOAM EU Group. Guideline for pesticide residue contamination for international trade in organic (2012). http://www.ifoam-eu.org/sites/default/files/page/files/ifoameu_reg_pesticide_residue_cont_guideline_201203.pdf
237. EPA. Pesticide registration manual: Chapter 3 – Additional considerations for biopesticide products. https://www.epa.gov/pesticide-registration/pesticide-registration-manual-chapter-3-additional-considerations
238. Chandler D, Bailey AS, Tatchell GM, Davidson G, Greaves J, Gran WP. The development, regulation and use of biopesticides for integrated pest management. Philos Trans R Soc Lond Ser B Biol Sci. 2011;366:1987–98.
239. Kumar S, Singh A. Biopesticides: present status and the future prospects. J Biofertil Biopestici. 2015;6:e129.
240. Fife J. Biopesticides vs. conventional pesticides: what's the difference? http://www.battelle.org/newsroom/the-battelle-insider/the-battelle-blog/2015/07/10/biopesticides-conventional-pesticides-what's-the-difference
241. Mnif I, Ghribi D. Potential of bacterial derived biopesticides in pest management. Crop Prot. 2015;77:52–64.
242. Thakore Y. The biopesticide market for global agricultural use. Ind Biotechnol. 2006;23:192–208.
243. Hansen BM, Salamitou S. Virulence of *Bacillus thuringiensis*. In: Charles JF, Delecluse A, Nielsen-Le RC, editors. Entomopathogenic bacteria: from laboratory to field application. Dodrecht: Kluwer Academic; 2000.
244. Jisha VN, Smitha RB, Benjamin S. An overview on the crystal toxins from *Bacillus thuringiensis*. Adv Microbiol. 2013;3:462–72.
245. Koul O. Microbial biopesticides: opportunities and challenges. CAB Rev Perspect Agric Vet Sci Nutr Nat Resour. 2011;6:1–26.
246. Nawaz M, Mabubu JI, Hua H. Current status and advancement of biopesticides: microbial and botanical pesticides. J Entomol Zool Stud. 2016;4:241–6.

Chapter 6
Biomaterials

Abstract The production of various biomaterials from bioresources and biowastes is a major industrial activity in bioeconomy. Providing markets with bio-based materials as replacements to the fossil-based ones is facing two main challenges. The first one is related to the wide range of materials to be replaced by bioproducts at a competitive basis (i.e., producing equal quantities of better quality products). The second challenge, which needs to be seriously taken into consideration in bioeconomy, is managing the expected competition over the available biomass between the industries involved in the production of materials and those in the biofuel and biochemical sectors.

In the present chapter, numerous materials derived from renewable biomass are presented, along with the involved mechanical, thermochemical, and biological production procedures. This includes pulp and paper, bioplastics from various biopolymers and microorganisms, as well as biochars and activated carbons with versatile applications such as energy storage, water and wastewater treatment, soil amendment and remediation, and CO_2 sequestration.

6.1 Introduction

More attention is being paid by researchers and industrials worldwide to replace fossil-based materials by greener alternatives from renewable natural resources. In this context, the sustainable conversion of bioresources and agro-industrial wastes into value-added materials is facing two serious challenges:

(i) The wide range and extensive amount of petroleum-derived materials to be replaced by biomass-derived ones at competitive costs
(ii) The increasing competition over biomass between the various bio-based industries mainly for bioenergy, biochemicals, and biomaterials production

In order to ensure sustainable industrial production schemes, scientists, industrials, and decision makers should be aware, well in advance, of the upcoming dilemmas in order to anticipate their occurrence and develop suitable plans to properly tackle those issues.

Although the first challenge currently constitutes the main task for bioeconomy within short- and medium-term planning, ensuring the sustainability of this

M. Sillanpää, C. Ncibi, *A Sustainable Bioeconomy*,
DOI 10.1007/978-3-319-55637-6_6

all-inclusive economic model relies on long-term planning. That's why anticipating serious conflicts is a key endeavor to lay a solid ground for bioeconomy. In the beginning, the food and non-food dilemma seemed to be the main issue regarding potential feedstocks. With numerous breakthroughs in the R&D field, related mainly to the fractionation and conversion of lignocellulosic biomass, the use of non-food feedstocks (natural resources or wastes) is gradually becoming the cost-effective choice for material, fuel, and chemical production.

Therefore, with the increasing recourse to non-food biomass for industrial applications, the competition over resources which caused many conflicts during the petroleum era might reemerge, but this time to control the supply chain of non-food feedstock. We are still far away from this scenario, but anticipating it could help avoiding serious problems for generations ahead, which is the core of sustainable development.

In the present chapter, the production of various biomaterials from natural resources and wastes will be presented including pulp and paper, various bio-plastics, as well as biochars and activated carbons for energy storage, water treatment, soil remediation, and CO_2 sequestration.

6.2 Pulp and Paper

The pulp and paper industry is mainly based on wood feedstock, and the still increasing growth of this important papermaking sector (including printing paper, paper and cardboard, and household/sanitary items [1]) would create a challenge for the countries with forested areas to ensure a long-term sustainable exploitation model of their resources. For the less-forested countries, the problem of the cost-effectiveness of any wood-based conversion process would be added. Therefore, the resource to non-wood bioresources as alternative feedstock to wood biomass gained momentum during the last decades. So far, although the commercial non-wood pulp production has been estimated to be around 6.5% of the global pulp production, it is expected to increase [2]. Nonetheless, in China and India, it was reported that more than 70% of feedstock for the pulp and paper industry is made of non-woody biomass, mainly agricultural by-products such as cereal straw, sugarcane bagasse, bamboo, and reeds [3].

In the present chapter, the discussion will be focused on the production of pulp and paper from renewable non-wood bioresources, by-products, and wastes. The recourse to wood species from fast-growing and hybrid trees remains a viable option to provide renewable feedstock for the pulp and paper industry [4, 5], provide a sustainable exploitation process from cultivation until pulping, based mainly on avoiding competition with food crops over land and water (through the exploitation of marginal lands sand the use of unconventional water resource), and generate minimum (at least mitigable) impact on the environment.

6.2.1 Conventional Pulping Technologies

Pulp is the product of wood or alternative lignocellulosic fibrous mechanically and/or chemically treated in order to detach the fibrous content (defibration), to be later bleached, dispersed in water, and reformed into a web [6]. Several pulping technologies ranging for well-established and newly developed ones were proposed to defibrate lignocellulosic biomass into pulp. In general, pulp could be produced by either mechanical or chemical pulping processes or a combination of both.

6.2.1.1 Mechanical Pulping Processes

Mechanical pulping is based on four main methods: (i) stone groundwood pulping (SGW) [7], (ii) refiner mechanical pulping (RMP) [8], (iii) thermomechanical pulping (TMP) [9], and (iv) chemi-thermomechanical pulping (CTMP) [10].

(i) *SGW*: The principle of this inexpensive technique is to press raw wood logs against a revolving wet grindstone made of silicon carbide or aluminum oxide grits. Papers produced from this pulp have high absorbing properties and good opacity making them suitable for newsprint and magazine paper. Although this mechanical method gives high yields, the resulting pulp is made of very short fibers leading the end products with poor strength properties [11]. To tackle this problem, the combination with long fibers mainly from the chemical processes is needed, hence the recourse to integrated pulping technologies.

(ii) *RMP:* In this technique the woody feedstocks including wood logs, chips, sawmill residues, and sawdust are ground between two rotating disks (or one rotating and one stationary). The produced pulp is made of fibers longer than the one for SGW which results in stronger and lighter weight papers.

(iii) *TMP:* During this process, woody feedstock is first chipped and steamed. Then, the soften biomass is fed into large refiners to be milled between two steel disks (same as in RMP) into a high-grade mechanical pulp. The TMP is the most common mechanical process to produce pulp from wood, despite its high energy requirements.

(iv) *CTMP*: This technique is based on a mild chemical pretreatment of the feedstock with sodium sulfite (2–5%) prior to refining (conventional TMP). The chemical pretreatment allows a less destructive extraction of fibers and the generation of high yields of long fibers, which increase the strength properties of pulp normally produced via TMP as well as improving its brightness. Nonetheless, like TMP, CTMP is a high energy-consuming process [12].

Overall, mechanical pulping has the advantage of converting up to 95% of the dry weight of the wood feedstock into pulp suitable to manufacture highly opaque papers with good absorbing and printing properties but with the drawback of being relatively weak and easy to discolor. The main drawback of this mechanical processing remains the needs for large energy supply.

6.2.1.2 Chemical Pulping Processes

The chemical treatment aims at dissolving the lignin content in order to facilitate the separation of the fiber content (mostly cellulose and some hemicellulose), with little to no mechanical action. The yields of chemical pulping range between 40 and 50% of the dry woody biomass.

In practice, and before the chemical treatment, the bark is removed and the logs are milled. Then, the wood chips are chemically cooked in the digester under controlled temperature and pressure conditions. Three main processes used in chemical pulping are (i) soda, (ii) sulfate (kraft), and (iii) sulfite processes.

(i) *Soda process*: This is the first chemical pulping process. It is based on the use of caustic soda as the cooking agent. The first commercial soda mills used poplar as raw material for soda pulping. Considering the large amounts of soda used in this process, recovering this chemical from the waste cooking liquor becomes necessary. Soda recovery was carried out via evaporating of the spent liquor and causticizing the formed sodium carbonate [13]. The causticizing step consists on converting chemically inactive sodium carbonate (Na_2CO_3) to the active sodium hydroxide (NaOH).

    The addition of anthraquinone (AQ) to soda pulping is a known procedure aiming at promoting the selective delignification, preserving the carbohydrate content and increasing the pulp yield. Nevertheless, the addition of this chemical produced pulp with lower bleachability and lower tear strength, compared to the soda pulping process [14].

(ii) *Sulfate (kraft) process*: Kraft process is based on the chemical treatment of wood materials (chips and sawdust) with sodium sulfide (Na_2S) and sodium hydroxide. The wood biomass is cooked in this highly alkaline solution in batch or continuous digesters for 1–3 h. With the kraft process, most of the lignin is dissolved along with a fraction of the hemicellulose, thus generating well-separated cellulose fibers. Indeed, it was reported that around 90% of the lignin content could be removed under such strong alkaline conditions [15]. Among the many advantageous features of kraft process is that it could accommodate various woody feedstocks (softwood and hardwood), as well as non-wood bioresources such as *Hibiscus* species [16], wheat straw [17], and bamboo [18]. Besides, the kraft pulp is of superior quality and the overall process is efficient in recovering cooking chemicals.

    Recently, with the advances in isolating kraft lignin from black liquor [19], new value-added materials and chemicals could be produced from the recovered lignin [20].

(iii) *Sulfite process*: To remove lignin, this process uses various chemicals including sulfur dioxide and calcium, sodium, magnesium, or ammonium bisulfite.

The woody biomass is mixed with the chemical solution and cooked in digesters at high temperature and pressure [21]. The main sulfite pulping processes were acid sulfite, bisulfite, neutral sulfite, and alkaline sulfite [22], giving various levels of acidity and alkalinity of the sulfite chemical solutions and leading to different degrees of delignification.

Among the advantages of sulfite pulping is that the process enables an efficient cellulose separation by removing lignin and hemicelluloses in the same step [23]. Regarding the quality of the produced pulp, the unbleached sulfite pulp is of a light color and could be used without bleaching (if high brightness is not required) and could be easily bleached to very bright pulps suitable for writing and printing paper. The unbleached pulp could also be blended with high-yielding mechanical pulping process to increase the mechanical strength for their pulps [24]. Besides, sulfite pulp is used for the production of dissolving grade pulps (also called dissolving cellulose) [25], provided further removal of hemicellulose [26]. Several studies reported the use of dissolving pulps as the raw material for the production of many textile fibers including viscose, rayon, and lyocell [26, 27], as well as other cellulose derivatives such as cellulose acetate, cellophane, and cellulose triacetate, a plastic-like material that could be manufactured into fibers or films [28].

Regarding the sulfite pulping, it has to be noted that it is only suitable for woody biomass with low extractive contents such as tannins, polyphenols, pigments, etc. because they tend to interfere with the sulfite pulping process [29].

Typically, chemical pulping requires large quantities of wood material due to the dissolution of the organic matter into the delignification media, which decreases the pulp yield to about half of the intimal amount of feedstock. Indeed, it was reported that, for the kraft process, for instance, pulp recovery from softwoods is between 44 and 47% for the bleached and unbleached kraft pulp, respectively. As far as hardwood is concerned, pulp recovery varies between 50 (bleached) and 52% (unbleached) [24], noting that most of the dissolved organic matter could be combusted to produce energy.

6.2.1.3 Organosolv Pulping Processes

Kraft pulping is the major process used for the production of pulp and paper based on its feedstock versatility and superior pulp quality.

Nonetheless, this process displays some drawbacks related to the pulp that needs an effective bleaching step [30], along with the emissions during the process of harmful residues and malodorous substances including sodium and calcium salts in addition to the emission of reduced sulfur compounds such as hydrogen sulfide, methyl mercaptan, dimethyl sulfide, and dimethyl disulfide, all with extremely low odor thresholds [31].

In order to mitigate those environmental problems, while producing a better quality pulp, researchers worked on developing alternative pulping processes, and the one that received much attention and then was widely adopted by industrials is the organosolv process, based on the use of organic solvents during the cooking stage [32]. Organosolv processes have been applied since the 1990s to hard- and softwood, as well as non-wood biomass [33], using various organic solvents including methanol. Two main processes use methanol in the pulping procedure: Organocell and ASAM (alkaline sulfite anthraquinone and methanol).

(i) *The Organocell process*: This pulping process is based on the utilization of methanol as organic solvent in addition to sodium hydroxide and AQ as catalyst. In practice, The Organocell process starts by impregnating the woody biomass with mixture of methanol (or ethanol) and water at a temperature ranging between 110 and 140 °C. The softened material is transferred to a digester where sodium hydroxide (18–22%) and catalytic amounts of AQ are added. The mixture is then cooked at 160–170 °C until reaching the desired degree of delignification [34].

(ii) *The ASAM*: The chemicals used in this process include methanol, sodium hydroxide (NaOH), sodium carbonate (Na_2CO_3), sodium sulfite (Na_2SO_3), as well as AQ. In the mixture AQ acts as a catalyst. Anthraquinone serves as a catalyst, and methanol assists in dissolving the lignin fraction as well as preventing it from condensation. Methanol helps also to improve the solubility of AQ [35]. The strength properties of the pulp produced by ASAM were found to be similar to kraft pulp derived from the same raw material. However, higher yield and lower residual lignin content were reported for ASAM, which unlike the kraft process do not generate reduced sulfur compounds. A previous study, using eucalyptus as feedstock, reported that the main advantage of ASAM process (comparted to kraft) is a better bleachability of the pulp, which enables the application of chlorine-free bleaching sequence, without excessively damaging pulp strength [36].

Other organic solvents were used in organosolv pulping such as acetic acid in the acetosolv process [37, 38] and peroxyformic acid in the Milox process [39, 40].

6.2.2 Emerging Pulping Technologies

Several new technologies are being developed (some in demonstration stage) to ensure lower energy and chemical consumption, in the one hand, and better quality paper products and minimal carbon footprint, on the other hand [41, 42]. The following Table 6.1 summarizes the main advantageous features of selected emerging pulping technologies with respect to production yields, energy requirements, and environmental impact.

Table 6.1 Overview of newly developed pulping processes

Emerging technology	Advantages	Commercial status
Directed green liquor utilization pulping [43]	Energy requirement is cut down by 25% Alkali consumption is reduced by 50% Lime kiln load is reduced by 30%, thus reducing the related fuel consumption H-factor[a] is reduced by 30%, at similar kappa number Pulp yield is increased by 1–3% Pulp strength is increased (10% gain in tear strength), along with bleachability	Demonstration phase
Membrane concentration of black liquor [44]	Energy cost is reduced for black liquor evaporation Less inorganic content in evaporators resulting in less fouling Active alkali concentrated in permeates for improved makeup liquor Eliminate evaporator or recovery boiler bottlenecks	Development phase
Borate auto-causticizing [45]	Energy efficiency in chemical recovery process is increased Lime demand is reduced CO_2 emissions are reduced from fuel burning and from calcining process in lime kiln Causticizing capacity and pulp production yield are increased without major investments Costs related to lime kiln operation and maintenance are reduced	Development phase (full auto-causticizing) Semicommercial phase (partial auto-causticizing)
Steam cycle washing [46]	Fuel/steam consumption is reduced by 40% Evaporative load is decreased by 50% Plant effluent and freshwater usage are reduced by 45% Fiber yield is increased by 1–2% Consumption of bleaching chemicals is reduced Operational costs could be reduced by $40–60 per air-dry ton of pulp	Demonstration phase
Recycled paper fractionation [47]	Energy consumption is lowered Efficiency of ink detachment is enhanced Pulp quality is improved Consumption of virgin fiber is reduced Production of de-inked pulp is promoted	Demonstration phase

[a]H-factor is a kinetic model for the rate of delignification in kraft pulping

6.2.3 *Pulp and Paper from Non-wood Bioresources*

Among the various potential non-wood feedstocks for the pulp and paper industry, bagasse and straw are considered the most promising residual biomass and were widely investigated by researchers worldwide [48]. Recently, the use of marine biomass for pulp and paper production has received a great deal of attention [49], considering their availability and high biomass productivity [50].

6.2.3.1 Papermaking from Terrestrial Bioresources

Several terrestrial non-wood biomasses were successfully tested for pulp and paper production. In Europe, the main sources of fibers from non-wood resources for papermaking and other industrial applications are flax (*Linum usitatissimum*) and hemp (*Cannabis sativa*) crops [51]. This is mainly due to the creation, in the early 1970s, of a common market organization for flax in the European Union. According to European experts in the field, this decision has widely contributed to maintaining the global competitiveness of flax and hemp fibers with other competing fibers (mainly cotton and synthetic fibers) [52].

In this context, a research group from Spain worked on identifying and quantifying the environmental impacts associated with the production of hemp and flax fibers for papermaking using the life cycle assessment methodology [53]. The environmental impacts associated to the tested crops were assessed via evaluating the potentials of global warming potential, acidification, eutrophication, and photochemical oxidant formation. In addition, two flow indicators were considered: energy and pesticide use. The assessment showed that the production of hemp fiber reported higher values for all tested environmental impact categories. On the contrary, flow indicators were more intensive in the flax scenario due to the irrigation and pesticide consumption (since hemp crop does not require pesticide application).

Other interesting studies reported the use of many other bioresources as feedstock for pulp and paper production. For instance, in Bangladesh, different sections of the jute plant (*Corchorus capsularis*) were investigated as feedstock for pulp production (bark, core, and whole plant) [54]. The results showed that, under identical cooking conditions, the bark gave higher pulp yield (50–52%) than core (41%), due to the higher α-cellulose content. The pulp yield for the whole jute plant was 44–46%. As well, the bark pulp reached 80% ISO brightness using less chemicals, compared to core pulp. For the latter, however, the tensile index was better mainly due to its shorter fiber content.

In Spain, the influences of the some key pulping conditions on the quality of *Miscanthus giganteus*-derived pulp were evaluated [55]. The pulping factors were NaOH percentage, digestion, and refining times. The assessment was based on a comparison with a pulp obtained from commercial fluting paper. The main findings showed that the fiber size distribution of the Miscanthus pulp was found to contain a higher fines (less than 0.2 mm) percentage than the CF pulp. In addition, the handsheets made from *Miscanthus* pulp showed better mechanical properties.

In India, two species of *Hibiscus* (*H. cannabinus* and *H. sabdariffa*) were subjected to optimized pulping conditions in order to enhance the quality of the produced pulp and paper [56]. The optimum cooking conditions for both *Hibiscus* species were found to be a 16% active alkali, 20% sulfidity, 160 °C temperature,

120 min cooking time, and a wood/liquor ratio of 1/4.5. The results also showed that an anthraquinone (AQ) dose of 0.05% at an active alkali dose of 13% produced kappa number similar to that obtained by using 15% active alkali.

In another study, the fibers of giant *Hesperaloe* (*Hesperaloe funifera*) were tested for pulp and paper production via different cooking methods [57]. It was shown that *Hesperaloe* pulp obtained by a cooking process with 10% NaOH and 1% AQ at 155 °C for 30 min exhibited interesting properties of yield (48.3%), viscosity (737 mL g^{-1}), kappa number (15.2), tensile index (83.6 Nm g^{-1}), stretch (3.8%), and burst index (7.34 kN g^{-1}).

6.2.3.2 Papermaking from Marine Bioresources

As far as R&D is concerned, the valorization of algae in pulp and paper production is relatively new. But, with the increasing demand for alternative feedstock, the subject gained momentum in the last decades with quite interesting findings. The following recent studies illustrate the worldwide extent of this research effort to produce pulp from various marine fibrous bioresources.

In South Korea, a papermaking process was developed using bleached pulps obtained from two red algae species: *Gelidium amansii* and *Gelidium corneum* [58]. The results revealed that the yield of bleached red algae pulp from both red algae was between 8 and 11%, with brightness over 80%. The produced handsheets had very high smoothness (Bekk method) and opacity, when compared to the one produced from commercial wood pulp, which are essential properties for high-quality printing paper.

In Tunisia, the utilization of the marine biomass *Posidonia oceanica* fibers for pulp production was assessed using the soda-AQ cooking procedure [59]. The operating conditions were 2 h of contact time, 20% NaOH alkali charge, and 0.1% AQ dose. The results showed that the values of cooking yield and kappa number both decreased (66–60% and 75–63%, respectively) when the cooking temperature was increased from 150 to170 °C. However, the transition from the pulping stage of the marine biomass to the actual papermaking was not possible mainly due to its lower degree of polymerization (around 500), reducing therefore the overall strength properties of the material.

A Taiwanese team investigated the feasibility of utilizing *Rhizoclonium* green algae in pulping and papermaking under several cooking procedures [60]. Cooking the algae pulp with 5–25% NaOH for 30–120 min at a temperature of 100 °C gave high algal pulp yields (between 70 and 80%). On the other hand, the best pulp mechanical strengths (breaking length 5.23 km, tensile strength 79.2 Nm g^{-1}, and bursting index 2.2 kpa m^2 g^{-1}) were obtained after cooking for 1 h with 20% NaOH. Nonetheless, the produced pulp lacked bursting, tearing, and folding strengths. Thus, blending the algae pulp with other kinds of pulps seems to be the best solution to (i) lower the cost and the environmental impact if blended with wood pulps and (ii) enhance the paper quality if blended with other non-wood bioresources.

The same problem (i.e., low mechanical strength) was also encountered for several other marine biomasses, hence the frequent tendency to blend them (raw or the derived pulp) with other pulps. In Greece, for instance, a research group assessed the potential use of various freshwater algae (filamentous and nonfilamentous green microalgae and diatoms from sewage treatment plants) as tissue paper pulp supplements [61]. The experiments showed that the addition of algal biomass to the conventional paper pulp (10% ratio) helped increasing its mechanical strength significantly. However, the brightness was negatively affected considering the relative high content in chlorophyll. Besides, an economic estimation in the study revealed that the cost of the algal biomaterials is about 45% lower than that of raw conventional pulp, corresponding to a reduction in the final paper price between 0.9 and 4.5% as long as they are supplemented between 2.5 and 10% in the tissue paper manufacture.

6.2.4 Pulp and Paper from Agro-Industrial Wastes

Many agro-industrial wastes were investigated for their aptitude to produce good quality pulp and paper. Sugarcane bagasse is among the most interesting alternatives to woody biomass based on its biochemical composition. Indeed, it consists of approximately 35–43% cellulose and 20–30% hemicelluloses and 20–27% lignin [62].

In a related study, depithed Sudanese bagasse was examined for its suitability for pulp production. Different pulping procedures were carried out with soda-AQ and alkaline sulfite-AQ (AS-AQ). It was revealed that soda pulping of bagasse at mild cooking conditions (12.4% active alkali, 60 min heating up time to the maximum temperature of 160 °C and 60 min cooking time) gave pulp yield of 55.8% and a relatively high kappa number of 14.3. Under harsher conditions (90 min cooking time at 165 °C), the yield dropped to 53.2%, but the pulp showed much higher tear strength. Overall, the AS-AQ was the bagasse-pulping process which gave the best results with respect to yield (57%), kappa number (6.2), pulp viscosity (1061 mL g^{-1}), and initial ISO brightness (35%) [63].

With quite similar biochemical composition (i.e., 60.7% holocellulose and 21.9% lignin), rice straw was also widely studied for pulp production. In this context, pulping tests on this agricultural residue were conducted using soda and soda-AQ at 1 wt%, potassium hydroxide, and sodium sulfate (kraft process) under two different sets of operating conditions (active alkali proportion, temperature, and time) [64]. The comparative analysis showed that the paper sheets made from pulp produced by cooking for 90 min with soda (15 wt%) and AQ (1 wt%) at 180 °C exhibited the best drainage index (23°SR), breaking length (3494 m), and stretch and burst index (3.34% and 2.51 kN g^{-1}, respectively).

Another agricultural by-product, wheat straw, was also investigated as feedstock for papermaking using soda, soda-AQ, AS-AQ methods to extract pulp, and a totally chlorine-free (TCF) procedure to bleach the produced pulp [65]. The analysis of the extracted pulp showed that soda-AQ and AS-AQ pulps had similar tensile strength, but the tear strength was better for the case of AS-AQ. As well, the latter pulping methods had much higher yield (52%) and ISO brightness (37%). Regarding the bleaching stage, it was demonstrated that wheat straw-derived AS-AQ pulp can be bleached in a TCF sequence and reach an interesting ISO brightness of 84%.

In Bangladesh, corn stalks and *Saccharum spontaneum* (kash), two agricultural wastes, were tested as raw biomaterials for pulp production [66]. Prior to the pulping phase, the lignified pith fraction was removed from a part of the studied samples via water pre-extraction at 150 °C for 1 h. The related results revealed that this pre-hydrolysis step dissolved 12.2% of biomass components (mainly pith and hemicelluloses) from corn stalks and 11.9% from kash, on a dry weight basis. Then, the soda-AQ pulping process was applied to both the non-hydrolyzed and pre-hydrolyzed corn stalks and kash. For the case of corn stalks, the pre-hydrolysis did not induce significant enhancement on the properties of the derived pulp. As for the kash biomass, it was shown that the pulp produced from the pre-hydrolyzed sample had higher tear index (11.6 vs. 8.3 mN m^2 g^{-1}) but lower tensile index (35.4 vs. 45.3 N m g^{-1}) than the non-hydrolyzed pulp.

6.3 Bioplastics

The worldwide pollution with plastic wastes has become a major environmental issue because the degradation of the widely used petroleum-based plastics takes several decades and produces many toxins, as well as macro- and micro-plastic fragments that persist in soil and water [67]. Nevertheless, the low production costs and high yields of petrochemical-derived plastics worsened the situation and delayed genuine initiatives to deal with this serious issue for a long time.

Taking into account the importance of plastic materials in our everyday life, the solution to this problem came with the production of alternative biodegradable plastics from various bioresources, commonly known as bioplastics [68, 69]. Bioplastics are made from natural biopolymers synthesized and accumulated by various organisms, generally in response to stress conditions [70, 71].

After decades of R&D, various kinds of biodegradable polymers were proven to be suitable for the production of biodegradable plastic that could be used for many applications including polyhydroxyalkanoates (PHAs), polyhydroxybutyrate (PHB), polylactide (PLA), polybutylene succinate (PBS), polycaprolactone (PCL), and poly(p-dioxanone) (PPDO). These biopolymers could be produced from the carbohydrates, lipids, and proteins of a wide range of terrestrial and marine organisms and microorganisms [72–75].

The following section illustrates the extent of the research effort in this field and the valuable results provided by researchers and inventors from all over the world.

6.3.1 Bioplastics from Carbohydrates

6.3.1.1 Starch-Based Bioplastics

Starch-based bioplastic is estimated at 20% of the total world bioplastics production [76]. Many research groups worked on this biopolymer to produce biodegradable plastics with interesting structural and thermal properties.

In Malaysia, a thermoplastic starch derived from sugar palm tree (*Arenga pinnata*) was developed in the presence of biodegradable glycerol as a plasticizer (15–40 wt%) [77]. The results revealed that the mechanical properties of plasticized starch increased with the increasing of glycerol until the optimum amount of 30 wt %. Meanwhile, its water absorption decreased as the glycerol was increased. It was also found that the addition of glycerol decreased the transition temperature of plasticized starch.

In Japan, a research team succeeded in the production of polyhydroxybutyrate (PHB) from soluble starch using the α-amylase cell-surface displaying *Corynebacterium glutamicum* [78]. The results from this single-step production process showed that the cells grown on starch accumulated higher PHB (6.4 wt%) than those grown on glucose (4.9 wt%). This indicated that the fermentable sugars resulting from starch hydrolysis by α-amylase induced PHB production. The productivity of PHB from starch (0.39 g L^{-1}) was slightly higher than that from glucose (0.35 g L^{-1}).

In another study, the thermal processing of starch-based polymers was investigated [79]. It was proven that the thermal properties of those polymers are much more complex than conventional polymers. Indeed, multiple chemical and physical reactions may occur during the processing, such as water diffusion, granular expansion, gelatinization, decomposition, melting, and crystallization. Among these phase transitions, the study focused on gelatinization because it is the basis of the conversion of starch to a thermoplastic.

6.3.1.2 Cellulose-Based Bioplastics

In China, the direct use of cellulose to fabricate bioplastic was investigated [80]. The plastic material was constructed by hot-pressing cellulose hydrogels, which were prepared from cellulose solution in an alkali hydroxide/urea aqueous system. The results showed that, due to the removal of cellulose molecules in the hydrogel state, the hot pressing induced a transition in the aggregated structure in the cellulose bioplastic, leading to a plastic deformation. The produced cellulose bioplastic was transparent, as a result of the uniformly orientated structure. Moreover, the cellulose-based bioplastic exhibited much higher tensile strength, flexural strength, and thermal stability, along with lower coefficient of thermal expansion than common plastics and regenerated cellulose films.

Another research team developed a cellulose acetate biopolymer suitable for bio-composite applications [81]. Plasticization of this biopolymer was carried out using triethylcitrate (TEC) as an eco-friendly plasticizer and under varying processing conditions. Hence, three types of processing were used to fabricate plasticized cellulose acetate: (i) compression molding, (ii) extrusion followed by compression molding, and (iii) extrusion followed by injection molding.

The analysis revealed that the processing mode affected the physico-mechanical and thermal properties of the cellulosic plastic. Indeed, compression-molded samples exhibited the highest impact strength, while samples that were extruded and then injection molded exhibited the highest tensile strength and modulus values. The results also showed that the coefficient of thermal expansion of the produced cellulose acetate increased with increasing amounts of plasticizer. Thus, plasticized cellulose acetate was found to be processable at 170–180 °C, approximately 50 °C below the melting point of neat cellulose acetate.

6.3.2 *Bioplastics from Lipids*

In Malaysia, the nonedible Jatropha oil was tested as a substitute to food-grade oils for bioplastic production [82]. The experiments were conducted in order to produce poly-3-hydroxybutyrate (P3HB), a biodegradable bioplastic, from Jatropha oil using the bacteria *Cupriavidus necator* H16. The related results showed that high (P3HB) accumulation of 87 wt% from 13.1 g L^{-1} of cell dry weight was obtained by the bacterial stain when 12.5 g L^{-1} of Jatropha oil and 0.54 g L^{-1} of urea were used. Besides, the bioplastic production in a 10 L lab-scale fermenter gave a yield of 0.78 g (P3HB) per gram of used Jatropha oil, after 48 h.

In the United States, a research group investigated the development of bio-thermoset plastics from several plant-based oils (e.g., linseed, soybean, cottonseed, oilseed radish, and peanut oils) using an optimal process of solvent-free epoxidation [83]. The study revealed that an epoxidation by hydrogen peroxide along with a catalyst was the most efficient method to epoxidize oils with a high conversion percentage. Indeed, fatty epoxides can be used directly as plasticizers to improve the flexibility, elasticity, and stability of polymer subjected to heat. As well, it was found that linseed oil containing linolenic acids provided the highest modulus and good impact resistance properties.

In the United Kingdom, a study compared the production yields of PHB by bacterium *Ralstonia eutropha* H16 using various feedstocks as carbon source including glucose, olive oil, and rapeseed oil [84]. The main result was that both oils gave the highest PHB yields in comparison with glucose, presumably due to a higher number of carbons per gram of oil (compared to glucose), which led to an enhanced microbial cell growth in oils. Indeed, the average cell dry weights were 3.4, 2.9, and 1.1 g L^{-1}, respectively, for olive oil, rapeseed oil, and glucose.

As well, it was reported that PHB from glucose had a higher molecular weight (7.35×10^5 g mol^{-1}) compared to PHB from rapeseed and olive oils (5.79×10^5

and 5.92×10^5 g mol^{-1}, respectively). The glycerol content in plant oils is believed to be the reason behind the lower molecular weight PHB from vegetable oils based on its ability to terminate PHA chain elongation [85].

6.3.3 *Bioplastics from Proteins*

6.3.3.1 Gluten-Based Bioplastics

Glycerol-plasticized wheat gluten was used to produce thermo-molded biodegradable plastics. The influence of incorporating aldehydes and L-cysteine was assessed regarding the morphology, moisture absorption, dynamic mechanical properties, tensile properties, and thermal degradation behavior of the produced plastics. The results showed that cross-linking through disulfide bonding led to a high degree of phase separation and a high glass transition temperature of the gluten-rich phase. Aldehyde-induced cross-linking seems to improve tensile strength, in one hand, and to lower elongation at break and Young's modulus in comparison with cross-linking via disulfide bonding in the cross-linker-free and the L-cysteine-containing plastics [86].

The influence of hydrophobic liquids (castor and dimethyl silicone oils) on the properties of glycerol-plasticized wheat gluten was monitored in another study. The analyses revealed that combining hydrophilic plasticizer and hydrophobic liquids is an effective method to improve the tensile strength of the wheat gluten-derived plastic. The improvement of mechanical properties was also observed in the produced plastics containing glycerol as plasticizer and dimethyl silicone oil as hydrophobic modifier [87].

6.3.3.2 Soy Protein-Based Bioplastic

Soy protein, a low-cost by-product of the soybean oil extraction industry, was proven to be a good precursor for bioplastic production, considering its rich composition in amino acids [88]. Nevertheless, the fabrication of soy protein-based plastics is limited for two main reasons: (i) high moisture absorption and (ii) low mechanical strength [89].

To overcome those problems, and benefit from this inexpensive source of proteins, several solutions were proposed. In China, a study was conducted in order to improve the flexibility and water resistance of soy protein isolate (SPI) using waterborne polyurethane (WPU). The plastic blend films were successfully prepared by casting the aqueous dispersions of SPI and WPU. The analyses revealed that the flexibility and water resistance of the soy protein-based films were improved significantly by the incorporation of WPU, which also helped enhancing the mechanical properties of the plastic films in water, leading to the possible applications under wet conditions [90].

The other effective solution was to blend soy protein with another biodegradable polymer. Several associations were investigated and some related studies will be discussed in Sect. 6.3.4.

6.3.3.3 Other Protein-Based Bioplastic

A recent research study focused on the thermomechanical polymerization of microalgal protein from *Chlorella sp.* and *Spirulina sp.* to develop algal-based bioplastics and thermoplastic blends [91]. The results showed that pressure, temperature, content of plasticizer, and processing time are major variables in polymerization and structure stabilization during the compression molding process of both algal protein biomass and thermoplastic blends. *Chlorella* showed better bioplastic elaboration potential than *Spirulina* microalgae, whereas the former showed better blend performance.

Besides, the algal bioplastics seem to provide biodegradability that can be tailored to have a wide range of material properties suitable for various applications including consumable and disposable plastic products and agricultural plastic products. Another interesting feature of this study is that the algal protein biomass can grow on nutrient-rich wastewater from livestock farms, municipal or industrial effluent sources, remediating (or at least mitigating) therefore the excess nitrogen and phosphorus.

Egg white protein (albumen), usually used in the food industry, has been proved to be an interesting source to produce bioplastics. Moreover, if compared to other common proteins like gluten, egg white was shown to be an adequate raw material to develop highly transparent bioplastics with suitable mechanical properties for the manufacture of biodegradable food packaging and other plastic products [92, 93].

But, like soy protein, the use of albumen on its own to produce plastics is not common for economic reasons (wide applications in the food industry). Thus, blends of this protein with other biodegradable and low-cost biopolymers have been investigated. Examples of those studies will be presented in the next section.

6.3.4 Bioplastics from Combined Sources

For the soy protein case, and in order to overcome the previously mentioned limits, a series of glycerol-plasticized soy protein plastics containing castor oil were prepared by intensive mixing and hot pressing. It was revealed that at high concentrations, phase separation occurred. But, the incorporation of low content of castor oil (glycerol/oil ratio = 9:1) resulted in a simultaneous enhancement of tensile strength, elongation at break, and Young's modulus, compared with neat glycerol-plasticized protein plastics. As well, increasing castor oil content enhanced the thermal stability of the protein plastics [94].

In Spain, a research team developed different blends of albumen protein and starch (from potato and corn) to be used as raw materials for bioplastics exhibiting high transparency and an improved mechanical behavior. Three different processing methods were investigated: (i) compression-molding-based manufacture, (ii) extrusion, and (iii) combination of both [95]. The results showed that cornstarch leads to bioplastics with higher values of tensile strength, while potato starch yields more transparent materials. Besides, materials with good mechanical properties and acceptable degree of transparency were obtained by compression molding, which seems to be a more effective way to prepare improved albumen/starch-based bioplastics than extrusion.

In Brazil, a research group combined rice flour (viz., starch) and cellulose fibers in order to develop new biodegradable plastic films. The elaboration was performed by direct mixing of the rice-derived starch with a plasticizer (glycerol or sorbitol) with and without cellulose fibers. The results revealed that the cellulose-reinforced plastic films presented lower water vapor permeability, compared with films without fibers. As for the films containing sorbitol, they were less permeable to water and more rigid. In addition, the incorporation of fibers mechanically reinforced the rice flour-based films, which presented higher tensile strength [96].

6.3.5 Bioplastics from Bacteria

6.3.5.1 Polyhydroxyalkanoates (PHAs)

PHAs are thermoplastic materials synthesized and accumulated by variety of bacteria (30–80% of cell dry weight) as intracellular energy and carbon storage materials under limited nutrient conditions [97]. PHA-derived plastics are considered among the best alternatives to replace the current petroleum-based plastics due to their durability in use, hydrophobicity (i.e., moisture resistance), biodegradability, and wide spectrum of applications [98].

The major issue regarding the PHAs is the high production cost. In order to overcome this obstacle and reach large-scale production levels, R&D efforts focused on two main approaches (generally combined): (i) enhancing the bioconversion rates of the bacteria via genetically engineered strains and (ii) using renewable and low-cost feedstock as carbon sources.

In this context, several agro-industrial oily wastes (oleic acid, soybean oil, and waste frying oil) were tested as substrates for PHA production by the new strain *Pseudomonas aeruginosa* NCIB 40045 [99]. Different PHA accumulation ratios were found, ranging from 29.4% for the waste frying oil up to 66.1% when waste-free fatty acids from soybean oil were used as carbon substrate.

In India, a research study targeted cardboard industry wastes in order to isolate PHA-accumulating bacteria and develop a cost-effective bioplastic production process [100]. The screening and isolation was performed in two coloration steps: (i) using Sudan Black B dye as a preliminary screening agent for lipophilic compounds and (ii) using Nile Blue A, a more specific dye for PHA granules. Thus, starting from 42 isolates, two bacterial stains, namely, *Enterococcus sp*. NAP11 and *Brevundimonas sp*. NAC1, showed maximum PHA production from cardboard industry wastewater with accumulation ratios of 79.3 and 77.6%, corresponding to polymer concentration of 5.24 and 4.04 g L^{-1}, respectively.

6.3.5.2 Polylactic Acid (PLA)

PLA is a type of thermoplastic polyester resulting from the chemical polymerization of the D- and L-lactic acids obtained from fermentation [101]. Bioplastics made from PLA are extensively used in biomedical applications as sutures, stents, dialysis devices, and drug capsules [102].

Basically, PLA is produced in a two-step process: first, by producing lactic acid after fermentation, followed by ring-opening polymerization of lactide, a cyclic dimer of lactic acid. In South Korea, a research team reported the possible production of PLA and its copolymers by direct fermentation using metabolically engineered *Escherichia coli* using glucose as a carbon source [103]. The results showed that the engineered bacterial strain (*E. coli* JLXF5) with the evolved propionate CoA-transferase and PHA synthase was able to produce poly (3-hydroxybutyrate-co-lactate) with a polymer content of 43 wt% in a chemically defined medium by the pH-stat fed-batch culture.

Other researchers investigated the use of PLA films for antimicrobial food packaging. In Turkey, for instance, olive leaf extract (from *Olea europaea* L.) was incorporated as antimicrobial agent into PLA films (using glycerol as plasticizer). Antimicrobial activities of the elaborated films were tested against *Staphylococcus aureus*. The main results revealed that increasing the amount of the olive leaf extract in the film disks from 0.9 to 5.4 mg caused a significant increase in the inhibitory zones from 9.10 to 16.20 mm, respectively. As well, the water solubility and the degradation rates of the films increased up to 19.3% and 22.4%, respectively [104].

6.3.5.3 Polybutylene Succinate (PBS)

Succinic acid, derived from the fermentation of renewable resources, is one of the intermediates in the metabolic pathway of various anaerobic and facultative microorganisms. *Actinobacillus succinogenes*, *Anaerobiospirillum succiniciproducens*, and *Mannheimia succiniciproducens* are among the most promising bacterial strains to produce succinic acid at high yields [105]. For instance, a succinic acid productivity of 3.9 g L^{-1} h^{-1} has been reported for *M. succiniciproducens* [106].

Several studies revealed that succinic acid is an abundant and inexpensive feedstock for bioplastics production with interesting properties, compared to petroleum-derived plastics. In the United States, a comparative study was conducted between succinic acid-based thermoplastic and polybutylene adipate

(PA), a common petrochemical-based polyester. The main results were that thermoplastic polyurethanes made using PBS exhibited higher glass transition temperatures and more hard-phase to soft-phase interaction than those with PA [107].

In order to improve the mechanical properties of the produced PBS-based plastics, several composites were developed. But, within an overall eco-friendly strategy, biocomposites are the most interesting subject of study. Thus, using the thermo-pressed molding technique, the PBS polymeric matrix was reinforced with different lignocellulosic fibers (coconut, sugarcane bagasse, curaua, and sisal) [108]. The results of this study showed that sisal and curaua fibers were the best reinforcing agents of PBS due to their high chemical compatibility with the aliphatic matrix as well as to their surface morphology. Sisal/PBS and curaua/PBS composites also showed greater resistance against water absorption, when compared with coconut/PBS and sugarcane bagasse/PBS composites.

6.4 Biochars and Activated Carbons

Consulting the literature related to carbonized biomaterials could be somehow confusing. Indeed, several designations are being used, some intermittently referring to different carbonaceous materials and others terming the same material with different nomenclatures based on specific production processes or application. This list includes biochar, hydrochar, biocarbon, charcoal, activated carbon, activated charcoal, and activated biochar.

Thus, in order to avoid unnecessary confusion, all carbonaceous materials presented in the present chapter will be categorized in two main groups: biochars and activated carbons. Biochar is the solid material obtained after carbonization of biomass in oxygen-limited environment. Activated carbons are biochars subjected to chemical and/or physical activation, either before or after carbonization.

6.4.1 Biochars from Bioresources and Organic Wastes

The increasing interest in producing biochars from a wide range of bioresources and organic wastes is mainly linked to the various application scenarios including a carbon sequestration, soil amendment and fertility improvement, and air and water depollution [109–111], along with the valorization of solid by-products and waste.

Overall, the main factors affecting the production yields and physiochemical properties of biochars are the nature of feedstock (i.e., its biochemical and elemental composition), pyrolysis temperature, heating rate, residence time, and initial moisture content. Such key parameters have a direct impact on the development of the porous structure, the distribution of pore size, as well as surface area and ion-exchange capacity [112]. One important factor to be taken into consideration for biomass pyrolysis is the behavior of the carbohydrates to the thermal treatment.

Indeed, the main biomass components (cellulose, hemicellulose, and lignin) have different thermal decomposition behavior. In this regard, it was reported for wood chips that the decomposition of hemicellulose and cellulose occurs in the respective temperature ranges of 200–380 °C and 250–380 °C. The thermal decomposition of lignin, on the other hand, starts from 180 °C and continued until 900 °C [113].

6.4.1.1 Production Procedures of Biochars

Pyrolysis

Pyrolysis is a thermochemical process during which biomass is decomposed under elevated temperature (300–800 °C) in oxygen-free environment such as kilns, furnaces, and reactors. Three main products are generated from the dry pyrolysis of organic matter: solid product (biochar), liquid product (bio-oil) from the partial condensation of the pyrolysis vapor, and a non-condensable volatile matter (syngas) made of various gases such as CO, CO_2, CH_4, H_2, and two-carbon hydrocarbons in varying proportions [114–116].

The following Fig. 6.1 illustrates a pyrolysis system with its main components.

Several pyrolysis processes were developed mainly based on the final temperatures and heating rates. This includes three main processes: slow, intermediate, and fast pyrolysis [118]. Very fast pyrolysis is generally referred to as flash pyrolysis [119, 120]. The key characteristics of the main pyrolysis processes are summarized in the following Table 6.2.

It is clear from this table that slow pyrolysis is the most suitable process for biochar production because of its higher solid yield (~35%), when compared with the other pyrolysis methods. In general, most of the slow pyrolysis processes are operated by exposing the biomass to a temperature ranging between 300 and 800 °C using low heating rates (10–30 °C min^{-1}) and for residence times varying from a couple of minutes to several hours [121, 122]. Temperature wise, slow and fast

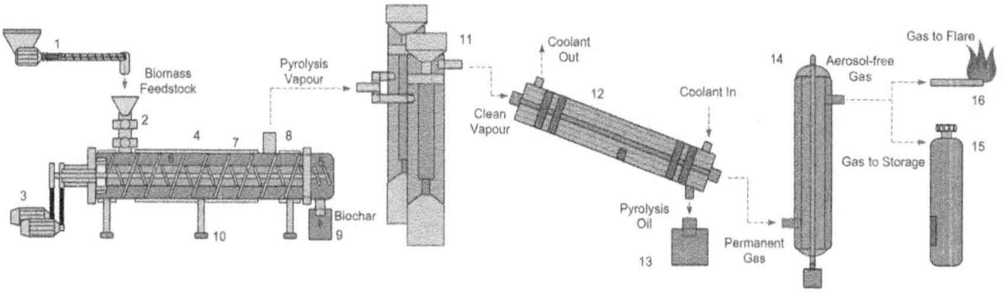

Fig. 6.1 Schematic diagram of a pyrolysis system with separate production of biochar, bio-oil, and syngas [117]. (1) Feeding system, (2) feed inlet, (3) electric motors, (4) the Pyroformer, (5) inner screw, (6) outer screw, (7) external heating jackets, (8) vapor outlet, (9) biochar pot, (10) stands, (11) hot gas filter, (12) shell and tube condenser, (13) oil vessel, (14) electrostatic precipitator, (15) gas vessel, and (16) gas flare

Table 6.2 Average yields of solid, liquid, and gaseous products from slow, intermediate, and fast pyrolysis

Process	Temperature (°C)	Residence time	Average yields (wt%) of pyrolysis products		
			Solid (biochar)	Liquid (bio-oil)	Gas (syngas)
Slow pyrolysis	300–800	Long (min to s)	35	30	35
Intermediate pyrolysis	450–500	Moderate (10–20 s)	25	50	25
Fast pyrolysis	300–1000	Short (<2 s)	12	75	13

pyrolysis processes are operated roughly within the same temperature range. Thus, the main operating parameter favoring high biochar yield is the low heating rate during slow pyrolysis.

In addition to the biochemical composition of the biomass, the pyrolysis operating conditions have a strong influence on the biochar's surface area, porous network, concentrations of fixed carbon, as well as heating value [123, 124]. As far as biochar production is concerned, the main technical issue with slow pyrolysis is that under typical operating conditions, the three solid, liquid, and gaseous products are roughly produced in the same ratio (cf. Table 6.2). In this context, simultaneous recovery of the liquid fraction (bio-oil) and the recirculation of the emitted gases (syngas) to provide heat to the process were reported to be viable options to enhance the overall efficiency and cost-effectiveness of slow pyrolysis [125].

Torrefaction, also known as mild pyrolysis, is a thermal process conducted between 150 °C and 300 °C under inert condition and slow heating rates in order to remove the moisture, CO_2, and oxygen contents in the treated biomass [126]. Compared to the biomass precursor, the solid product from torrefaction has a lower moisture content, lower O/C ratio, and higher energy content, which makes it a better feedstock for further processing including pyrolysis, gasification, and combustion or co-combustion with coal [127, 128]. Torrefaction was also reported to improve the grindability and reactivity of wood biomass [129].

Hydrothermal Carbonization (HTC)

During this process, the organic matter in biomass is thermochemically decomposed in oxygen-free environment with the presence of water and under autogenous pressure, hence the name hydrothermal carbonization or hydrous pyrolysis [130].

During the hydrothermal process, water is added to the feedstock and the mixture is heated in a sealed reactor, and like dry pyrolysis, the main products of hydrothermal decomposition of biomass are biochar [131], bio-oil [132], and syngas [133]. Based on the end products, the hydrothermal treatment could be categorized into three main processes: hydrothermal carbonization (temperatures below 250 °C), hydrothermal liquefaction (temperatures between 250 and 400 °C), and hydrothermal gasification (temperature above 400 °C).

In general, HTC process yields more solid compounds and more water-soluble organic compounds and emits less gas (mostly CO_2) than dry pyrolysis [134]. Overall, several reactions could occur simultaneously during HTC including dehydration, decarboxylation, condensation polymerization, and aromatization [135]. Nonetheless, the precise reaction network is still not yet fully understood [136].

The influence of the various operating parameters (temperatures, residence times, and biomass/water ratios) on the solid yields of hydrothermal carbonization of tomato peel wastes is given in Table 6.3.

Furthermore, many studies proved that the structure and chemical composition of biochars from HTC were substantially different from the ones produced via dry pyrolysis [138]. Indeed, it was reported that, from a chemical perspective (i.e., kind of chemical bonds and their quantity), HTC biochars were much closer to natural coal than charcoal, hence the term "artificial coalification" [139].

Besides, due to the evolution of H_2O and CO_2 in the dehydration and decarboxylation reactions, both biochars from dry pyrolysis and HTC exhibit lower hydrogen to carbon (H/C) and oxygen to carbon (O/C) ratios than the raw precursor. Nonetheless, biochars from HTC tend to have higher H/C and O/C ratios [140], mainly due to a higher ratio of decarboxylation to dehydration reaction rates in HTC than in dry pyrolysis [141].

In practice, and for large-scale production schemes, HTC is still confronted with serious challenges, responsible for increasing the capital cost and operating and maintenance charges [142]. Heat recovery is one of the most important issues to be dealt with in order to enable competitive production schemes of biochars. In HTC, heat could be recovered from the hot process water, which poses some technical

Table 6.3 Average solid yields under various HTC operating parameters [137]

Temperature (°C)	Residence time (h)	Biomass/water ratio	Solid yield (wt%)
150	10	6.7	65.0
170	5	3.3	64.6
170	5	10	68.0
170	15	3.3	60.8
170	15	10	61.3
200	10	1.1	49.8
200	10	6.7	62.0
200	10	12.3	61.5
230	5	3.3	49.6
230	5	10	62.2
230	15	3.3	27.6
230	15	10	35.4
250	10	6.7	29.4

problems. As for dry pyrolysis, the produced gas could be burnt to supply heat either within the reactor or by heat exchangers, which makes the energetic requirements for pyrolysis less than for hydrothermal process. In addition, post-production separation step could be necessary to recover the solid biochar from water, which also needs to be processed [134].

6.4.1.2 Characteristics of Biochars

Biomass composition, reaction temperature, pressure, residence time, heating rate, and moisture content are the main parameters determining the characteristics of the produced biochars via dry or wet pyrolysis [143, 144].

The variation of one single parameter will generate biochars with distinct physicochemical characteristics related to their physical structure such as surface area and porous network, as well as their chemical properties including pH and elemental composition (carbon, oxygen, nitrogen, phosphorus, potassium, and calcium). For instance, scientists reported that biochars produced from seaweeds, manures, and crop residues have richer nutrient content, higher pH, and less stable carbon, when compared with biochars from lignocellulosic biomass [145–147].

The following Table 6.4 gives the main characteristics of biochars produced by dry and wet pyrolysis for a wide range of precursor including natural terrestrial and marine resources and agro-industrial by-products and wastes.

6.4.2 Activated Carbons: Activation and Characteristics

6.4.2.1 Activation Procedures

Biochars generated from the thermal treatment of biomass (pyrolysis or HTC) could be activated mainly to increase the surface area pore volume and add more functional groups or graft new ones. Two main activation methods are used: physical and chemical activation [159].

Physical Activation

In order to produce physically activated carbons, the biomass feedstock is carbonized and activated with an oxidizing gas including air [160], carbon dioxide [161], steam [162, 163], and ozone [164], separately or combined.

The physical activation process occurs in two stages. First, the unstructured carbonized material is decomposed during the thermal treatment, and a porous network starts to form as fine pores enclosed in the carbon structure start to open. Thus, at this stage, the biomass is basically converted to a fixed carbon mass with an undeveloped porous network. Then, during the second stage, the crystalloid carbon

Table 6.4 Elemental and physicochemical characteristics of biochars produced from various bioresources and wastes

Biomass	Reaction temperature (°C)	Biochar yield (%)	Surface area (m^2 g^{-1})	Pore volume (cm^3 g^{-1})	pH	C (%)	O (%)	N (%)	Ash (%)
Orange peel [148]	200	61.6	7.8	0.01	–	57.90	34.40	1.88	0.3
	300	37.2	32.3	0.03	–	69.30	22.20	2.36	1.6
	400	30.0	34.0	0.01	–	71.70	20.80	1.92	2.1
	500	26.9	42.4	0.02	–	71.40	20.30	1.83	4.3
Pine needles [149]	200	75.3	6.2	–	–	57.10	36.31	0.88	0.9
	400	30.0	112.4	0.04	–	77.85	18.04	1.16	2.3
	600	20.4	206.7	0.07	–	85.36	11.81	0.98	2.8
Peanut shell [150]	300	36.9	3.1	–	7.8	68.27	25.89	1.91	1.2
	700	21.9	448.2	0.20	10.6	83.76	13.34	1.14	8.9
Fescue grass (*Festuca arundinacea*) [151]	200	96.6	3.3	–	–	47.20	45.10	0.61	5.7
	400	37.2	8.7	–	–	77.30	16.70	1.24	16.3
	600	29.8	75.0	–	–	89.0	7.60	0.99	18.9
Buffalo weed [152]	300	50.0	4.0	0.01	8.7	78.09	7.44	10.21	20.4
	700	29.0	9.3	0.02	12.3	84.96	6.56	7.40	32.3
Rapeseed (straw/stalk) [153]	400	39.4	16.0	1.24	–	71.34	10.84	1.43	12.2
	600	32.2	17.6	1.26	–	78.48	3.94	1.53	13.9
	800	28.2	19.0	1.15	–	79.51	2.61	1.45	15.3
Apple tree branch [154]	400	28.3	11.9	–	7.0	70.18	20.56	0.76	–
	600	16.6	208.7	–	10.0	81.46	13.63	0.46	–
	800	15.5	545.4	–	10.0	84.84	5.81	0.34	–
Oak tree branch [154]	400	35.8	5.6	–	6.4	70.52	21.47	0.69	–
	600	22.0	288.6	–	8.8	81.22	15.96	0.48	–
	800	19.1	398.1	–	9.7	82.85	17.29	0.32	–
Rice straw [154]	400	39.3	46.6	–	8.6	49.92	12.02	1.22	–
	600	23.4	129.0	–	10.2	33.78	13.68	0.41	–
	800	18.3	256.9	–	10.5	29.17	3.71	0.25	–

(continued)

Table 6.4 (continued)

Biomass	Reaction temperature (°C)	Biochar yield (%)	Surface area (m^2 g^{-1})	Pore volume (cm^3 g^{-1})	pH	C (%)	O (%)	N (%)	Ash (%)
Cottonseed hull [155]	200	83.4	–	–	–	51.90	40.50	0.60	3.1
	350	36.8	4.7	–	–	77.00	15.70	1.90	7.9
	650	25.4	34.0	–	–	91.00	5.90	1.60	8.3
Sewage sludge [156]	500	–	35.6	0.06	7.2	26.59	4.29	3.95	64.1
	600	–	19.1	0.05	8.0	27.68	3.89	3.76	93.8
	700	–	18.1	0.05	13.1	27.84	0.79	2.92	68.0
Poultry litter [157]	350	54.3	3.9	–	8.7	51.07	15.63	4.45	30.7
Poultry manure [158]	450	–	15.4	0.05	8.1	43.84	12.78	1.8	37.9
Cattle manure [158]	450	–	13.5	0.08	8.9	55.55	14.93	2.16	22.3

fraction and the carbon structure with fine pores are further destructed by the activation reactions, which generate larger pores [165].

During the activation, the carbon will be partially combusted according to the following reactions [166]:

$$C + H_2O \rightarrow CO + H_2 \qquad (\Delta H_{298K} = +117\ \text{kJ mol}^{-1}) \qquad \text{(Eq.6.1)}$$

$$C + CO_2 \rightarrow 2CO \qquad (\Delta H_{298K} = +159\ \text{kJ mol}^{-1}) \qquad \text{(Eq.6.2)}$$

Around 800 °C the following equilibrium involving steam may occur:

$$CO + H_2O \leftrightarrow CO_2 + H_2 \qquad (\Delta H_{298K} = +41\ \text{kJ mol}^{-1}) \qquad \text{(Eq.6.3)}$$

The reactions of carbon with water steam and carbon dioxide are endothermic and occur at moderate rates. However, the reactions of carbon with oxygen are exothermic and proceed at a rapid rate. Besides, it was reported that the physical activation with steam and carbon dioxide generated highly porous carbonaceous materials [167].

Chemical Activation

For the chemical activation, there are three main approaches:

- The most common is the impregnation of biomass in chemical solutions (pretreatment), followed by a carbonization stage [168, 169].
- The second method is the opposite as it starts by carbonizing the biomass (pyrolysis or HTC) then soaking the produced biochar in chemical solutions [170].
- The third approach is to thermally carbonize the biomass, chemically treat the generated biochar, and then subject the activated material to another thermal treatment under carbonizing conditions [171].

In general, the porosity of chemically activated carbons is generated by dehydration reactions occurring at low temperatures using chemicals such as phosphoric acid (H_3PO_4), nitric acid (HNO_3), potassium hydroxide (KOH), sodium hydroxide (NaOH), zinc chloride ($ZnCl_2$), calcium chloride ($CaCl_2$), etc. [172–174]. The utilization of those chemicals usually results in activation efficiencies higher than those of physical activation. Nonetheless, several drawbacks are registered for chemical activation including the corrosive effect of the chemical agents and the need to recover those chemicals via costly separation and purification methods [175].

During the chemical pretreatment of biomass (aka impregnation), several phenomena could occur and thus modify various characteristics of the natural precursor such as depolymerization (mainly lignin) and hydrolysis reactions which weaken the structure, eliminate the volatile matter, and increase the elasticity and swelling of the treated biomass [176]. The kind and concentration of the chemical agents and the pretreatment or impregnation time are the main factors affecting the porous structure of the produced activated carbon.

In an interesting study, the influence of different activation methods (CO_2 for physical activation, and H_3PO_4, NaOH, and KOH for chemical activation) was investigated using sewage sludge as precursor [177]. The main results showed in this work are the inefficiency of CO_2 and H_3PO_4 to generate a higher surface area and pore volume. However, the utilization of NaOH or KOH as chemical activating agents produced activated carbons from sewage sludge with high specific surface area and pore volume. As well, it was proven that the generated surface and volume values tend to increase as hydroxide/precursor ratios were increased. For instance, regarding the case of NaOH, increasing the ratio from 1/1 to 1/3 led to the increase of both BET surface and pore volume from 689 to 1224 $m^2\ g^{-1}$ and from 0.29 to 0.44 $cm^3\ g^{-1}$, respectively. The chemical agent responsible for the production of the activated carbon from sewage sludge with the highest specific surface area (1686 $m^2\ g^{-1}$) and pore volume (0.64 $cm^3\ g^{-1}$) was KOH at a 3/1 hydroxide/precursor ratio.

6.4.2.2 Characteristics of Activated Carbons

The combined effect of chemical and thermal treatments (or vice versa) during the conversion of bioresources and wastes to activated carbons produces carbonaceous materials with various structural characteristics, which are usually expressed using key parameters such as specific surface area, pore volume, and pore size distribution. The potential application fields of those porous carbonaceous materials are highly dependent on those characteristics.

According to the classification promoted by the International Union of Pure and Applied Chemistry (IUPAC), porous materials such as activated carbons could contain three main kinds of pores: micropores (diameters less than 2 nm), mesopores (2–50 nm), and macropores (more than 50 nm) [178].

Regarding the specific surface area, it reflects the total surface covered by the pores. In particle, it is estimated from the equilibrium adsorption of a gas (generally nitrogen) measured in a range of relative pressures between 0.01 and 0.3. There are two main procedures to determine the surface area of a porous material via gas adsorption isotherm data, the commonly used Brunauer-Emmett-Teller (BET) method [179] and the I-point method [180]. A new method to evaluate specific surface area from the Freundlich model was also proposed [181].

The following Table 6.5 gives the structural characteristics (specific surface area and total pore volume) of selected activated carbons produced via various chemical activation methods, and Fig. 6.2 shows the porous structure of one of the select ACs.

Table 6.5 Structural characteristics of chemically activated carbons produced from various bioresources and wastes

Biomass	Chemical activation conditions	BET surface area ($m^2\ g^{-1}$)	Pore volume ($cm^3\ g^{-1}$)
Pear peel [182]	H_3PO_4 at 400 °C for 3 h	1025	2.16
Broccoli stems [182]		1177	1.52
Soybean oil cake [183]	KOH at 600 °C for 1 h	600	0.30
	KOH at 800 °C for 1 h	618	0.29
	K_2CO_3 at 600 °C for 1 h	643	0.34
	K_2CO_3 at 800 °C for 1 h	1353	0.68
Rice husk [184]	NaOH at 800 °C for 1.5 h	1015	0.75
Rice husk [185]	$ZnCl_2$ at 700 °C for 1 h	750	0.38
Sugarcane bagasse [185]		674	0.34
Sewage sludge [186]	H_3PO_4 at 650 °C for 1 h	289	0.43
	H_2SO_4 at 650 °C for 1 h	408	0.52
	$ZnCl_2$ at 650 °C for 1 h	555	0.75
Coconut shell [187]	$ZnCl_2$ at 800 °C for 2 h	1510	0.75
	KOH at 800 °C for 4 h	2309	0.90
Palm stone [187]	$ZnCl_2$ at 800 °C for 2 h	1291	0.78
Artichoke [188]	H_3PO_4 at 300 °C for 1 h	2038	2.47
Parkinsonia aculeata wood sawdust [189]	H_3PO_4 at 450 °C for 0.5 h	968	0.70
	KOH at 300 °C for 2 h	768	0.37
Safflower seed press cake[a] [190]	$ZnCl_2$ at 600 °C for 1 h	249	0.15
	$ZnCl_2$ at 700 °C for 1 h	492	0.25
	$ZnCl_2$ at 800 °C for 1 h	772	0.36
	$ZnCl_2$ at 900 °C for 1 h	801	0.39
Posidonia oceanica seagrass fibers [191]	H_3PO_4 at 600 °C for 1 h	946	0.23
	KOH at 600 °C for 1 h	763	0.13
	$ZnCl_2$ at 600 °C for 1 h	503	0.06
	H_2O_2 at 600 °C for 1 h	60	0.02
Lotus stalk [192]	H_3PO_4 at 450 °C for 1 h	1220	1.19

(continued)

Table 6.5 (continued)

Biomass	Chemical activation conditions	BET surface area (m^2 g^{-1})	Pore volume (cm^3 g^{-1})
Vetiver roots [193]	H_3PO_4 at 600 °C for 1 h	1272	1.19

[a]This biowaste was pyrolyzed at 500 °C prior to the chemical activation and the second pyrolysis

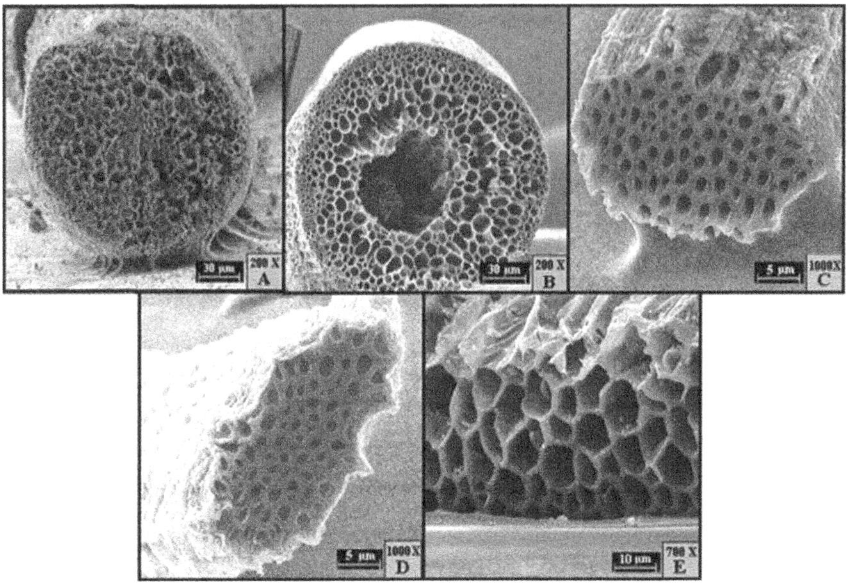

Fig. 6.2 SEM images of chemically activated carbons (CAC) from *P. oceanica* fibers: activation free sample (**a**), CAC-$ZnCl_2$ (**b**), CAC-H_3PO_4 (**c**), CA-H_2O_2 (**d**), and CAC-KOH (**e**) [191]

6.4.3 Applications of Biochars and Activated Carbons

The physical structure and chemical composition of biochars and activated carbons have a direct impact on their application potentialities. For instance, as far as biochars are concerned, pyrolysis reactions conducted under high temperatures are generally aimed at producing biochars effective in adsorbing organic contaminants by increasing surface area, pore volume, carbon content, and hydrophobicity [154, 194]. On the other hand, biochars generated at low temperatures are believed to be more effective in adsorbing, and thus removing, inorganic and polar organic pollutants by oxygen-containing functional groups and electrostatic attraction [195].

Carbonaceous materials (raw biochars or activated carbons) from diverse natural resources and wastes have already found numerous applications in strategic fields such as energy storage, drinking water purification, wastewater treatment, as well as carbon dioxide sequestration.

6.4.3.1 Electrical Energy Storage: Case of Supercapacitors

The production of porous carbonaceous materials from low-cost bioresources or wastes with specific structural and chemical properties is key for optimized gas and energy storage including electrical energy [196, 197] and hydrogen [198, 199]. The increasing recourse to this kind of materials in the energy sector is mainly related to the rapid depletion of fossil fuel reserves and many environmental considerations, in the one hand, and the still expensive technologies to produce renewable energy. Indeed, the prerequisite for the success and durability of any technology producing renewable energy (solar, wind, etc.) is an efficient storage system.

Matching the electricity supply to demand is one of the main challenges facing grid operators in any nation. With the increasing tendency to include electricity supplies from renewable but intermittent sources, scientists are predicting that this balancing issue along with managing peak demand will become more challenging [200]. Currently, many research studies are working on developing and improving the performance of SCs and other energy storage devices such as lithium-ion batteries [201].

Supercapacitors (SCs) are devices capable of managing higher power rates than batteries. In the same volume, it was reported that SCs can deliver higher power (hundred- to thousandfold), but they are not able to store the same amount of charge as batteries (usually 3–30 times lower) [202]. Based on those characteristics, SCs are suitable devices for any applications requiring power bursts without high energy storage capacity, which is the case for portable electrical devices and hybrid vehicles. On the other hand, compared to secondary or rechargeable batteries, SCs have been reported to have higher power, higher energy density, better electrochemical stability, and longer life cycle [203].

The following Table 6.6 illustrates the characteristics of the main electrochemical energy storage technologies, including carbon-based SCs.

Several carbonaceous materials were investigated as the electrodes of supercapacitors based on the properties of those materials to exert high electric conductivity and to exhibit a large pseudo-capacitance [205]. But most carbon-based electrodes tend to show low capacitance values [206]. One of the methods developed to overcome this issue was to graft nitrogen-containing functional groups onto the carbon materials [207, 208].

Numerous studies reported the use of biochars and activated carbons derived from biomass as electrodes in supercapacitors. Overall, it was revealed that the capacitive performance of the carbon-based electrodes is highly influenced by the types of functional groups on the surface, the electric conductivity, specific surface area, as well as the porous structure and pore distribution [209].

Table 6.6 Specifications of the main electrochemical energy storage devises [204]

Characteristics	Electrolytic capacitor	Carbon supercapacitor	Battery
Specific energy (Wh kg^{-1})	<0.1	1–10	10–100
Specific power (W kg^{-1})	≫10,000	500–10,000	<1000
Discharge time	10^{-6}–10^{-3}	s to min	0.3–3 h
Charging time	10^{-6}–10^{-3} s	s to min	1–5 h
Charge/discharge efficiency (%)	~100	85–98	70–85
Cycle life (cycles)	Quasi-infinite	>500,000	~1000

Table 6.7 Specific capacitance of activated carbons derived from various agro-industrial biowastes

Biowaste	Activation method	BET surface area (m^2 g^{-1})	Specific capacitance (F g^{-1})	Electrolyte
Camellia oleifera shell [210]	$ZnCl_2$	1935	374	1 M H_2SO_4
Waste coffee beans [211]	$ZnCl_2$	1019	368	1 M H_2SO_4
Argan seed shells [212]	KOH	2062	355	1 M H_2SO_4
Waste tea leaves [213]	KOH	2841	330	2 M KOH
Sugarcane bagasse [214]	$ZnCl_2$	1788	300	1 M H_2SO_4
Distillers dried grains[a] [215]	KOH	2959	260	6 M KOH
Cornstalk core [216]	KOH	2495	260	6 M KOH
Sunflower seed shell [217]	KOH	1162	244	30 wt% KOH
Banana peel [218]	$ZnNO_3$	1650	206	6 M KOH
Peanut shell [219]	$ZnCl_2$	1552	199	1 M $TEABF_4$/PC[b]
Rubber wood saw-dust [220]	CO_2	912	138	1 M H_2SO_4
Oil palm kernel shell [221]	Steam	727	123	1 M KOH
Cotton stalk [222]	H_3PO_4	1481	114	1 M $TEABF_4$
Rice straw [223]	H_3PO_4	396	112	1 M H_2SO_4

[a]By-product of the corn-to-ethanol industry
[b]Tetraethylammonium tetrafluoroborate ($TEABF_4$) in propylene carbonate (PC)

The following Table 6.7 details the activation conditions, structural properties, and specific capacitance of various activated carbons derived from natural resources and biowastes.

6.4.3.2 Soil Amendment

Extensive research studies focused (and still do) on assessing the short- and long-term impact of returning carbon to the soil via biochar amendment. The main objectives for this important R&D effort are to increase soil fertility, valorize degraded or marginal lands, and sequester carbon.

Alongside its contribution to mitigate global warming and climate change by sequestrating carbon and reducing greenhouse gas emissions [224–226], the incorporation of biochar to soil was proven to increase its pH (useful to neutralize acidic soils) [227] and cation exchange capacity (CEC) [228]. Indeed, the presence of phenolic, carboxyl, and hydroxyl functional groups enables biochars to interact with protons in the soil, reduce their concentration, and therefore increase the soil pH. As well, silicates, carbonates, and bicarbonates in biochars were also shown to react with H^+ in acidic soils [229].

In a related study, biochars from nine crop residues (straws from canola, wheat, corn, rice, soybean, peanut, faba bean and mung bean, and rice hull) were amended to a soil at a loading rate of 10 g kg^{-1}, and after 60 days, the soil pH increased by 0.59–1.05, with respect to the used biochar [230]. Incorporation of biochars to soil also results in the discharge of its basic cations, leading to the replacement of Al and H^+ and therefore the enhancement of the CEC of soil [228].

On the other hand, improvement in the retention of soil nutrients and water was also reported after the amendment of soils with biochars. For instance, the utilization of biochar produced from pecan shells with dried and ground switchgrass reduced nitrate leaching from soil within 25 days [231]. The same study also noticed that the application of biochar can cause a short-term immobilization of nitrogen which results in a temporary reduction of NO_3–N concentration potentially available for crops. Regarding another important macronutrient, phosphorus (P), scientists showed that the availability of P is generally enhanced with biochar amendment [232]. For instance, it was revealed that the maximum P retention capacity of the soil increased proportionally with the quantity of added biochars, derived from hardwoods and poultry litter [233].

As a consequence of those chemical properties, many scientists reported the beneficial impact of biochar addition on improving crop yields [234, 235], especially in degraded or low-fertility soils [236]. For example, the incorporation of cow manure biochar at loading rates of 15 and 20 t ha^{-1}, respectively, improved the production yield of maize grain by 150 and 98%, when compared with the non-amended soil. Such enhancement in plant growth was partly related to the improved soil properties after the biochar amendment including pH and organic carbon content [237].

Nonetheless, considering the various and complex physicochemical properties of both soils and biochars, the direct correlation between crop yield improvement and biochar amendment is not always valid. As well, the crop species plays an influential role in the trilateral relationship (i.e., soil, plant, and biochar). Indeed, in one study, and for the same biochar addition and in the same soil, an increase in soybean biomass production was registered but also a reduction in that of wheat and radish [238].

6.4.3.3 Water and Soil Decontamination

In general, biochars and activated carbon are characterized by high specific surface area and pore fraction and a surface chemistry-rich oxygenated functional groups and aromatic compounds [239, 240]. These characteristics enabled many raw biochars and especially activated carbons derived from biomass to effectively adsorb various kinds of organic and inorganic pollutant including heavy metals, phenolic compounds, dyes, surfactants, pharmaceutical drugs, pesticides, and many other toxics compounds.

An illustrative summary of the main mechanisms governing the interactions between biochars and organic/inorganic contaminants is depicted on Fig. 6.3.

The following Table 6.8 shows the utilization of biochars produced from various bioresources and organic wastes for the decontamination of pesticide-polluted soils.

Even though adsorption is the major phenomenon affecting the bioavailability, and thus the efficacy, of pesticides in biochar-amended soils, potential release due to desorption has to be assessed and taken into consideration in order to have a more accurate prediction about the impact of biochar amendment [246, 247]. In this regard, several studies confirmed the release of pesticides after being adsorbed onto biochars in amended soil. The mechanisms that are believed to be responsible for such reversible phenomenon include the deformation the macroporous network as a result to the swelling of the adsorbent, as well as the weak binding between certain pesticides and the available functional groups of biochars [248].

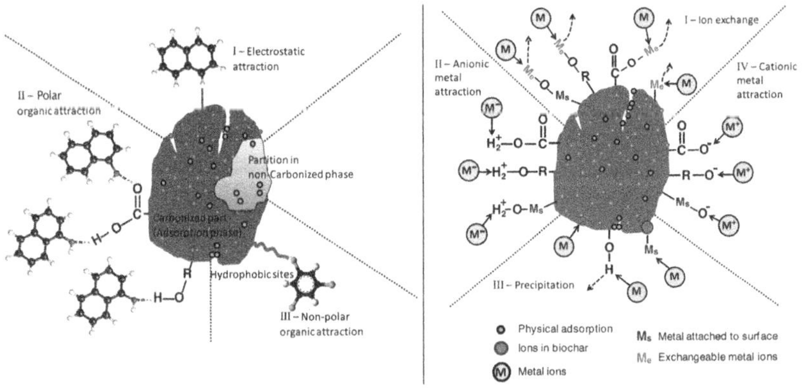

Fig. 6.3 Main mechanisms occurring during the interactions of biochar with organic (*left side*) and inorganic (*right side*) contaminants. Adapted from [152] (*circles* on biochar particle showing partition or adsorption)

Table 6.8 Biochars for various bioresources and their adsorption capacities in pesticide-contaminated soils

Biochar precursor	Pyrolysis temperature (°C)	Soil type	Amendment rate (%)	Targeted pesticide	Adsorption (%)[a]
Paper mill sludge [241]	550	Ferrosol	1	Atrazine	515
				Diuron	448
Composted alperujo[b] [242]	500	Sandy loam	2	Tricyclazole	400–500
Hardwood sawdust [243]	500	Sandy loam	2	Fluometuron	340–365
Hardwood [243]	540	Sandy loam	2	Fluometuron	300–310
Poultry manure [241]	550	Ferrosol	1	Atrazine	270
				Diuron	220
Wood chip pellets [244]	500	Silt loam	10	Aminocyclopyrachlor	95–240
Hardwood [244]	540	Silt loam	10	Aminocyclopyrachlor	50
				Bentazone	40
Macadamia nut shells [244]	850	Silt loam	10	Aminocyclopyrachlor	18–25
				Bentazone	13
Beechwood [245]	550	Sandy loam	1.5	Imazamox	5

[a]Biochar amendment increased the adsorption of pesticides in soils by:
[b]Solid olive-mill by-product

The following Table 6.9 illustrates many study cases for water depollution using biochars and activated carbons derived from different natural resources and biowastes.

After describing the various production procedures to convert bioresources and derived wastes into several added-value biomaterials in the present chapter, and biofuels and biochemicals in the previous two chapters, the next chapter will deal with the industrial core of bioeconomy, integrated biorefineries, from historical, technological, and R&D perspectives.

Table 6.9 Biochars and activated carbons for different bioresources and their application in various water decontamination scenarios

Precursor	Activation conditions	Targeted pollutants
Biochars		
Dairy manure [249]	200 °C	Lead and atrazine
Mixture of wood chips and barks [250]	450 °C	Atrazine and simazine (herbicides)
Cotton straw [251]	450–850 °C	Chlorpyrifos and fipronil (insecticides)
Soybean stover and peanut shell [252]	300–700 °C	Trichloroethylene
Ramie residues [253]	300–600 °C	Hexavalent chromium
Bamboo, pepperwood, sugarcane bagasse, and hickory wood [254]	450–600 °C	Sulfamethoxazole (antibiotic)
Switchgrass [255]	450 °C	Reactive red 195 A dye
Bamboo [256]	7000 °C	Nitrobenzenes, phenols, anilines
Cow bones [257]	450 °C	Uranium (U^{VI})
Rice husk [258]	450–500 °C	Tetracycline (antibiotic)
Activated carbons		
Sisal waste [259]	K_2CO_3/ 600–800 °C	Paracetamol, ibuprofen (pharmaceuticals)
Japanese cedar [260]	H_2O_2/350 °C	Phenol
Turbinaria turbinata alga [261]	Steam/800 °C H_3PO_4/ 600 °C	Methylene blue dye
Raspberry leaves [262]	H_2SO_4/ 600 °C	Ibuprofen, naproxen, clofibric acid (pharmaceuticals)
Coconut tree sawdust, silk cotton hull, banana pith, maize cob [263]	Concentrated H_2SO_4	Rhodamine-B, Congo red, methyl violet, malachite green (dyes), Hg (II), Ni (II)
Bean pod waste [264]	Steam/600 °C	Arsenic and manganese ions
Textile sewage sludge [265]	KOH/ 400–700 °C	Strontium
Arundo donax Linn and pomelo peel [266]	H_3PO_4/ 450 °C	Ciprofloxacin (antibiotic)
Pine tree cones [267]	K_2CO_3–KOH/750 °C	Sodium dodecylbenzene sulfonate (surfactant)
Olive-mill waste [268]	KOH/ 300–800 °C	Bisphenol A and chromium (III)

References

1. McCarthy P, Lei LJ. Regional demands for pulp and paper products. For Econ. 2010;16: 127–44.
2. Laftah WA, Abdul Rahman WAW. Pulping process and the potential of using non-wood pineapple leaves fiber for pulp and paper production: a review. J Nat Fibers. 2016;13:85–102.

3. Saijonkari-Pahkala K. Non-wood plants as raw material for pulp and paper. Academic Disseration. Available online at: https://helda.helsinki.fi/bitstream/handle/10138/20756/nonwoodp.pdf?...1
4. Ali F, Sarma TC, Saikia CN. Pulp and paper from certain fast-growing plant species. Bioresour Technol. 1993;45:65–7.
5. Ai J, Tschirner U. Fiber length and pulping characteristics of switchgrass, alfalfa stems, hybrid poplar and willow biomasses. Bioresour Technol. 2010;101:215–21.
6. Passas R. Natural fibres for paper and packaging. In: Kozłowski RM, editor. Handbook of natural fibres: processing and applications, vol. 2. Cambridge: Woodhead; 2012. p. 367–400.
7. Oliver-Ortega H, Granda LA, Espinach FX, Mendez JA, Julian F, Mutjé P. Tensile properties and micromechanical analysis of stone groundwood from softwood reinforced bio-based polyamide11 composites. Compos Sci Technol. 2016;132:123–30.
8. Sabharwal HS, Akhtar M, Blanchette RA, Young RA. Refiner mechanical and biomechanical pulping of jute. Holzforschung. 1995;49:537–44.
9. Harinath E, Biegler LT, Dumont GA. Predictive optimal control for thermo-mechanical pulping processes with multi-stage low consistency refining. J Process Control. 2013;23: 1001–11.
10. Hou Q, Wang Y, Liu W, Liu L, Xu N, Li Y. An application study of autohydrolysis pretreatment prior to poplar chemi-thermomechanical pulping. Bioresour Technol. 2014;169: 155–61.
11. Fernando D, Rosenberg P, Persson E, Daniel G. Fibre development during stone grinding: ultrastructural characterisation for understanding derived properties. Holzforschung. 2007; 61:532–8.
12. Martin N, Anglani N, Einstein D, Khrushch M, Worrell E, Price LK. Opportunities to improve energy efficiency and reduce greenhouse gas emissions in the U.S. pulp and paper industry. Lawrence Berkeley National Laboratory, CA. Report LBNL-46141. 2000. Available online at: https://publications.lbl.gov/islandora/object/ir%3A115962/spms_tab
13. Sixta H. Chemical pulping. In: Sixta H, editor. Handbook of pulp. Weinheim: Wiley-VCH; 2006.
14. Francis RC, Bolton TS, Abdoulmoumine N, Lavrykova N, Bose SK. Positive and negative aspects of soda/anthraquinone pulping of hardwoods. Bioresour Technol. 2008;99:8453–7.
15. Gierer J. Chemical aspects of Kraft pulping. Wood Sci Technol. 1980;14:241–66.
16. Dutt D, Upadhyay JS, Singh B, Tyagi CH. Studies on *Hibiscus cannabinus* and *Hibiscus sabdariffa* as an alternative pulp blend for softwood: an optimization of Kraft delignification process. Ind Crop Prod. 2009;29:16–26.
17. Hou Q, Yang B, Liu W, Liu H, Hong Y, Zhang R. Co-refining of wheat straw pulp and hardwood Kraft pulp. Carbohydr Polym. 2011;86:255–9.
18. Yuan Z, Kapu NS, Beatson R, Chang XF, Martinez DM. Effect of alkaline pre-extraction of hemicelluloses and silica on Kraft pulping of bamboo (*Neosinocalamus affinis* Keng). Ind Crop Prod. 2016;91:66–75.
19. Tomani P. The LignoBoost process. Cellul Chem Technol. 2010;44:53–8.
20. Gellerstedt G. Softwood Kraft lignin: raw material for the future. Ind Crop Prod. 2015;77: 845–54.
21. Bryce JRG. Sulfite pulping. In: Casey JP, editor. Pulp and paper: chemistry and chemical technology. 3rd ed. New York: Wiley; 1980. p. 291–376.
22. Ingruber O. Sulfite science part I: sulfite pulping cooking liquor and the four bases. In: Ingruber O, Kocurek M, Wong A, editors. Sulfite science and technology. The joint textbook committee of the paper industry. 3rd ed. Atlanta: TAPPI/CPPA; 1985. p. 3–23.
23. Schild G, Sixta H, Estova L. Multifunctional alkaline pulping, delignification and hemicellulose extraction. Cellul Chem Technol. 2010;4:35–45.
24. Office of Technology Assessment. Technologies for reducing dioxin in the manufacture of bleached wood Pulp (OTA-BP-O-54). Washington, DC: U.S. Government Printing Office; 1989.

25. Christov L, Biely P, Kalogeris E, Christakopoulos P, Prior BA, Bhat MK. Effects of purified endo-β-1,4-xylanases of family 10 and 11 and acetyl xylan esterases on eucalypt sulfite dissolving pulp. J Biotechnol. 2000;83:231–44.
26. Wang H, Pang B, Wu K, Kong F, Li B, Mu X. Two stages of treatments for upgrading bleached softwood paper grade pulp to dissolving pulp for viscose production. Biochem Eng J. 2014;82:183–7.
27. Schild G, Sixta H. Sulfur-free dissolving pulps and their application for viscose and lyocell. Cellulose. 2011;18:1113–28.
28. Saka S, Takahashi T. Effects of solvent addition to acetylation medium on cellulose triacetate prepared from low-grade dissolving pulp. Cellul Cellul Deriv. 1995;219–226
29. Biermann CJ. Handbook of pulping and papermaking. Cambridge, MA: Academic Press; 1996.
30. Zhao X, van der Heide E, Zhang T, Liu D. Single-stage pulping of sugarcane bagasse with peracetic acid. J Wood Chem Technol. 2011;31:1–25.
31. US EPA. Chemical Wood Pulping. Report. Available online at https://www3.epa.gov/ttnchie1/ap42/ch10/final/c10s02.pdf
32. Johansson A, Aaltonen O, Ylinen P. Organosolv pulping: methods and pulp properties. Biomass. 1987;13:45–65.
33. Hergert HL. Developments in organosolv pulping. An overview. In: Young RA, Akhtar M, editors. Environmentally friendly technologies for the pulp and paper industry. New York: Wiley; 1998.
34. Kinstrey RB. An overview of strategies for reducing the environmental impact of bleach-plant effluents. TAPPI J. 1993;76:105–13.
35. Muurinen E. Organosolv pulping—A review and distillation study related to peroxyacid pulping. Academic Dissertation. Available online at: http://jultika.oulu.fi/files/isbn9514256611.pdf
36. Kordsachia O, Wandinger B, Patt R. Some investigations on ASAM pulping and chlorine free bleaching of eucalyptus from Spain. Holz Roh Werkst. 1992;50:85–91.
37. Abad S, Santos V, Parajó JC. Two-stage acetosolv pulping of eucalyptus wood. Cellul Chem Technol. 2003;35:333–43.
38. Ferrer A, Vega A, Rodríguez A, Jiménez L. Acetosolv pulping for the fractionation of empty fruit bunches from palm oil industry. Bioresour Technol. 2013;132:115–20.
39. Dapía S, Santos V, Parajó JC. Carboxymethylcellulose from totally chlorine-free-bleached milox pulps. Bioresour Technol. 2003;89:289–96.
40. Ligero P, Villaverde JJ, Vega A, Bao M. Pulping cardoon (*Cynara cardunculus*) with peroxyformic acid (MILOX) in one single stage. Bioresour Technol. 2008;99:5687–93.
41. Kramer KJ, Masanet E, Xu T, Worrell E. Energy efficiency improvement and cost saving opportunities for the pulp and pulp industry. Ernest Orlando Lawrence Berkeley National Laboratory. Report LBNL-2268E. 2009. Available online at: https://www.energystar.gov/ia/business/industry/downloads/Pulp_and_Paper_Energy_Guide.pdf
42. Kong L, Hasanbeigi A, Price L. Assessment of emerging energy-efficiency technologies for the pulp and paper industry: a technical review. J Clean Prod. 2016;122:5–28.
43. Industrial Technologies Program (ITP). Highly energy efficient D-GLU (Directed Green Liquor Utilization) pulping. U.S. Department of Energy, Washington, DC. 2011. Available online at: http://energy.gov/sites/prod/files/2014/05/f16/dglu_pulping.pdf
44. Adnan S, Hoang M, Wang HT, Bolto B, Xie ZL. Recent trends in research, development and application of membrane technology in the pulp and paper industry. Appita. 2010;63:235–41.
45. Bjork M, Sjogren T, Lundin T, Rickards H, Kochesfahani S. Partial borate autocausticizing trial increases capacity at Swedish mill. TAPPI. 2005;88:15–9.
46. Muehlethaler E, Starkey Y, Salminen R, Harding D. Steam cycle washer for unbleached pulp. Final report. 21st Century Pulp and Paper and Idaho National Laboratory, Port Townsend, WA. 2008. Available online at: http://www.osti.gov/scitech/servlets/purl/937487

47. Kemppainen K, Körkkö M, Niinimäki J. Fractional pulping of toner and pigment-based inkjet ink printed papers—ink and dirt behavior. Bioresources. 2011;6:2977–89.
48. Carvalho DMD, Perez A, García JC, Colodette JL, López F, Diaz MJ. Ethanol-soda pulping of sugarcane bagasse and straw. Cellul Chem Technol. 2014;48:355–64.
49. Seo YB, Lee YW, Lee CH, You HC. Red algae and their use in papermaking. Bioresour Technol. 2010;101:2549–53.
50. Jung KA, Lim SR, Kim Y, Park JM. Potentials of macroalgae as feedstocks for biorefinery. Bioresour Technol. 2013;135:182–90.
51. Lloveras J, Santiveri F, Gorchs G. Hemp and flax biomass and fibre production and linseed yield in irrigated Mediterranean conditions. J Ind Hemp. 2006;11:3–15.
52. Evaluation of the common market organization for flax and hemp. Summary of the final report. Available online at: http://ec.europa.eu/agriculture/eval/reports/lin/sum_en.pdf
53. Gonzalez-Garcia S, Hospido A, Feijoo G, Moreira MT. Life cycle assessment of raw materials for non-wood pulp mills: hemp and flax. Resour Conserv Recycl. 2010;54:923–30.
54. Jahan MS, Kanna GH, Mun SP, Chowdhury DA. Variations in chemical characteristics and pulpability within jute plant (*Chorcorus capsularis*). Ind Crop Prod. 2008;28:199–205.
55. Marin F, Sanchez JL, Arauzo J, et al. Semichemical pulping of *Miscanthus giganteus*. Effect of pulping conditions on some pulp and paper properties. Bioresour Technol. 2009;100: 3933–40.
56. Dutta D, Upadhyay JS, Singh B, Tyagi CH. Studies on *Hibiscus cannabinus* and *Hibiscus sabdariffa* as an alternative pulp blend for softwood: an optimization of Kraft delignification process. Ind Crop Prod. 2009;29:16–26.
57. Sanchez R, Rodriguez A, Navarro E, et al. Use of *Hesperaloe funifera* for the production of paper and extraction of lignin for synthesis and fuel gases. Biomass Bioenergy. 2010;34: 1471–80.
58. Seo YB, Lee YW, Lee CH, You HC. Red algae and their use in papermaking. Bioresour Technol. 2010;101:2549–53.
59. Khiari R, Mhenni MF, Belgacem MN, Mauret E. Chemical composition and pulping of date palm rachis and *Posidonia oceanica*—a comparison with other wood and non-wood fibre sources. Bioresour Technol. 2010;101:775–80.
60. Chao KP, Su YC, Chen CS. Feasibility of utilizing Rhizoclonium in pulping and papermaking. J Appl Phycol. 2000;12:53–62.
61. Ververis C, Georghiou K, Danielidis D, et al. Cellulose, hemicelluloses, lignin and ash content of some organic materials and their suitability for use as paper pulp supplements. Bioresour Technol. 2007;98:296–301.
62. Jain A, Wei Y, Tietje A. Biochemical conversion of sugarcane bagasse into bioproducts. Biomass Bioenergy. 2016;93:227–42.
63. Khristova P, Kordsachia O, Patt R, et al. Environmentally friendly pulping and bleaching of bagasse. Ind Crop Prod. 2006;23:131–9.
64. Rodriguez A, Moral A, Serrano L, et al. Rice straw pulp obtained by using various methods. Bioresour Technol. 2008;99:2881–6.
65. Hedjazi S, Kordsachia O, Patt R, et al. Alkaline sulfite–anthraquinone (AS/AQ) pulping of wheat straw and totally chlorine free (TCF) bleaching of pulps. Ind Crop Prod. 2009;29: 27–36.
66. Jahan MS, Rahman MM. Effect of pre-hydrolysis on the soda-anthraquinone pulping of corn stalks and *Saccharum spontaneum* (kash). Carbohydr Polym. 2012;88:583–8.
67. Cole M, Lindeque P, Halsband C, Galloway ST. Microplastics as contaminants in the marine environment: a review. Mar Pollut Bull. 2011;62:2588–97.
68. Gross RA, Kalra B. Biodegradable polymers for the environment. Science. 2002;297:803–7.
69. Jabeen N, Majid I, Nayik GA. Bioplastics and food packaging: a review. Cogent Food Agric. 2015;1:1117749.
70. Kadouri D, Jurkevitch E, Okon Y, Castro-Sowinski S. Ecological and agricultural significance of bacterial polyhydroxyalkanoates. Crit Rev Microbiol. 2005;31:55–67.

71. Osanai T, Numata K, Oikawa A, et al. Increased bioplastic production with an RNA polymerase sigma factor SigE during nitrogen starvation in *synechocystis* sp. PCC 6803. DNA Res. 2013;20:525–35.
72. Sagnelli D, Hebelstrup KH, Leroy E, et al. Plant-crafted starches for bioplastics production. Carbohydr Polym. 2016;152:398–408.
73. Morone P, Tartiu VE, Falcone P. Assessing the potential of biowaste for bioplastics production through social network analysis. J Clean Prod. 2015;90:43–54.
74. Hossain ABMS, Ibrahim NA, AlEissa MS. Nano-cellulose derived bioplastic biomaterial data for vehicle bio-bumper from banana peel waste biomass. Data Brief. 2016;8:286–94.
75. Hempel F, Bozarth AS, Lindenkamp N, et al. Microalgae as bioreactors for bioplastic production. Microb Cell Factories. 2011;10(81):1–6.
76. Queiroz AU, Collares FP. Innovation and industrial trends in bioplastics. J Macromol Sci C Polym Rev. 2009;49:65–78.
77. Saharia J, Sapuan SM, Zainudin ES, Maleque MA. Thermo-mechanical behaviors of thermoplastic starch derived from sugar palm tree (*Arenga pinnata*). Carbohydr Polym. 2013;92: 1711–6.
78. Song Y, Matsumoto K, Tanaka T, Kondo A, Taguchi S. Single-step production of polyhydroxybutyrate from starch by using α-amylase cell-surface displaying system of *Corynebacterium glutamicum*. J Biosci Bioeng. 2013;115:12–4.
79. Liu H, Xie F, Yu L, et al. Thermal processing of starch-based polymers. Prog Polym Sci. 2009;34:1348–68.
80. Wang Q, Cai J, Zhang L, et al. A bioplastic with high strength constructed from a cellulose hydrogel by changing the aggregated structure. J Mater Chem A. 2013;1:6678–86.
81. Mohanty AK, Wibowo A, Misra M, Drzal LT. Development of renewable resource-based cellulose acetate bioplastic: effect of process engineering on the performance of cellulosic plastics. Polym Eng Sci. 2003;43:1151–61.
82. Ng KS, Ooi WY, Goh LK, et al. Evaluation of jatropha oil to produce poly (3-hydroxybutyrate) by *Cupriavidus necator* H16. Polym Degrad Stab. 2010;95:1365–9.
83. Kim JR, Sharma S. The development and comparison of bio-thermoset plastics from epoxidized plant oils. Ind Crop Prod. 2012;36:485–99.
84. Irorere VU, Bagherias S, Blevins M, Kwiecień I, Stamboulis A, Radecka I. Electrospun fibres of polyhydroxybutyrate synthesized by *Ralstonia eutropha* from different carbon sources. Int J Polym Sci. 2014;2014:705359.
85. Hyakutake M, Saito Y, Tomizawa S, Mizuno K, Tsuge T. Polyhydroxyalkanoate (PHA) synthesis by class IV PHA synthases employing *Ralstonia eutropha* PHB-4 as host strain. Biosci Biotechnol Biochem. 2011;75:1615–7.
86. Sun S, Song Y, Zheng Q. Morphologies and properties of thermo-molded biodegradable plastics based on glycerol-plasticized wheat gluten. Food Hydrocoll. 2007;21:1005–13.
87. Song Y, Zheng Q. Improved tensile strength of glycerol-plasticized gluten bioplastic containing hydrophobic liquids. Bioresour Technol. 2008;99:7665–71.
88. Sun XS, Kim HR, Mo X. Plastic performance of soybean protein components. J Am Oil Chem Soc. 1999;76:119–23.
89. Lodha P, Netravali AN. Thermal and mechanical properties of environment-friendly ‘green’ plastics from stearic acid modified-soy protein isolate. Ind Crop Prod. 2005;21:49–64.
90. Tian H, Wang Y, Zhang L, et al. Improved flexibility and water resistance of soy protein thermoplastics containing waterborne polyurethane. Ind Crop Prod. 2010;32:13–20.
91. Zeller MA, Hunt R, Jones A, Sharma S. Bioplastics and their thermoplastic blends from Spirulina and chlorella microalgae. J Appl Polym Sci. 2013;130:3263–75.
92. Jerez A, Partal P, Martinez I, et al. Egg white-based bioplastics developed by thermomechanical processing. J Food Eng. 2007;82:608–17.
93. Lee JY, Li P, Lee J, et al. Ethanol production from *Saccharina japonica* using an optimized extremely low acid pretreatment followed by simultaneous saccharification and fermentation. Bioresour Technol. 2013;127:119–25.

94. Tian H, Wu W, Guo G, et al. Microstructure and properties of glycerol plasticized soy protein plastics containing castor oil. J Food Eng. 2012;109:496–500.
95. Gutierrez JG, Partal P, Morales MG, Gallegos C. Effect of processing on the viscoelastic, tensile and optical properties of albumen/starch-based bioplastics. Carbohydr Polym. 2011;84:308–15.
96. Dias AB, Muller CM, Larotonda FD, Laurindo JB. Mechanical and barrier properties of composite films based on rice flour and cellulose fibers. Food Sci Technol. 2011;44:535–42.
97. Kim YB, Lenz RW. Polyesters from microorganisms. In: Babel W, Steinbuchel A, editors. Advances in Biochemical Engineering Biotechnology. Berlin: Springer; 2000. p. 51–79.
98. de Koning G. Physical properties of bacterial poly((R)-3-hydroxyalkanoates). Can J Microbiol. 1995;41:303–9.
99. Fernandez D, Rodriguez E, Bassas M, et al. Agro-industrial oily wastes as substrates for PHA production by the new strain *Pseudomonas aeruginosa* NCIB 40045: effect of culture conditions. Biochem Eng J. 2005;26:159–67.
100. Bhuwal AK, Singh G, Aggarwal NK, et al. Isolation and screening of polyhydroxyalkanoates producing bacteria from pulp, paper, and cardboard industry wastes. Int J Biomater. 2013; 2013:1–10.
101. Nampoothiri KM, Nair NR, John RP. An overview of the recent developments in polylactide (PLA) research. Bioresour Technol. 2010;101:8493–501.
102. Shi X, Sun L, Jiang J, et al. Biodegradable polymeric microcarriers with controllable porous structure for tissue engineering. Macromol Biosci. 2009;9:1211–8.
103. Jung YK, Lee SY. Efficient production of polylactic acid and its copolymers by metabolically engineered *Escherichia coli*. J Biotechnol. 2011;151:94–101.
104. Erdohan ZO, Çama B, Turhan KN. Characterization of antimicrobial polylactic acid based films. J Food Eng. 2013;119:308–15.
105. Kaneuchi C, Seki M, Komagata K. Production of succinic acid from citric acid and related acids by lactobacillus strains. Appl Environ Microbiol. 1988;54:3053–6.
106. Lee PC, Lee SY, Hong SH, Chang HN. Isolation and characterization of a new succinic acid-producing bacterium, *Mannheimia succiniciproducens* MBEL55E, from bovine rumen. Appl Microbiol Biotechnol. 2002;58:663–8.
107. Sonnenschei MF, Guillaudeu SJ, Landes BG, Wendt BL. Comparison of adipate and succinate polyesters in thermoplastic polyurethanes. Polymer. 2010;51:3685–92.
108. Frollini E, Bartolucci N, Sisti L, Celli A. Poly(butylene succinate) reinforced with different lignocellulosic fibers. Ind Crop Prod. 2013;45:160–9.
109. Abdel-Fattah TM, Mahmoud ME, Ahmed SB, Huff MD, Lee JW, Kumar S. Biochar from woody biomass for removing metal contaminants and carbon sequestration. J Ind Eng Chem. 2015;22:103–9.
110. Windeatt JH, Ross AB, Williams PT, Forster PM, Nahil MA, Singh S. Characteristics of biochars from crop residues: potential for carbon sequestration and soil amendment. J Environ Manag. 2014;146:189–97.
111. Ahmed MB, Zhou JL, Ngo HH, Guo W, Chen M. Progress in the preparation and application of modified biochar for improved contaminant removal from water and wastewater. Bioresour Technol. 2016;214:836–51.
112. Gai X, Wang H, Liu J, et al. Effects of feedstock and pyrolysis temperature on biochar adsorption of ammonium and nitrate. PLoS One. 2014;9:e113888.
113. Gasparovic L, Korenova Z, Jelemensky L. Kinetic study of wood chips decomposition by TGA. Chem Pap. 2010;64:174–81.
114. Demirbas A, Arin G. An overview of biomass pyrolysis. Energy Sources. 2002;24:471–82.
115. Laird DA, Brown RC, Amonette JE, Lehmann J. Review of the pyrolysis platform for coproducing bio-oil and biochar. Biofuels Bioprod Biorefin. 2009;3:547–62.
116. Nguyen TL, Hermansen JE, Nielsen RG. Environmental assessment of gasification technology for biomass conversion to energy in comparison with other alternatives: the case of wheat straw. J Clean Prod. 2013;53:138–48.

117. Yang Y, Brammer JG, Mahmood ASN, Hornung A. Intermediate pyrolysis of biomass energy pellets for producing sustainable liquid, gaseous and solid fuels. Bioresour Technol. 2014; 169:794–9.
118. Brewer CE, Schmidt-Rohr K, Satrio JA, Brown RC. Characterization of biochar from fast pyrolysis and gasification systems. Environ Prog Sustain Energy. 2009;28:386–96.
119. Horne PA, Williams PT. Influence of temperature on the products from the flash pyrolysis of biomass. Fuel. 1996;75:1051–9.
120. Onay O, Kockar OM. Slow, fast and flash pyrolysis of rapeseed. Renew Energy. 2003;28: 2417–33.
121. Ncibi MC, Jeanne-Rose V, Mahjoub B, et al. Preparation and characterisation of raw chars and physically activated carbons derived from marine *Posidonia oceanica* (L.) fibres. J Hazard Mater. 2009;165:240–9.
122. Gómez N, Rosas JG, Cara J, Martínez O, Alburquerque JA, Sánchez ME. Slow pyrolysis of relevant biomasses in the Mediterranean basin. Part 1. Effect of temperature on process performance on a pilot scale. J Clean Prod. 2016;120:181–90.
123. Idris J, Shirai Y, Anduo Y, et al. Improved yield and higher heating value of biochar from oil palm biomass at low retention time under self-sustained carbonization. J Clean Prod. 2015;104:475–9.
124. Crombie K, Masek O. Pyrolysis biochar systems, balance between bioenergy and carbon sequestration. GCB Bioenergy. 2015;7:349–61.
125. Zhang L, Xu C, Champagne P. Overview of recent advances in thermochemical conversion of biomass. Energy Convers Manag. 2010;51:969–82.
126. Benavente V, Fullana A. Torrefaction of olive mill waste. Biomass Bioenergy. 2015;73: 186–94.
127. Sadaka S, Negi S. Improvements of biomass physical and thermochemical characteristics via torrefaction process. Environ Prog Sustain Energy. 2009;28:427–34.
128. Couhert C, Salvador S, Commandre J. Impact of torrefaction on syngas production from wood. Fuel. 2009;88:2286–90.
129. Arias B, Pevida C, Fermoso J, Plaza M, Rubiera F, Pis J. Influence of torrefaction on the grindability and reactivity of woody biomass. Fuel Process Technol. 2008;89:169–75.
130. Liu Z, Balasubramanian R. Upgrading of waste biomass by hydrothermal carbonization (HTC) and low temperature pyrolysis (LTP): a comparative evaluation. Appl Energy. 2014;114:857–64.
131. Liu Z, Quek A, Hoekman SK, Balasubramanian R. Production of solid biochar fuel from waste biomass by hydrothermal carbonization. Fuel. 2013;103:943–9.
132. Kumar D, Pant KK. Production and characterization of biocrude and biochar obtained from non-edible de-oiled seed cakes hydrothermal conversion. J Anal Appl Pyrolysis. 2015;115: 77–86.
133. Palumbo AW, Sorli JC, Weimer AW. High temperature thermochemical processing of biomass and methane for high conversion and selectivity to H2-enriched syngas. Appl Energy. 2015;157:13–24.
134. Libra JA, Ro KS, Kammann C, et al. Hydrothermal carbonization of biomass residuals: a comparative review of the chemistry, processes and applications of wet and dry pyrolysis. Biofuels. 2011;2:89–124.
135. Alatalo SM, Repo E, Mäkilä E, Salonen J, Vakkilainen E, Sillanpää M. Adsorption behavior of hydrothermally treated municipal sludge & pulp and paper industry sludge. Bioresour Technol. 2013;147:71–6.
136. Funke A, Ziegler F. Hydrothermal carbonization of biomass: a summary and discussion of chemical mechanisms for process engineering. Biofuels Bioprod Biorefin. 2010;4:160–77.
137. Sabio E, Alvarez-Murillo A, Roman S, Ledesma B. Conversion of tomato-peel waste into solid fuel by hydrothermal carbonization: influence of the processing variables. Waste Manag. 2016;47:122–32.

138. Cao X, Ro KS, Libra JA, et al. Effects of biomass types and carbonization conditions on the chemical characteristics of Hydrochars. J Agric Food Chem. 2013;61:9401–11.
139. Schuhmacher JP, Huntjens FJ, van Krevelen DW. Chemical structure and properties of coal XXVI—studies on artificial coalification. Fuel. 1960;39:223–34.
140. Antal MJ, Gronli M. The art, science, and technology of charcoal production. Ind Eng Chem Res. 2003;42:1619–40.
141. Ruyter HP. Coalification model. Fuel. 1982;61:1182–7.
142. Pratt K, Moran D. Evaluating the cost-effectiveness of global biochar mitigation potential. Biomass Bioenergy. 2010;34:1149–58.
143. Gul S, Whalen JK, Thomas BW, Sachdeva V, Deng H. Physico-chemical properties and microbial responses in biochar-amended soils: mechanisms and future directions. Agric Ecosyst Environ. 2015;206:46–59.
144. Intani K, Latif S, Rafayatul Kabir AKM, Müller J. Effect of self-purging pyrolysis on yield of biochar from maize cobs, husks and leaves. Bioresour Technol. 2016;218:541–51.
145. Brewer CE, Brown RC. Biochar. In: Sayigh A, editor. Comprehensive renewable energy. Oxford: Elsevier; 2012. p. 357–84.
146. Novak JM, Cantrell KB, Watts DW. Compositional and thermal evaluation of lignocellulosic and poultry litter chars via high and low temperature pyrolysis. Bioenergy Res. 2013;6: 114–30.
147. Huff MD, Kumar S, Lee JW. Comparative analysis of pinewood, peanut shell, and bamboo biomass derived biochars produced via hydrothermal conversion and pyrolysis. J Environ Manag. 2014;146:303–8.
148. Chen B, Chen Z. Sorption of naphthalene and 1-naphthol by biochars of orange peels with different pyrolytic temperatures. Chemosphere. 2009;76:127–33.
149. Chen B, Zhou D, Zhu L. Transitional adsorption and partition on nonpolar and polar aromatic contaminants by biochars of pine needles with different pyrolytic temperatures. Environ Sci Technol. 2008;42:5137–43.
150. Ahmad M, Lee SS, Dou X, et al. Effects of pyrolysis temperature on soybean Stover- and peanut shell-derived biochar properties and TCE adsorption in water. Bioresour Technol. 2012;118:536–44.
151. Keiluweit M, Nico PS, Johnson MG, Kleber M. Dynamic molecular structure of plant - biomass-derived black carbon (biochar). Environ Sci Technol. 2010;44:1247–53.
152. Ahmad M, Rajapaksha AU, Lim JE. Biochar as a sorbent for contaminant management in soil and water: a review. Chemosphere. 2014;99:19–33.
153. Karaosmanoglu F, Ergudenler AI, Sever A. Biochar from the straw–stalk of rapeseed plant. Energy Fuel. 2000;14:336–9.
154. Jindo K, Mizumoto H, Sawada Y, Sanchez-Monedero MA, Sonoki T. Physical and chemical characterization of biochars derived from different agricultural residues. Biogeosciences. 2014;11:6613–21.
155. Uchimiya M, Chang S, Klasson KT. Screening biochars for heavy metal retention in soil: role of oxygen functional groups. J Hazard Mater. 2011;190:432–41.
156. Zielinska A, Oleszczuk P. Evaluation of sewage sludge and slow pyrolyzed sewage sludge-derived biochar for adsorption of phenanthrene and pyrene. Bioresour Technol. 2015;192: 618–26.
157. Cantrell KB, Hunt PG, Uchimiya M, Novak JM, Ro KS. Impact of pyrolysis temperature and manure source on physicochemical characteristics of biochar. Bioresour Technol. 2012;107: 419–28.
158. Liu N, Charrua AB, Weng CH, Yuan X, Ding F. Characterization of biochars derived from agriculture wastes and their adsorptive removal of atrazine from aqueous solution: a comparative study. Bioresour Technol. 2015;198:55–62.
159. Bhatnagar A, Hogland W, Marques M, Sillanpää M. An overview of the modification methods of activated carbon for its water treatment applications. Chem Eng J. 2013;219: 499–511.

160. Ould-Idriss A, Stitou M, Cuerda-Correa EM. Preparation of activated carbons from olive-tree wood revisited. II. Physical activation with air. Fuel Process Technol. 2011;92:266–70.
161. Jung SH, Kim JS. Production of biochars by intermediate pyrolysis and activated carbons from oak by three activation methods using CO_2. J Anal Appl Pyrolysis. 2014;107:116–22.
162. Shim TY, Yoo JS, Ryu C, Park YK, Jung J. Effect of steam activation of biochar produced from a giant Miscanthus on copper sorption and toxicity. Bioresour Technol. 2015;197: 85–90.
163. Demiral H, Demiral I, Karabacakoglu B, Tümsek F. Production of activated carbon from olive bagasse by physical activation. Chem Eng Res Des. 2011;89:206–13.
164. Jimenez-Cordero D, Heras F, Alonso-Morales N, Gilarranz MA, Rodriguez JJ. Ozone as oxidation agent in cyclic activation of biochar. Fuel Process Technol. 2015;139:42–8.
165. Cha JS, Park SH, Jung SC. Production and utilization of biochar: a review. J Ind Eng Chem. 2016;40:1–15.
166. Bansal RC, Donnet JB, Stoeckli F. Active carbon. New York: Marcel Dekker; 1998.
167. Buchel KH, Moretto HH, Woditsch P. Industrial inorganic chemistry. 2nd ed. Weinheim: Wiley-VCH; 2000.
168. Ncibi MC, Altenor S, Seffen M, Brouers F, Gaspard S. Modelling single compound adsorption onto porous and non-porous sorbents using a deformed Weibull exponential isotherm. Chem Eng J. 2008;145:196–202.
169. Meryemoglu B, Irmak S, Hasanoglu A. Production of activated carbon materials from kenaf biomass to be used as catalyst support in aqueous-phase reforming process. Fuel Process Technol. 2016;151:59–63.
170. Pak SH, Jeon MJ, Jeon YW. Study of sulfuric acid treatment of activated carbon used to enhance mixed VOC removal. Int Biodeter Biodegrad. 2016;113:195–200.
171. Park J, Hung I, Gan Z, Rojas OJ, Lim KH, Park S. Activated carbon from biochar: influence of its physicochemical properties on the sorption characteristics of phenanthrene. Bioresour Technol. 2013;149:383–9.
172. Menéndez-Diaz JA, Martin-Gullon I. Types of carbon adsorbents and their production, in activated carbon surfaces in environmental remediation. In: Bandosz TJ, editor. Interface science and technology, vol. 7. Oxford: Academic Press, Elsevier; 2006.
173. Kumar A, Jena HM. High surface area microporous activated carbons prepared from fox nut (*Euryale ferox*) shell by zinc chloride activation. Appl Surf Sci. 2015;356:753–61.
174. Gokce Y, Aktas Z. Nitric acid modification of activated carbon produced from waste tea and adsorption of methylene blue and phenol. Appl Surf Sci. 2014;313:352–9.
175. Prahas D, Kartika Y, Indraswati N, Ismadji S. Activated carbon from jackfruit peel waste by H_3PO_4 chemical activation. Chem Eng J. 2008;140:32–42.
176. Wyman CE, Dale BE, Elander RT, Holtzappl M, Ladisch MR, Lee YY. Coordinated development of leading biomass pretreatment technologies. Bioresour Technol. 2005;96: 1959–66.
177. Ros A, Lillo-Rodenas MA, Fuente E, Montes-Moran MA, Martin MJ, Linares-Solano A. High surface area materials prepared from sewage sludge-based precursors. Chemosphere. 2006;65:132–40.
178. Rouquerol F, Rouquerol J, Sing K. Adsorption by powders and porous solids: principles, methodology and applications. San Diego: Academic Press; 1999.
179. Brunauer S, Emmett PH, Teller EJ. Adsorption of gases in multimolecular layers. Am Chem Soc. 1938;60:309–19.
180. Pomonis PJ, Petrakis DE, Ladavos AK, et al. A novel method for estimating the C-values of the BET equation in the whole range $0 < P/Po < 1$ using a Scatchard-type treatment of it. Microporous Mesoporous Mater. 2004;69:97–107.
181. Passe-Coutrin N, Altenor S, Cossement D, Jean-Marius C, Gaspard S. Comparison of parameters calculated from the BET and Freundlich isotherms obtained by nitrogen adsorption on activated carbons: A new method for calculating the specific surface area. Microporous Mesoporous Mater. 2008;111:517–22.

182. Pelaez-Cid AA, Herrera-Gonzalez AM, Salazar-Villanueva M, Bautista-Hernandez A. Elimination of textile dyes using activated carbons prepared from vegetable residues and their characterization. J Environ Manag. 2016;181:269–78.
183. Tay T, Ucar S, Karagoz S. Preparation and characterization of activated carbon from waste biomass. J Hazard Mater. 2009;165:481–5.
184. Lin L, Zhai SR, Xiao ZY, Song Y, An QD, Song XW. Dye adsorption of mesoporous activated carbons produced from NaOH-pretreated rice husks. Bioresour Technol. 2013; 136:437–43.
185. Kalderis D, Bethanis S, Paraskeva P, Diamadopoulos E. Production of activated carbon from bagasse and rice husk by a single-stage chemical activation method at low retention times. Bioresour Technol. 2008;99:6809–16.
186. Zhang FS, Nriagu JO, Itoh H. Mercury removal from water using activated carbons derived from organic sewage sludge. Water Res. 2005;39:389–95.
187. Hu Z, Guo H, Srinivasan MP, Yaming N. A simple method for developing mesoporosity in activated carbon. Sep Purif Technol. 2003;31:47–52.
188. Benadjemia M, Millière L, Reinert L, Benderdouche N, Duclaux L. Preparation, characterization and methylene blue adsorption of phosphoric acid activated carbons from globe artichoke leaves. Fuel Process Technol. 2011;92:1203–12.
189. Nunell GV, Fernandez ME, Bonelli PR, Cukierman AL. Conversion of biomass from an invasive species into activated carbons for removal of nitrate from wastewater. Biomass Bioenergy. 2012;44:87–95.
190. Angin D, Altintig E, Köse TE. Influence of process parameters on the surface and chemical properties of activated carbon obtained from biochar by chemical activation. Bioresour Technol. 2013;148:542–9.
191. Ncibi MC, Ranguin R, Pintor MJ, Jeanne-Rose V, Sillanpää M, Gaspard S. Preparation and characterization of chemically activated carbons derived from Mediterranean *Posidonia oceanica* (L.) fibres. J Anal Appl Pyrol. 2014;109:205–14.
192. Huang LH, Sun YY, Yang T, Li L. Adsorption behavior of Ni (II) on lotus stalks derived active carbon by phosphoric acid activation. Desalination. 2011;268:12–9.
193. Altenor S, Carene B, Emmanuel E, Lambert J, Ehrhardt JJ, Gaspard S. Adsorption studies of methylene blue and phenol onto vetiver roots activated carbon prepared by chemical activation. J Hazard Mater. 2009;165:1029–39.
194. Gai X, Wang H, Liu J, et al. Effects of feedstock and pyrolysis temperature on biochar adsorption of ammonium and nitrate. PLoS One. 2014;9:e113888.
195. Li J, Li Y, Wu M, Zhang Z, Lu J. Effectiveness of low-temperature biochar in controlling the release and leaching of herbicides in soil. Plant Soil. 2013;370:333–44.
196. Jiang J, Zhang L, Wang X. Highly ordered macroporous woody biochar with ultra-high carbon content as supercapacitor electrodes. Electrochim Acta. 2013;113:481–9.
197. Sun W, Lipka SM, Swartz C, Williams D, Yang F. Hemp-derived activated carbons for supercapacitors. Carbon. 2016;103:181–92.
198. Sethia G, Sayari A. Activated carbon with optimum pore size distribution for hydrogen storage. Carbon. 2016;99:289–94.
199. Heo YJ, Park SJ. Synthesis of activated carbon derived from rice husks for improving hydrogen storage capacity. J Ind Eng Chem. 2015;31:330–4.
200. Pierpoint LM. Harnessing electricity storage for systems with intermittent sources of power: policy and R&D needs. Energy Policy. 2016;96:751–7.
201. Agarkar S, Yadav P, Fernandes R, Kothari D, Suryawanshi A, Ogale S. Minute-made activated porous carbon from agro-waste for Li-ion battery anode using a low power microwave oven. Electrochim Acta. 2016;212:535–44.
202. Miller JR, Simon P. Electrochemical capacitors for energy management. Science. 2008;321: 651–2.

203. Inal IIG, Holmes SM, Banford A, Aktas Z. The performance of supercapacitor electrodes developed from chemically activated carbon produced from waste tea. Appl Surf Sci. 2015; 357:696–703.
204. Pandolfo A, Hollenkamp A. Carbon properties and their role in supercapacitors. J. Power Sources. 2006;157:11–27.
205. Cho EA, Lee SY, Park SJ. Effect of microporosity on nitrogen-doped microporous carbons for electrode of supercapacitor. Carbon Lett. 2014;15:210–3.
206. Chen XY, Chen C, Zhang ZJ, et al. Nitrogen-doped porous carbon for supercapacitor with long-term electrochemical stability. J Power Sources. 2013;230:50–8.
207. Kim KS, Park SJ. Synthesis and high electrochemical capacitance of N-doped microporous carbon/carbon nanotubes for supercapacitor. J Electroanal Chem. 2012;673:58–64.
208. Wu K, Liu Q. Nitrogen-doped mesoporous carbons for high performance supercapacitors. Appl Surf Sci. 2016;379:132–9.
209. Abioye AM, Ani FN. Recent development in the production of activated carbon electrodes from agricultural waste biomass for supercapacitors: a review. Renew Sust Energ Rev. 2015; 52:1282–93.
210. Zhang J, Gong L, Sun K, Jiang J, Zhang X. Preparation of activated carbon from waste *Camellia oleifera* shell for supercapacitor application. J Solid State Electrochem. 2012;16: 2179–86.
211. Rufford TE, Hulicova-Jurcakova D, Zhu Z, Lu GQ. Nanoporous carbon electrode from waste coffee beans for high performance supercapacitors. Electrochem Commun. 2008;10:1594–7.
212. Elmouwahidi A, Zapata-Benabithe Z, Carrasco-Marin F, Moreno-Castilla C. Activated carbons from KOH-activation of argan (*Argania spinosa*) seed shells as supercapacitor electrodes. Bioresour Technol. 2012;111:185–90.
213. Peng C, Yan XB, Wang RT, Lang JW, Ou YJ, Xue QJ. Promising activated carbons derived from waste tea-leaves and their application in high performance supercapacitors electrodes. Electrochim Acta. 2013;87:401–8.
214. Rufford TE, Hulicova-Jurcakova D, Khosla K, Zhu Z, Lu GQ. Microstructure and electrochemical double-layer capacitance of carbon electrodes prepared by zinc chloride activation of sugar cane bagasse. J Power Sources. 2010;195:912–8.
215. Jin H, Wang X, Gu Z, Polin J. Carbon materials from high ash biochar for supercapacitor and improvement of capacitance with HNO_3 surface oxidation. J Power Sources. 2013;236: 285–92.
216. Cao Y, Wang K, Wang X, et al. Hierarchical porous activated carbon for supercapacitor derived from corn stalk core by potassium hydroxide activation. Electrochim Acta. 2016;212: 839–47.
217. Li X, Xing W, Zhuo S, et al. Preparation of capacitor's electrode from sunflower seed shell. Bioresour Technol. 2011;102:1118–23.
218. Lv Y, Gan L, Liu M, et al. A self-template synthesis of hierarchical porous carbon foams based on banana peel for supercapacitor electrodes. J. Power Sources. 2012;209:152–7.
219. He X, Ling P, Qiu J, et al. Efficient preparation of biomass-based mesoporous carbons for supercapacitors with both high energy density and high power density. J Power Sources. 2013;240:109–13.
220. Taer E, Deraman M, Talib IA, Awildrus A, Hashmi SA, Umar AA. Preparation of highly porous binderless activated carbon monolith from rubber wood sawdust by a multi-step activation process for application in supercapacitors. Int J Electrochem Sci. 2011;6:3301–15.
221. Misnon II, Zain NKM, Abdul AR, Vidyaharan B, Jose R. Electrochemical properties of carbon from oil palm kernel shell for high performance supercapacitors. Electrochim Acta. 2015;174:78–86.
222. Chen M, Kang X, Wumaier T, et al. Preparation of activated carbon from cotton stalk and its application in supercapacitor. J Solid State Electrochem. 2013;17:1005–12.

223. Adinaveen T, Kennedy LJ, Vijaya JJ, Sekaran G. Surface and porous characterization of activated carbon prepared from pyrolysis (rice straw) by two-stage procedure and its applications in supercapacitor electrodes. J Mater Cycles Waste Manage. 2014;17:736–47.
224. Meyer S, Bright RM, Fischer D, Schulz H, Glaser B. Albedo impact on the suitability of biochar systems to mitigate global warming. Environ Sci Technol. 2012;46:12726–34.
225. Smith P. Soil carbon sequestration and biochar as negative emission technologies. Glob Chang Biol. 2016;22:1315–24.
226. Woolf D, Amonette JE, Street-Perrott A, Lehmann J, Joseph S. Sustainable biochar to mitigate global climate change. Nature Commun. 2010;1(56):1–9.
227. Yuan JH, Xu RK. The amelioration effects of low temperature biochar generated from nine crop residues on an acidic Ultisol. Soil Use Manag. 2011;27:110–5.
228. Liang B, Lehmann J, Solomon D, et al. Black carbon increases cation exchange capacity in soils. Soil Sci Soc Am J. 2006;70:1719–30.
229. Gul S, Whalen JK, Thomas BW, Sachdeva V, Deng H. Physico-chemical properties and microbial responses in biochar-amended soils: mechanisms and future directions. Agric Ecosyst Environ. 2015;206:46–59.
230. Yuan JH, Xu RK, Wang N, Li JY. Amendment of acid soils with crop residues and biochars. Pedosphere. 2011;21:302–8.
231. Novak JM, Busscher WJ, Watts DW, Laird DA, Ahemdna MA, Niandou MAS. Short-term CO_2 mineralization after additions of biochar and switchgrass to a Typic Kandiudult. Geoderma. 2010;154:281–8.
232. Kloss S, Zehetner F, Wimmer B, Buecker J, Rempt F, Soja G. Biochar application to temperate soils: effects on soil fertility and crop growth under greenhouse conditions. J Plant Nutr Soil Sci. 2014;177:3–15.
233. Dari B, Nair VD, Harris WG, Nair PKR, Sollenberger L, Mylavarapu R. Relative influence of soil- vs. biochar properties on soil phosphorus retention. Geoderma. 2016;280:82–7.
234. Pandey V, Patel A, Patra DD. Biochar ameliorates crop productivity, soil fertility, essential oil yield and aroma profiling in basil (*Ocimum basilicum* L.). Ecol Eng. 2016;90:361–6.
235. Liu Y, Lu H, Yang S, Wang Y. Impacts of biochar addition on rice yield and soil properties in a cold waterlogged paddy for two crop seasons. Field Crop Res. 2016;191:161–7.
236. Kimetu J, Lehmann J, Ngoze S, et al. Reversibility of soil productivity decline with organic matter of differing quality along a degradation gradient. Ecosystems. 2008;11(5):726–39.
237. Uzoma KC, Inoue M, Andry H, Fujimaki H, Zahoor A, Nishihara E. Effect of cow manure biochar on maize productivity under sandy soil condition. Soil Use Manag. 2011;27:205–12.
238. Van Zwieten L, Kimber S, Morris S, et al. Effects of biochar from slow pyrolysis of papermill waste on agronomic performance and soil fertility. Plant Soil. 2010;327:235–46.
239. Peng P, Lang YH, Wang XM. Adsorption behavior and mechanism of pentachlorophenol on reed biochars: pH effect, pyrolysis temperature, hydrochloric acid treatment and isotherms. Ecol Eng. 2016;90:225–33.
240. Moon HS, Kim IS, Kang SJ, Ryu SK. Adsorption of volatile organic compounds using activated carbon fiber filter in the automobiles. Carbon Lett. 2014;15:203–9.
241. Martin SM, Kookana RS, Van Zwieten L, Krull E. Marked changes in herbicide sorption–desorption upon ageing of biochars in soil. J Hazard Mater. 2012;231:70–8.
242. Garcia-Jaramillo M, Cox L, Cornejo J, Hermosin MC. Effect of soil organic amendments on the behavior of bentazone and tricyclazole. Sci Total Environ. 2014;466:906–13.
243. Cabrera A, Cox L, Spokas KA, et al. Comparative sorption and leaching study of the herbicides fluometuron and 4-chloro-2-methylphenoxyacetic acid (MCPA) in a soil amended with biochars and other sorbents. J Agric Food Chem. 2011;59:12550–60.
244. Cabrera A, Cox L, Spokas L, Hermosin MC, Cornejo J, Koskinen WC. Influence of biochar amendments on the sorption–desorption of aminocyclopyrachlor, bentazone and pyraclostrobin pesticides to an agricultural soil. Sci Total Environ. 2014;470:438–43.
245. Dechene A, Rosendahl I, Laabs V, Amelung W. Sorption of polar herbicides and herbicide metabolites by biochar-amended soil. Chemosphere. 2014;109:180–6.

246. Tatarkova V, Hiller E, Vaculik M. Impact of wheat straw biochar addition to soil on the sorption, leaching, dissipation of the herbicide (4-chloro-2-methylphenoxy) acetic acid and the growth of sunflower (*Helianthus annuus* L.). Ecotoxicol Environ Saf. 2013;92:215–21.
247. Delwiche KB, Lehmann J, Walter MT. Atrazine leaching from biochar-amended soils. Chemosphere. 2014;95:346–52.
248. Khorram MS, Wang Y, Jin X, Fang H, Yu Y. Reduced mobility of fomesafen through enhanced adsorption in biochar-amended soil. Environ Toxicol Chem. 2015;34:1258–66.
249. Cao X, Ma L, Liang Y, Gao B, Harris W. Simultaneous immobilization of lead and atrazine in contaminated soils using dairy-manure biochar. Environ Sci Technol. 2011;45:4884–9.
250. Zheng W, Guo M, Chow T, Bennett DN, Rajagopalan N. Sorption properties of greenwaste biochar for two triazine pesticides. J Hazard Mater. 2010;181:121–6.
251. Yang XB, Ying GG, Peng PA, et al. Influence of biochars on plant uptake and dissipation of two pesticides in an agricultural soil. J Agric Food Chem. 2010;58:7915–21.
252. Ahmad M, Lee SS, Dou X, et al. Effects of pyrolysis temperature on soybean Stover- and peanut shell-derived biochar properties and TCE adsorption in water. Bioresour Technol. 2012;118:536–44.
253. Zhou L, Liu Y, Liu S, et al. Investigation of the adsorption-reduction mechanisms of hexavalent chromium by ramie biochars of different pyrolytic temperatures. Bioresour Technol. 2016;218:351–9.
254. Yao Y, Gao B, Chen H, et al. Adsorption of sulfamethoxazole on biochar and its impact on reclaimed water irrigation. J Hazard Mater. 2012;209:408–13.
255. Mahmoud ME, Nabil GM, El-Mallah NM, Bassiouny HI, Kumar S, Abdel-Fattah TM. Kinetics, isotherm, and thermodynamic studies of the adsorption of reactive red 195 a dye from water by modified switchgrass biochar adsorbent. J Ind Eng Chem. 2016;37: 156–67.
256. Yang K, Yang J, Jiang Y, Wu W, Lin D. Correlations and adsorption mechanisms of aromatic compounds on a high heat temperature treated bamboo biochar. Environ Pollut. 2016;210:57–64.
257. Ashry A, Bailey EH, Chenery SRN, Young SD. Kinetic study of time-dependent fixation of U^{VI} on biochar. J Hazard Mater. 2016;320:55–66.
258. Liu P, Liu WJ, Jiang H, Chen JJ, Li WW, Yu HQ. Modification of bio-char derived from fast pyrolysis of biomass and its application in removal of tetracycline from aqueous solution. Bioresour Technol. 2012;121:235–40.
259. Mestre AS, Bexiga AA, Proenca M, et al. Activated carbons from sisal waste by chemical activation with K_2CO_3: kinetics of paracetamol and ibuprofen removal from aqueous solution. Bioresour Technol. 2011;102:8253–60.
260. Mochidzuki K, Sato N, Sakoda A. Production and characterization of carbonaceous adsorbents from biomass wastes by aqueous phase carbonization. Adsorption. 2005;11:669–73.
261. Altenor S, Ncibi MC, Emmanuel E, Gaspard S. Textural characteristics, physiochemical properties and adsorption efficiencies of Caribbean alga *Turbinaria turbinata* and its derived carbonaceous materials for water treatment application. Biochem Eng J. 2012;67:35–44.
262. Dubey SP, Dwivedi AD, Lee C, Kwon YN, Sillanpaa M, Ma LQ. Raspberry derived mesoporous carbon-tubules and fixed-bed adsorption of pharmaceutical drugs. J Ind Eng Chem. 2014;20:1126–32.
263. Kadirvelu K, Kavipriya M, Karthika C, et al. Utilization of various agricultural wastes for activated carbon preparation and application for the removal of dyes and metal ions from aqueous solutions. Bioresour Technol. 2003;87:129–32.
264. Budinova T, Savova D, Tsyntsarski B, et al. Biomass waste-derived activated carbon for the removal of arsenic and manganese ions from aqueous solutions. Appl Surf Sci. 2009;255: 4650–7.
265. Kaçan E, Kütahyali C. Adsorption of strontium from aqueous solution using activated carbon produced from textile sewage sludges. J Anal Appl Pyrolysis. 2012;97:149–57.

266. Sun Y, Li H, Li G, Gao B, Yue Q, Li X. Characterization and ciprofloxacin adsorption properties of activated carbons prepared from biomass wastes by H_3PO_4 activation. Bioresour Technol. 2016;217:239–44.
267. Valizadeh S, Younesi H, Bahramifar N. Highly mesoporous K_2CO_3 and KOH/activated carbon for SDBS removal from water samples: batch and fixed-bed column adsorption process. Environ Nanotechnol Monit Manage. 2016;6:1–13.
268. Bautista-Toledo MI, Rivera-Utrilla J, Ocampo-Perez R, Carrasco-Marin F, Sanchez-Polo M. Cooperative adsorption of bisphenol-a and chromium(III) ions from water on activated carbons prepared from olive-mill waste. Carbon. 2014;73:338–50.

Chapter 7
Biorefineries: Industrial-Scale Production Paving the Way for Bioeconomy

Abstract The development and application of industrial-scale conversion procedures in high-yielding and cost-efficient production facilities is a vital endeavor to successfully implement bioeconomy and produce valuable products such as biofuels and platform chemicals from biomass and derived wastes in a sustainable manner. Such effort is expected to benefit various strategic sectors mainly related to energy security by gradually reducing the dependency on fossil fuels (which will appease many geopolitical tensions around the world) and the mitigation of various environmental issues such as global warming and toxic wastes and emissions.

The facilities where such sustainable bio-based production processes are operated are referred to as biorefineries. In this chapter, various categories of biorefineries are reported (green, whole grain, lignocellulosic biomass, oleochemical, and marine biorefineries), along with the related strategies and technologies. Considering the strategic importance of such production facilities in the bioeconomy concept, several issues and challenges are also presented and discussed with respect to both the design and operation of biorefineries. The urgent need for a wider expansion plan of biorefineries well implemented in their local environments (sustainable management of biomass, water, and energy resources) is also highlighted. Besides, two commercially available biorefining technologies are reported in order to represent the biorefining technological know-how in Europe (Borregaard, Norway) and Northern America (Envergent Technologies, Canada/United States).

7.1 Introduction

A biorefinery is an industrial facility assembled and operated to produce one or many commodities from renewable biomass and wastes, including biofuels, heat and power, biochemicals, and biomaterials [1, 2]. Over the years, biorefining was defined in several ways, but the most authoritative definition remains the one elaborated by the International Energy Agency (IEA Bioenergy Task 42—Biorefineries), stating that "*Biorefining is the sustainable processing of biomass into a spectrum of bio-based products (food, feed, chemicals, and/or materials) and bioenergy (biofuels, power, and/or heat)*" [3].

M. Sillanpää, C. Ncibi, *A Sustainable Bioeconomy*,
DOI 10.1007/978-3-319-55637-6_7

At a conceptual level, biorefineries are similar to the petrochemical refineries, which produce a wide range of fuels (gasoline, diesel, kerosene, etc.), products (asphalt, plastics, synthetic fibers, wires, etc.), and chemicals (paints, food and fuel additives, fertilizers, detergents, etc.), but from fossil resources. From a practical perspective, a biorefinery is the engine propelling bioeconomy and its sustainable dimension. At first, the gradual worldwide implementation of biorefineries will help to mitigate many serious issues including the recurrent conflicts over fossil resources, food crisis, global warming and other environmental threats. At maturity, which will probably be reached by the time petroleum is depleted, biorefineries will be able to compete with refineries relying on still-available fossil resources, coal and natural gas, as feedstocks for the production of diverse products.

According to some scientists, the transition to a biorefinery-based economy is difficult as it would require large investments in new infrastructure to produce, store, and deliver the end products to customers [4]. That would be true if we are starting from scratch, which is not the case. Despite the heavy legacy of the "petroleum era," a century-long expertise was gained by industrialists, and tremendous breakthroughs were achieved by researchers in many related fields, which could and will benefit the effort to facilitate and speed up the implementation of bioeconomy.

As well, many of the already operating infrastructure and logistics including plants, storage facilities, and transportation means could be readapted or upgraded, with respect to the renewable feedstocks to be used and the commodities to be produced. Furthermore, biorefinery concept is flexible enough to allow various implementation scenarios, from small-scale plants in rural areas using locally available bioresources such as agricultural residues to large biorefineries using wastes from other industries or municipalities [5].

7.2 Biorefineries: Green Production Facilities

7.2.1 Historical Background

Contrary to the widespread current conception, history tells us that the production of energy, fuels, and products from biomass is the norm and that the one from petroleum or coal is the exception. Nobody will deny the fact that fossil fuels, coal and petroleum, literally "fueled" the industrial revolution. So, yes it was a highly influential episode in the history of mankind, but still it remains a very short one and it has to be perceived that way.

Biorefining, or biomass conversion into consumable and/or marketable products, has its roots deep into mankind's history. At first, biorefining was mostly orientated towards the transformation of biomass into foodstuff. The contemporary innovations related to this ongoing biorefining activity include the industrial production of refined sugar (early nineteenth century), potato starch (mid-nineteenth century), wheat and corn starch, and soy and palm oils (early twentieth century).

With the increase and diversification of industrial activities, especially among the developed countries, biorefining took a new dimension as the demand for fuel started to increase substantially, especially for the thriving car industry. In this context, the first spark-ignition piston engine using alcohol was invented by Samuel Morey in 1826 [6]. Then, during the 1860s, the German engineer Nikolaus Otto [7] invented the four-stroke internal combustion engine, i.e., Otto engine, running on ethanol. In the early twentieth century, the German engineer Rudolf Diesel invented the diesel engine and firstly used peanut oil as automotive fuel [8]. While Henry Ford originally designed his famous Model T car to run on bioethanol, the engine was later accommodated to run also on petroleum-derived fuels such as gasoline and kerosene [9].

In 1925, the famous American entrepreneur and car manufacturer, Henry Ford, said that "*The fuel of the future is going to come from fruit like that sumach out by the road, or from apples, weeds, sawdust—almost anything. There is fuel in every bit of vegetable matter that can be fermented. There's enough alcohol in one year's yield of an acre of potatoes to drive the machinery necessary to cultivate the fields for a hundred years*" [10]. Almost a century later, research centers and universities are still working on biofuel or Ford's "fuel of the future."

In the meantime, and starting from the1930s, the large-scale discoveries and exploitation of petroleum reserves marked the beginning of a new industrial era, heavily relying on cheap fuels derived from fossil crude oil. This has directly impacted the biofuel sector as the use of ethanol substantially declined. Nowadays, despite all the problems related to fossil fuels, they remain the unchallenged resources for heating, automotive fuel, and energy. During this petroleum era, biofuels managed to resurface couple of times, not in response to the environmental issues induced by petroleum, but to temporarily compensate the shortages in petroleum supply.

The first episode was during the Second World War, where the shortage of fuels led to the increased production of alcohol from grains and potatoes, which was blended with gasoline to meet military demand to fuel. In Japan, for instance, between 1937 and 1944, an alcohol monopoly system (i.e., governmental monopoly on manufacturing and/or retailing alcoholic beverages) was established to produce ethanol from potatoes and provide fuel (bioethanol and gasoline blend) for the airplanes. The production yields were around 170 million liters of bioethanol per year [11]. By the end of the war in 1945, the share of bioethanol in the total liquid fuels was estimated at 26.7% [12].

The second period was the oil embargo crisis between 1973 and 1974 which started as a geopolitical tension in the Middle East to become a worldwide crisis shortly afterwards. Oil prices increased and severe shortages in fuel supplies started to directly and indirectly affect the economies of many countries in Europe, North America, and South Africa. By the end of the embargo in March 1974, the oil price quadrupled from US$3 to around $12 per barrel [13]. Besides its global geopolitical impact, which still can be easily observed nowadays, the oil embargo crisis incited governments, and forced others, to quickly take measures to downsize the fuel consumption in order to mitigate the impact of this crisis [14], and then to look for alternative resources to avoid the dependency on foreign oil, as a medium term objective. This strategic orientation gained a new momentum with the increasing awareness about the phenomena of global warming and climate change (generally after some pressure from NGOs, environmental lobbies, and the general public).

Thus, researchers and scientists from around the world started to refocus their effort on biofuels to increase their countries' national energy autonomy. Promising results justified and facilitated the introduction of regional, national, and internationals legislations and the establishment of panels and funding bodies aiming at pushing forward the industrialization of biorefinery concept.

The first steps of the biorefining industry were made by small and medium-sized enterprises mainly involved in the paper pulping sectors. Recently, an increasing number of large multinational companies and multi-billion dollar firms started to invest, some cautiously, in sustainable fuels and chemicals such as BP (ethanol from sugarcane), BASF (chemicals), DuPont (cellulosic ethanol), Honeywell's UOP (diesel and jet fuel), and UPM (paper and biodiesel). For example, since 2006, BP has announced the investment of more than US$2 billion in biofuels research, development, and operations in Brazil and Europe [15]. The Chinese company Sunshine Kaidi New Energy Group is planning to build a billion-euro biorefinery in Kemi (Finland) for the production of 200,000 t of biofuels per year [16]. Also in Finland, UPM, one of the world leading forest industry company, has invested 175 million euros to build its biorefinery in Lappeenranta to produce biodiesel from wood. The Spanish company Abengoa officially opened its cellulosic biorefinery in Hugoton, Kansas, and it's the first facility to apply, at a commercial level, enzymatic hydrolysis process to convert cellulosic biomass into fermentable sugars that are then converted into bioethanol or other transportation fuels. The total investment in the project was estimated at around US$350 million [17].

After this brief historical account, let's analyze the various biorefining processes and first start by comparing conventional petrochemical and biomass-based refineries.

7.2.2 *Biorefineries Versus Petroleum Refineries*

In the recent past, humanity witnessed several shifts in raw materials aiming at creating wealth and prosperity, mainly through the industrial sector. The first shift was during the industrial revolution from biomass to coal. Later, another shift occurred and coal was gradually replaced by petroleum and natural gas.

Nowadays, the world is preparing itself for another shift, this time back to biomass. So, we started with renewable resources, shifted to a fossil resource, traded one fossil resource for two others, and now going back to the only resource that will never run out, biomass. Therefore, substituting fossil resources by renewable ones should be seen from this angle. Indeed, contextualizing this expected shift will lay a solid ground for its proper implementation, benefit for the previous advantageous aspects, and avoid the disadvantageous ones.

In practice, there are many analogies between petroleum refineries and biorefineries. Indeed, for both processes, refining includes three main operations to convert petroleum or biomass into various end products:

1. *Separation/fractionation*: aiming at separating the various components either in crude oil (distillation) or biomass (fractionation) into streams for convertible intermediates.
2. *Conversion*: for the case of crude oil, this step helped reducing the length of some hydrocarbon chains to desired compounds (catalytic reforming). As for the biomass, the conversion aims at transforming the biomass components into the end products via the fermentation of simple sugars (ethanol, butanol, organic acids, etc.) or the transesterification of vegetable and algal oils (biodiesel).
3. *Purification*: this final stage helps recovering the end chemicals or products from the conversion reaction mixture. Several processes could be applied including distillation, membrane separation, adsorption, solvent extraction, etc.

Despite these similarities, biomass, on one hand, and crude oil and natural gas, on the other hand, differ considerably in composition, hence the need to operate some modification to "convert" conventional refineries into biorefineries. Table 7.1 illustrates the main analogies and differences between the two refineries.

Thus, the main difference between petroleum refineries and biorefineries is the composition of the respective of the feedstocks. Indeed, petroleum is rich in hydrocarbons with almost no oxygen content. Biomass, on the other hand, has a smaller fraction of hydrocarbons and a much higher content [19]. The complex and heterogeneous composition of biomass used to be a major obstacle for the development of biorefining. Several R&D breakthroughs in biomass pretreatment [20–22] and fractionation [23–25] helped, to a large extent, overcoming this issue, especially for the recalcitrant lignocellulosic resources. Nonetheless, further optimization for large-scale applications is still needed (i.e., higher production yield, lower residues, and at lower cost).

From the beginning, the possible application of biomass in "upgraded" conventional refineries attracted the attention of both researchers and industrialists. Among the various scenarios, scientists are recommending the use of biomass or biomass-derived compounds in the oil and syngas production platforms and benefit from the already operational facilities including oil cracking, hydrotreating, and gasification. The expected products from this straightforward integration of biomass in conventional refineries could include biodiesel, bioalcohols, acids, and other added-value chemicals derivable from biosyngas [26].

Table 7.1 Comparative analysis between petroleum refineries and biorefineries [18]

Factors	Petroleum refineries	Biorefineries
Feedstock	Relatively homogeneous	• Heterogeneous composition (carbohydrates, lignin, proteins, oils, etc.) • Most components are in polymeric configuration (cellulose, starch, proteins, and lignin)
	• Low oxygen content • The weight of the product (mole/mole) tends to increase after the process	• High oxygen content • The weight of the product tends to decrease after the process
Main building blocks	Ethylene, propylene, methane, benzene, toluene, and xylene isomers	Glucose (C6 sugar), xylose (C5 sugar), fatty acids (oleic and stearic acids)
Conversion processes	• Mostly chemical processes • Relatively homogeneous processes to produce building blocks • Various chemical conversion routes	• Combination of chemical and biotechnological processes • Relatively heterogeneous processes to produce building blocks • Limited (bio)chemical routes

7.2.3 *Major Categories of Biorefineries*

Biorefineries are designed and operated to produce various end products via various conversion routes that form a wide variety of renewable resources. In order to fully benefit from the various potential feedstocks (terrestrial and marine biological resources and biowastes), different biorefining configurations were developed. With respect to the selected feedstock(s), different handling equipments, thermochemical and biochemical conversion procedures, and separation and purification technologies could be integrated and applied.

In this context, several classification of biorefineries were reported according to the nature of the feedstocks, the production technologies (first to fourth generation), and the adopted conversion process. Tables 7.2 and 7.3 summarize the various classifications.

In the following section, a brief description of each class of biorefinery will be presented. The classification based on the kind of feedstocks will be followed, because it is the availability and composition of the feedstock that are the main factors to target specific end product(s) and therefore apply proper conversion technologies.

7.2.3.1 Green Biorefineries

Green biorefineries are based on the processing of green biomass, including green grass or crops (alfalfa, stover, immature cereals, etc.) for the production of a wide range of bioproducts such as bioethanol, amino acids, fibers, fertilizers, and biogas [30]. This processing relies mainly on the efficiency of the mechanical fractionation of the fresh green biomass, which results in the production of a fiber-rich cake and a

Table 7.2 Classification of biorefineries based on the used feedstock and the generation technology [4, 27, 28]

	Classification	Feedstocks	End products
Classification by feedstock	Green biorefineries	Grasses and green plants	Bioethanol
	Cereal (whole grain) biorefineries	Starch crops, sugar crops, and grains	Bioethanol
	Oleo-chemical (oilseed) biorefineries	Oilseed crops and oil plants	Vegetable oils, biodiesels
	Forest biorefineries	Forest harvesting residues, barks, wood sawdust, pulping liquors, and fibers	Biofuels, biochemicals, and biomaterials
	Lignocellulosic biomass biorefineries	Agricultural wastes, crop residues, urban wood wastes, industrial organic wastes	Lignocellulosic bioethanol, bio-oil, bio-syngas, biochar
	Algal biorefineries	Algae, seagrass	Bioethanol, bio-oil, biodiesel
Classification by generation	First-generation biorefineries	Sugar, starch, vegetable oils, or animal fats	Bioalcohols, vegetable oil, biodiesel, bio-syngas, biomethane
	Second -generation biorefineries	Nonfood crops, energy crop, and agro-industrial residues (wheat straw, corn stover, and stalk, etc.)	Bioalcohols, bio-oil, biohydrogen, bio-Fischer-Tropsch diesel, biochar
	Third-generation biorefineries	Algae	Bioethanol, bio-oil, biodiesel
	Fourth-generation biorefineries	Vegetable oil, biodiesel	Biogasoline

Table 7.3 Classification of biorefineries based on the conversion technology [4, 18, 29]

Conversion technology	End products
Fermentation-based biorefineries	Bioethanol, biobutanol, organic acids
Pyrolysis-based biorefineries	Bio-oil, biochemicals, biohydrogen, biochar
Hydrothermal process-based biorefineries	Bio-oil, biochar, biochemicals, biohydrogen
Gasification-based biorefineries	Bio-syngas, biohydrogen, methanol, dimethyl ether

biomass juice rich in water-soluble compounds (nutrients, lactic acid, amino acids) [31, 32].

The green juice is then treated, mainly via fermentative routes, for the production of bioethanol, lactic acid, amino acids, and proteins [33]. As for the press cake, it can be valorized as green feed pellets or further processed (thermochemical and/or biological procedures) for the production of various organic acids such as levulinic acid (a high added-value platform chemical [34]) or bio-syngas [35]. The

Table 7.4 Chemical composition (g/kg) of dry matter in the juices of different green biomasses [40]

Components	Green juice from Italian rye-grass	Green juice from clover grass	Green juice from alfalfa
Water soluble carbohydrates	449.4	330.8	137.0
Free carbohydrates	283.1	219.5	135.8
Fructans	166.3	111.3	0
Succinic acid	15.2	5.7	3.2
Malonic acid	17.7	5.7	53.5
Citric acid	8.9	14.6	8.3
Malic acid	42.8	36.9	33.7
Acetic acid	0	0	0
Lactic acid	3.3	0	0
Formic acid	4.5	0	0
Total organic acids	92.4	62.9	98.7
Proteins (N × 6.25)[a]	174.0	264.2	349.0
Dry matter (%)	5.4	5.9	6.0

[a]Protein content calculated from the nitrogen (N) content

waste stream (by-product) of this processing is a nutrient-rich juice commonly known as brown juice.

The implementation of the green biorefinery technology is at an advanced stage of development in many European countries including Germany [36], Austria [37], Switzerland [38], and Denmark [39]. Table 7.4 illustrates the composition of the green juices extracted from various green grass and crops.

7.2.3.2 Whole Grain Biorefineries

Whole grain biorefineries (*aka.* cereals biorefineries) are using raw biomass of cereals including rye, wheat, triticale, sweet sorghum, and corn, as feedstock to produce a wide range of end products. The first step is the separation, mainly by mechanical means, of the grains (cereals or corn) rich in starch, cellulose, and proteins, and the lignocellulosic straw.

The grains processing is generally conducted via dry or wet milling. During the dry milling, the whole grains are mechanically grinded. Water is added to the produced flour, and the resulting mixture or mash is subjected to chemical or enzymatic hydrolysis treatment and cooked in order to breakdown the starch content [41, 42]. As for the wet milling, the grains are first soaked in water with sulfur dioxide in order to soften the kernels, loosen the hulls, and break the bonds between germ and endosperm [43]; then the mixture is milled. Several separation technologies could be applied to recover pure streams of starch, cellulose, proteins, as well as oil [44]. Considering the high content of polysaccharidic components in the raw feedstock, the main conversion procedure in whole grain biorefineries is fermentation [45–47].

It has to be noted that the hydrolysis phase could be integrated with the fermentation phase by applying a simultaneous saccharification and fermentation process, i.e., simultaneously incorporating saccharifying enzymes and fermenting yeasts in the same reactor [48, 49]. The fractionation of the various grains components prior to conversion process is also a promising configuration for whole grain biorefinery to diversity their end products portfolio. With such configuration, bioethanol, biobutanol, organic acids, and other added-value fine chemicals could be produced from the starch and cellulose fraction via fermentative routes. The oil content, especially from corn, is used for the production of biodiesel via the well-established transesterification procedure [50, 51].

The straw residues generated during the first separation stage could be further processed in another biorefining system using lignocellulosic biomass as feedstock.

7.2.3.3 Lignocellulosic Biomass Biorefineries

Lignocellulosic bioresources and wastes from forestry, wood-processing industries (mainly pulp and paper), agro-industrial activities, and municipal solid wastes are considered to be one of the major feedstock sources for thriving industries in the circular and bio-based economies including the energy chemicals, pharmaceuticals, and materials production sectors. Indeed, lignocellulosic biomasses are widely available around the world and could be acquired at moderate cost. As well, their application as feedstock for the production of various commodities helped mitigating the serious conflict over the use of food crops for fuel production [52, 53].

The composition of lignocellulosic biomass consists of cellulose and hemicellulose (polysaccharides) and the complex and recalcitrant lignin. The relative content of each component varies from one species to another and within the same species with respect to the age and the growing conditions. In general, the holocellulosic content (cellulose and hemicellulose) is estimated at 50–75% of total dry weight, and the lignin fraction at 10–25% [54]. The presence of the aromatic polymer lignin necessitates a pretreatment step in order to remove it and make holocellulosic fraction more accessible for the conversion stage hydrolysis.

Several pretreatments could be applied, individually or combined, including mechanical, chemical, or biological (enzymatic) procedures [55–57]. Then, a chemical or enzymatic hydrolysis (saccharification) is applied to convert the polysaccharides into simple fermentable C5 and C6 sugars. After this stage, a wide range of bio-based products including biofuels and fine chemicals could be produced via diverse fermentative routes [58].

The forest and pulp and paper industries are facing serious challenges due the recurrent worldwide economic crisis, even in developed countries, and the declining demand for paper and paperboard production. The coproduction of pulp and paper and other added-value bioproducts, especially from side streams (wastes and by-products), was first applied as a "survival strategy." Then, the winning strategy opened the door for more elaborate conversion schemes and production facilities that could be qualified as the first implementations of the forest or lignocellulose biorefineries' concept.

The Forest-based sector Technology Platform (FTP) formed the "Biorefinery Taskforce" in order to investigate the full potentialities of wood-based biorefineries to produce a wide range of added-value products from both forest raw materials and by-products of the forest-based industries [59]. The related study revealed that the pulp mills could produce various biofuels and bioproducts from forest biomass and wood-processing residues via the application of advanced fractionation and conversion technologies. Besides, lignin, the most important residue from the wood industry, was also confirmed to be a promising feedstock for the production of various platform chemicals [60], which could be used to produce fine chemicals [61], biodegradable polymers [62], carbon nanofibers [63], and other polymers and biomaterials [64–66].

7.2.3.4 Oleo-Chemical Biorefineries

Industries in the oleo-chemistry sector have been using renewable oil and fat-containing raw materials including many oilseed plants or the by-products of meat-processing industries (animal fats) for a long time. Therefore, oleo-chemical industries are among the most predisposed facilities to work with renewable resources for the production of biofuels, biochemicals, and biomaterials.

In practice, after mechanical and solvent-based chemical processing of the oilseeds in order to extract the oil content, it is estimated that fatty acids and esters make up around 85–90% of the recovered oily material. From chemical and structural perspectives, the components in oilseeds resemble the hydrocarbons in fossil resources. That is why the incorporation of oil-containing biomass in already operating refineries with, or instead of, nonrenewable feedstocks is one of the smoothest transitions towards bio-based economy [67].

In the R&D field, the direct utilization of raw vegetable oils from oilseeds (*aka.* straight vegetable oil, SVO) to run diesel engines seemed an interesting endeavor in the scientific research community [68, 69], either exclusively or in combination with conventional diesel fuels. However, the high boiling point and viscosity of SVO, compared to diesel fuel [70], was the main obstacle for such straightforward application, hence the need for further processing of the vegetable oils to produce biodiesel [71, 72], modifying the engines [73], and mixing with conventional fossil diesel [74] or renewable biodiesel [75]. In this context, several methods were investigated and applied to reduce the viscosity of vegetable oils to the required levels including transesterification, micro-emulsion, catalytic cracking, pyrolysis, and enzymatic hydro-esterification [76, 77].

The production of biodiesel for vegetable oils via transesterification is currently the main process for large-scale production of biodiesel, and it is based on the incorporation of methanol with alkaline compounds such as sodium and potassium hydroxides [78]. Nonetheless, the content of free fatty acids (FFAs) in those oils has a major impact on both the properties and yield of the derived biodiesel. Indeed, the alkaline transesterification process applied to oils, with FFAs content superior to 2.5%, results in the formation of soap formation, which substantially reduces the biodiesel yield and complicates the purification and separation stage [79, 80]. To remediate this issue, several acid catalysts were applied to vegetable oils with high FFAs including sulfuric acid, hydrochloric acid, and sulfonic acid [81]. Nonetheless, the high cost related to the acquisition of the catalyst and then its separation from the produced biodiesel constituted a new set of challenges.

To address these issues, researchers worked on the development of enzymatic catalysts, mainly lipase [82], for the production of biodiesel from a wide range of vegetable oils. The related findings revealed that the application of specific enzymes in the conversion process of oils (with low or high FFAs content) to biodiesel reduced the impact of the issues related to the incorporation of alkali or acid chemicals for the production of fatty acid methyl ester (FAME) or fatty acid ethyl ester (FAEE) [83], avoided the saponification phenomenon, generated pure glycerol stream (by-product), and enabled a lower energy consumption [84]. On the other hand, however, the high cost of enzymes and the easy inhibition of their activity by organic solvents constitute the main disadvantages of using enzymatic catalysts in the biodiesel production process [85].

Other recent research studies are promoting the application of new processes in oleo-chemical biorefineries for the conversion of renewable oils from oilseeds, waste cooking oils, and animal fats to biodiesel [86]. Pyrolysis is one of the alternative processes for the commonly used transesterification. Indeed, several research teams carried out comparative analyses between the two processes, and the general consensus is that pyrolysis has more advantageous features than transesterification. Indeed, it was reported that pyrolyzed vegetable oils tend to generate liquid biofuels with chemical compositions comparable to the ones of conventional diesel fuel, and thus compatible with engines and fuel standards [87].

For the previously mentioned biorefineries, all the feedstocks are from terrestrial resources (green biomass, grains, lignocellulosic biomass, and oilseeds). Figure 7.1 gives a schematic illustration of the possibility to combine various conversion platforms. Such combination scenarios are mainly aimed at diversifying the end products and also to enable operating petrochemical refineries to coproduce various bio-based products including renewable fuels and added-value specialty chemicals.

Along with the terrestrial feedstocks, marine bioresources including algae and seagrasses are quickly gaining the interest of researchers and industrialists in the bio-based industry sector as renewable and highly available nonfood feedstock for biorefineries, *aka*. Algal or fourth-generation biorefineries.

7.2.3.5 Marine Biorefineries

Marine biorefineries, also specifically referred to as algal biorefineries, are refining facilities operated to produce different biofuels, biochemicals, and biomaterials for various marines bioresources including microalgae, macroalgae (seaweeds),

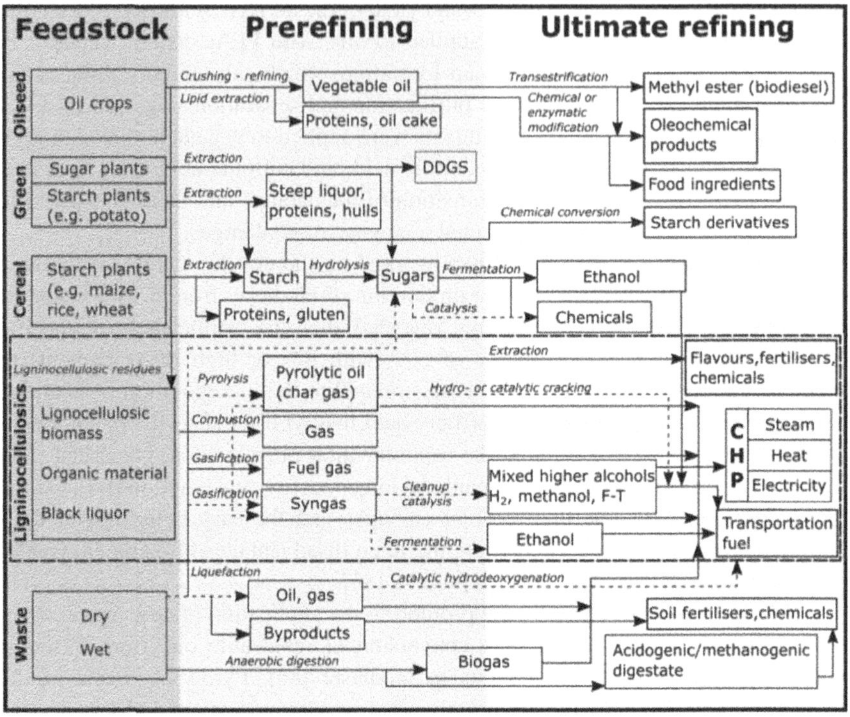

Fig. 7.1 The concept of an integrated biorefinery combining various terrestrial feedstocks and multiple conversion platforms [88]

seagrasses, and other aquatic plants growing or being cultivated in seas and oceans. From an industrial perspective, marine biomass has a considerable potential to meet a substantial part of the rising worldwide requirement for biomass. Indeed, besides being available from their native ecosystems, marine bioresources could be cultivated for industrial utilization without any competition with food crops over water and soil [89], with possible cultivation in marginal waters such as municipal wastewaters [90]. In addition to their high availability, marine algae and seagrasses are also characterized by high biomass production yields, wide range of synthesized biochemicals, and an important contribution in carbon capture and global warming mitigation [91].

Marine bioresources can accumulate substantial amounts of oils, carbohydrates, and vitamins, depending on species and growing (or cultivation) conditions. As a consequence, several studies worked on valorizing such promising biomass for the production of many commodities, mainly biofuels and platform chemicals. In this thriving research field, the applicability of oleaginous algae as raw biomass for biodiesel production was already demonstrated [92–94] and its viability verified through life cycle assessment [95]. Other marine bioresources, mainly seaweeds and seagrasses, have a rich carbohydrates content, which can be fractionated and/or converted into various added-value bioproducts such as bioethanol from microalgae [96], seaweeds [97], and seagrasses [98], as well as acetone and butanol [99] and organic acids [100]. Biomethane and bio-syngas are also among the various end products of marine biorefineries [101, 102].

Currently, the industrial valorization of marine biomasses into biofuels is gaining most on the interest among scientists, industrialists, and decision makers, which is understandable considering the urgency to make available renewable alternative to fossil fuels. However, this should not get the focus of researchers around the world away from other added-value potential products, including omega-3 fatty acids, biopolymers, biofertilizers, bioplastics, proteins, and also natural pigments [103]. In this context, many scientists assert that the sole production of biofuel from algae raises serious question about the sustainability of such production scheme and even its economic feasibility [104]. The integrated biorefining concept, by definition, implies the production of biofuels with commodities, which will help to improve the economics of such sustainable production model [105, 106].

Figure 7.2 presents an integrated biorefining process to convert algal feedstock into biomass via hydrothermal liquefaction (HTL) into bio-oil, biochar, syngas, biochemicals, and biofertilizers. Other conversion procedures could be applied to marine bioresources according to their respective composition. This includes fermentation [107], transesterification [108], pyrolysis [109], along with HTL [110].

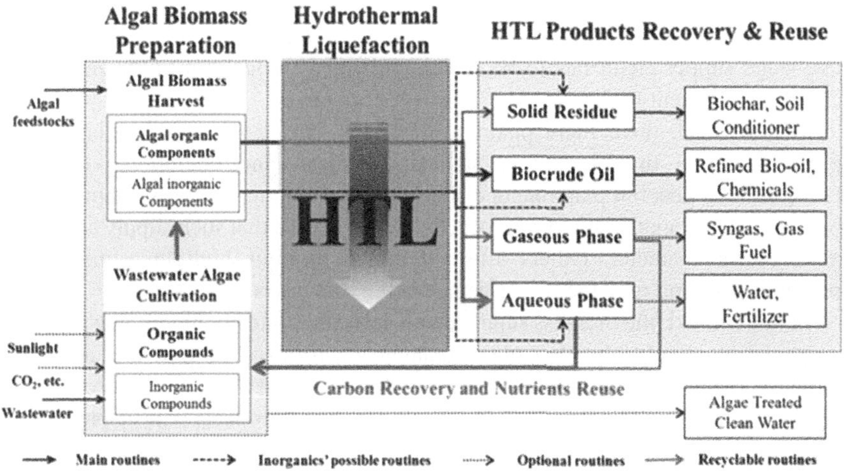

Fig. 7.2 An integrated biorefining concept of algal biomass based on the hydrothermal liquefaction process [111]

7.3 Implementation of Integrated Biorefineries

7.3.1 Implementation Designs of Biorefineries

Industrial integrated biorefineries are processing facilities enabling an optimized conversion of various bioresources into added-value fuels and chemicals, with limited wastes, thus generating profit in a sustainable manner.

In general, biorefineries are designed based on a common engineering-based design approach that includes the following four phases: (1) defining the problem, (2) elaborating a conceptual design, (3) detailing the design specification, and (4) finally its implementation [112].

With various R&D breakthroughs, multiple biomass conversion pathways are available for industrialists to be applied [113], depending on the availability of feedstock and the targeted end product(s). Therefore, making strategic choices regarding those two aspects (i.e., feedstock supply chain and conversion process engineering) are the key factors to be taken into consideration when designing an integrated biorefinery.

7.3.1.1 Biomass Supply Chain Network Design

A supply chain network involves all the participants operating in the flow of raw material from producers to conversion and processing units, and then the end products to consumers. Basically, for biorefining facilities, biomass supply chain includes five phases: (1) feedstock production, (2) logistics, (3) conversion process, (4) products distribution, and (5) end-use [114]. In order to reach the last stage of products' end-use in the most efficient way (highest quality and minimum cost), the five-stage supply chain has to be well coordinated through an optimized supply chain management (SCM) [115].

Within SCM, three main planning decisions can be taken depending on the planning frame time including strategic, tactical, and operational decisions [116, 117]. In general, planning decisions related to biomass supply chain network belong to the strategic ones. Experts in the field reported that such supply chain is of a long-term character. Others affirmed that it requires high investments, and potential planning revision might be needed in 3–8 years [118].

In this context, the biomass supply chain network related to biorefining activities is primarily affected by two main strategic decisions. The first one about the biomass (what will be used as feedstock) and the second one related to the production facilities (where and how the targeted end products will be produced).

Strategic Decisions in Biomass Supply

Regarding the choice of feedstock, there are two major scenarios: the single feedstock and multiple feedstocks models. A related bibliometric analysis (from January 1997 until July 2016) revealed that 54.8% of the research studies on this matter was conducted based on the multi-feedstocks supply chain and the remaining 45.2% on single feedstock models [119]. The main advantages of multiple feedstocks biorefining scenarios are the possibility to cope with fluctuations in the supply of any of the feedstocks in terms of both price and availability. This model also promotes the competition between the suppliers to provide biomass at lower cost. However, the diverse composition of biomass necessitates the incorporation of different pretreatment and conversion technologies, which means a higher investment, and operations and maintenance costs. Single feedstock model, on the other hand, has lesser technological requirements, but the availability of biomass for industrial processing is directly affected by any issue disturbing the supply chain. This short-term unavailability could be related to variation on seasonal productivity, climatic incidents, logistics issues, etc.

But, the most important matter is the possible establishment of a monopoly over this single feedstock, which will directly affect the prices of the end products. In general, few end products are produced by a single feedstock model (primarily biofuels with, in some cases, biochemicals, biofertilizers, etc.). Currently, with the R&D breakthroughs in biomass fractionation, new set of bioproducts could be produced in single feedstock biorefineries [120].

In the general context of biorefineries and biomass supply chain, the notions of supply chain and value chain are sometimes wrongly used to describe the same concept although they refer to two different notions [121]. Supply chain is related to the flow of materials from the supplier of feedstock to the customer (end product). Value chain, on the other hand, is the set of activities aiming at creating value from this chain, including marketing and innovations [122, 123].

Strategic Decisions in Facilities-Related Matters

This includes a set of decisions on matters related to the location, the choice of the biomass conversion and separation technologies, daily load of feedstock, and the incorporation of different other components in the facility including handling, blending, pretreatment, and storage. For instance, the decision to build and operate the biorefinery in a specific location is of strategic importance and has direct impact on both the investment and operating costs and therefore on the profitability of those green and sustainable businesses [124, 125]. Several aspects need to be taken into consideration including the vicinity to feedstock sources and the distribution network. In this specific context, since the biomass feedstock is much more bulky (compared to the end products), transportation costs are higher for the feedstock, especially in long distances. Thus, the vicinity to the feedstock source is generally prioritized over the distribution chain [126].

Overall, it is the set of choices about the processing unit in the biorefinery that generates most the decision variables, along with the site location. That is why most of the mathematical models are based on binary variables, expressing the foundation of a processing facility in a specific location, for a specific time frame [127].

7.3.1.2 Biomass Conversion Process Design

In integrated biorefineries, numerous biomass conversion routes could be applied for the production of various bioproducts. This wide range of conversion pathways and end products complicated the design of integrated biorefineries, which necessitated the development of different systematic screening tools to address the process design of those facilities [128].

Research studies on process design are mainly focusing on the development of the process and then the possible integration and optimization scenarios. For the development phase, superstructure and sub-processes are synthesized, generally via a hierarchical decomposition of the process or algorithms such as the ones for heat exchange and reactor networks [129]. As for integration and optimization, the design is mainly related to feedstock and energy integration and process intensification, as well as repeated simulations or mathematical modeling [27, 130].

In this regard, several detailed process models were developed using reliable simulation engines such as Aspen Plus, ChemCAD, Aspen HYSYS, Pro/II, SuperPro Designer, etc. [131]. The use of such process simulation tools helped elaborating reliable process designs by providing complete flowsheets, optimized operating factors, and useful information about the input amount and the optimal size of the processing units. At first, and in order to generate a simulated process design of a biorefinery, the targeted bioproducts and any recognized process constraints are sent to the process design module. Consequently, different processes enabling the production of the desired bioproducts are modeled using one of the well-established simulation softwares. More advanced process design tools, such as computer-aided molecular design (CAMD) and thermodynamics-based pinch analyses [132], are applied to carry out mass and energy integration regarding the possible integration of the various processes to generate a fully optimized and highly efficient process. For the more complicated combination scenarios, large-block analysis (LBA), hierarchical analysis, supply chain management (SCM), etc. were reported to be of greater assistance [114].

For instance, the LBA of processes was developed as a comparative tool to assess the applicability of different retrofit cases, using the same basis, such as the scenario of a biofuel production retrofit at a Kraft mill [133]. This methodology enables the establishment of common assumptions through the combination of various simulated data related to the production yield, mass, and energy balances, as well as capital investments [134].

On another important matter, one of the main decisions in process designing, is, first, to identify the "best" bioproduct or platform molecule and (i.e., high yield and value) then design the optimal conversion routes to produce the targeted products.

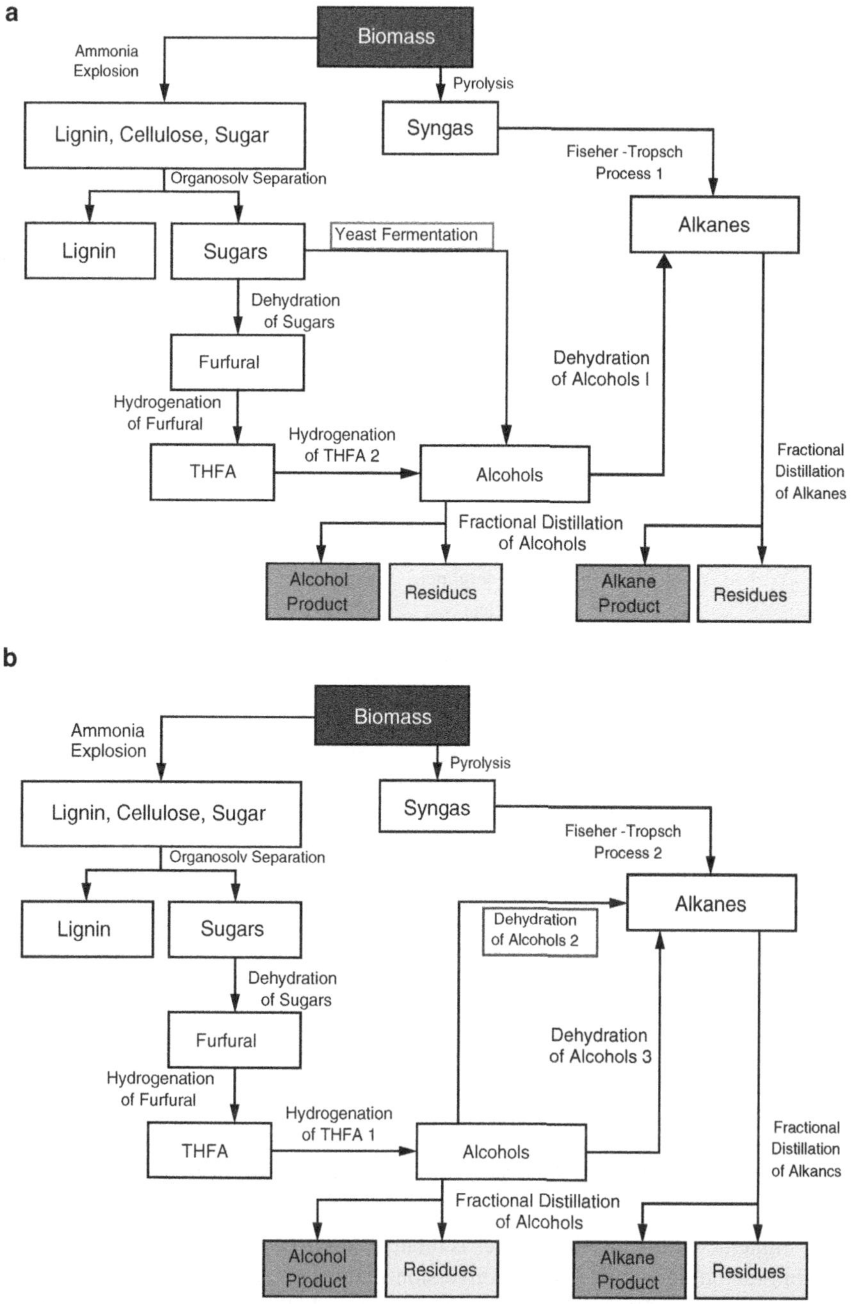

Fig. 7.3 Flow diagrams of two integrated biorefining scenarios: (**a**) for maximum production yield and (**b**) for maximum economic value. Adapted from [135]

Table 7.5 Estimated prices of several feedstock materials for the biorefinery industry [18, 66, 136–141]

Raw biomass	Price (€/t dry matter)
Jatropha oil	465[a]
Safflower	419[a]
Corn	326[a]
Sorghum	236[a]
Jatropha fruit	128[a]
Beech wood	100
Grass	70–80
Wood chips	79[a]
Wheat	73
Poplar wood	71
Leaves (alfalfa, beet, etc.)	50–70
Birch wood	66
African palm	60[a]
Eucalyptus	55[a]
Wheat straw	45[a]
Olive pruning	41[a]
Rice	37
Sweet sorghum	36[a]
Sugarcane	25
Sugarcane straw	18

[a]Prices originally given in US dollars

Figure 7.3 depicts the flow diagrams of two integrated biorefinery scenarios: the first one to maximize the production yield and the second to increase the economic value of the biorefining process. Further details about the respective production yields and values of each biorefining scenario could be found in [135].

Several studies reported the values of several bioproducts, which could be produced after biomass refining, as a tool to select the most suitable feedstock (s) and end product(s) that would generate a higher profitability for such sustainable production system. In the related literature, the available cost data is reported either from real or simulated-case scenarios. A wide collection of average prices from several feedstocks and end products is given in Tables 7.5 and 7.6, respectively.

7.3.2 Obstacles Facing the Implementation of Biorefineries

The implementation of biorefineries is related to many technological and economic factors including supply chain, conversion pathways, and overall economic efficiency. Each one of those factors is posing a set of challenges to the development of fully operational biorefining facilities. As well, the implementation of any biorefinery is highly influenced by the space in which its units are operating and its end products stored and distributed.

Table 7.6 Estimated prices of several end products (platform chemicals, biofuels, polymers, sugars, adhesives, plastics, etc.) from the biorefinery industry [136, 137, 142]

End product	Price (€/t)
1,2,4-Butanetrio-trinitrate	90,000
Xylonic acid	61,370
Difurfuryl disocyanate	15,000
Vanillin	12,000
XOS	8500
Polyester	7523
HMF	5000
Lignin castor oil	4200
Polyamide2	4000
Xylitol	3500
APP	2500
Activated carbon	2093
Polyacrylate	1971
Hydrogel	1887
XB polyester	1750
Isopropanol	1667
Aromatic polyols	1620
n-Butanol (chemical)	1505[a]
Bio polyester	1483
PEF	1079
Pf resin	1076
Polyamide1	1000
Polypropylene	970
n-Butanol (biofuel)	940[a]
Propylene	930
Ethylene	906
Carbon black	835
Ethylene glycol	800
Bio PVC	681
Furfural	597
Sorbitol	583
Ethanol	560
Pulp	550
Isosorbite	502
Dichloroethane	433
Glucose	373
Wood adhesive	360
Bio-oil	340[a]
Biochar	109

[a]Prices originally given in US dollars

The social dimension of implementing sustainable production facilities in specific regions is becoming a key factor. Many research studies in social sciences are providing new valuable perspectives, in addition to the technological and economic ones [143].

Overall, the wider implementation and expansion of biorefineries is mainly restricted by the complexity of such facilities, especially the integrated ones designed, built, and operated in order to process various feedstocks and produce a wide range of bioproducts using different pretreatment, conversion, and purification technologies. Thus, each biorefining process scenario (single–multiple/input–output) has its own degree of complexity, which is a valuable insight for its subsequent implementation. In this context, the IEA Bioenergy Task 42 has published a working document on the notion of biorefinery complexity index (BCI) and the assessment of this index for different biorefining scenarios [144].

Considering the various factors challenging the implementation of a biorefinery, we will discuss two study cases: forest biorefinery (terrestrial bioresources) and algal biorefinery (marine bioresources).

7.3.2.1 Challenges Facing Forest Biorefineries

According to the World Bank data, forests are contributing to the livelihoods of about 1.6 billion people around the world. Both raw and industrially processed forest products are sources for economic growth and employment, with an estimated global value around US$270 billion, of which developed countries account for 80%. Worldwide, the various forest industries provide direct and indirect employment for roughly 50 million people [145].

Lately, the forest industry, *aka.* the cradle of the biorefining concept, is facing several challenges, some of them are specific and other commonly found in mature industrial activities. In general, forest industries, mostly concentrated in developed countries (Northern Europe and America), are benefiting from substantial capital assets and large domestic markets. However, latest investigations are reporting higher production costs for slowly shrinking markets [146]. In response to this escalating difficult situation and its inevitable impact on conventional forest-based industrial activities such as the pulp and paper industry, several structural and technological changes were both necessary and urgent to ensure the continuity and profitability of this strategic sector in many Scandinavian and North American countries [147].

The most promising long-term strategy to transform the struggling industries into sustainable business models with improved financial performance was the implementation of the biorefinery concept. Such transformation helped diversifying the products' portfolio and including new and cheaper feedstock materials, but it also generated new set of technological challenges, as well as many market and policy obstacles. In an interesting study, Canadian experts from the Natural Sciences and Engineering Research Council (NSERC) have put forward different potential obstacles to a wider implementation of forest biorefineries. The analysis was conducted using the multiple-criteria decision-making (MCDM) analytical tool via the analytic hierarchy process [148]. According to this study, the main barriers to the implementation of a forest biorefinery are summarized in the following four points:

- The governmental policies could be responsible for delaying the implementation of biorefineries due to the lack of promoting legislations, and in some countries, the impact of the influential fossil fuels lobby.
- The previously implemented biorefineries encountered many technological problems, especially that most biorefining scenarios were based on inefficient early generation technologies.
- The limited cooperation with industrial partners from outside the forest sector via collaborative networking was a critical factor that affected the early stages of the implementation of biorefineries.
- The forest industry was described as a risk-averse industry with marginal financial performance. This "risk-averse culture," according to the scientists conducting this analysis, limited the orientation of decision makers towards emerging and cost-effective biorefining strategies, in favor of short-term decisions [148].

Overall, industrial companies in the forest sector willing to invest in the thriving biorefining business need to be ready to operate serious structural changes and adopt new business models [149]. According to some experts in the field, the general understanding of these fundamental structural changes is still limited [150], hence the need for a joint effort to facilitate these changes at both conceptual and implementation levels.

7.3.2.2 Challenges Facing Algal Biorefineries

Many research studies and expertise reports are highlighting the benefits of a wider implementation of algal biorefineries considering the renewability and high availability of those marine feedstocks [151, 152] and the fact that the biochemical composition of many micro and macroalgae enables the production of various added-value bioproducts for the energy, chemicals, medical, food and feed sectors [153] and even new compounds for new business ideas [154].

Nonetheless, the implementation of such promising biorefining facilities is still facing some serious challenges, mainly technical and economical ones. On the technical side, the large-scale cultivation and harvesting of algal biomass is the main obstacle to a wide expansion of the algal biorefining strategy. For instance, many scientists are stating that, for the case of producing biofuels from algae, the cost of the algal biomass was the decisive factor in the determination of the selling prices of the produced biofuel [155]. In other studies, the economic assessment, the energy and carbon balance, as well as the environmental impact of many cultivation systems were conducted for various algal bioresources, including open raceway ponds (ORP) and photobioreactors (PBR) [156].

A comparative analysis between those two cultivation systems revealed that the cost of producing algae in ORP is lower than the cost of the same algal production in PBR. A detailed cost breakdown showed that the production cost in ORP is related to the raw materials and operation and maintenance costs. On the other hand, the cost of producing algal biomass in PBR is mainly associated with the capital cost of the PBR system [157].

Nonetheless, the productivity and energy efficiency of both cultivation systems still need to be improved in order to further reduce the production cost of algae and meet the increasing requirement for biomass from the biorefinery sector. The same study revealed that the production cost could be reduced by more than 50% if the main cultivation inputs, water, nutrients, and carbon dioxide could be acquired at a lower cost. In this regard, the cultivation of algae in marginal wastewater seems a promising solution to avoid the use of freshwater resources and provide the necessary nutrients to the algae for a higher biomass productivity. Several research studies explored this option and promising results were reported using municipal [158], dairy [159], and slaughterhouse [160] wastewaters.

Recently, other cultivation systems are being investigated for large-scale production of algae, including algal turf scrubbers [161] and various configurations of biofilm-based algal cultivation systems [162] such as the rotating algal biofilm reactor where the algae are grown as biofilm attached on a surface. With this technology, the algal biomass was more easily harvested in comparison with the suspended cultures in ORPs and PBRs. Such practical advantage was particularly suitable to harvest algal biomass cultivated in wastewater media. Besides, the biofilm-based algae were reported to require less expensive downstream processing during their conversion into biofuels and other algae-derived products [163].

On a related matter, but from a biotechnological perspective, many research investigations worked on improving the biomass productivity and biochemical content (mainly polysaccharides and lipid) of algae via genetic modifications. Several scenarios of strain selection and genetic modifications of algae could be applied. The main goal of such research endeavor is to select native stains with improved biomass productivity or to engineer their genetic material in order to develop new algal stains with a simultaneous improvement of both the lipid or carbohydrate content and biomass yield [164]. Other genetic modification aimed at improving the resistance to potential contamination by wild organisms and reducing the pigment content to minimize the negative impact of self-shading and optimize the photosynthetic reaction [165, 166].

However, despite the fact that the advances in sequencing technologies helped identifying the genes involved in metabolic reactions, applying genetically modified algae in large-scale cultivation systems remain a risky decision, especially in open pond systems. Even PRBs, which are closed bioreactors, are also vulnerable as the leakages are inevitable [167], which could lead to serious environmental complications.

Besides the main challenge related to the large-scale cultivation and harvesting of algal biomass, the lack of full evaluation of the economic feasibility of algal biorefinery is posing another challenge delaying the wider implementation of such biorefining concept [168]. Indeed, the examination of the literature related to algal biorefineries shows a wide range of conversion pathways and design scenarios for the production of biofuels and biochemicals from algal biomass, but limited numbers of studies addressed the subject of assessing the cost-effectiveness of those various biorefining configurations [169, 170]. Some researchers even related the slow implementation of algal biorefineries to the inconsistent and uncertain technological and economic data [171], upon which interested entrepreneurs and decision makers will base their investment decisions.

Overall, the implementation of a biorefining facility, forest, algal, or any category of biorefinery implies several influential technological challenges and marketing risks. Therefore, it is both critical and urgent for legislative and executive authorities, industrialists, and researchers to join their efforts in order to endorse and ratify legislations facilitating the implementation of biorefineries and make the necessary funds available for the promising R&D activities and pro-sustainability support group.

Such joint effort will help to mitigate most challenges currently faced by biorefineries and reduce the risks related to the implementation of new technologies and even new biorefining concepts in the future to tolerable levels. In the meanwhile, measures such as the carbon emission trading system and other policies favoring sustainability will enable a better competitive potential to the various bioproducts of biorefineries.

7.4 Biorefining Technologies: Green Production Processes

The technological configuration of a biorefinery is mainly related to the nature, availability, and quantity of the feedstock material(s) to be processed, on the one hand, and the type and amount of the end product(s) to be produced and sold, on the other hand. These two main factors will dictate all the operation conditions to be combined and applied to pretreat and convert the feedstock and separate and purify the end product. Then, the developed technology needs to be subjected to extensive techno-economic and environmental analyses in order to identify the most robust, sustainable, and profitable configuration. In this regard, experts in the biorefineries sector are affirmed that the development of an efficient, cost-effective, and marketable biorefining technology relies on one major factor: full process optimization [172, 173].

Currently, several commercially available biorefining technologies are being applied worldwide for the production of added-value biofuels, electricity, fine chemicals, and other valuable commodities. The global market for this thriving technological sector is expected to grow from US$466.6 billion in 2016 to US $714.6 billion by 2021, corresponding to an annual growth rate of 8.9% within the same period [174].

In the following section, two commercially available biorefining technologies are selected to represent the biorefining technological know-how in the two leading regions in the world in the biorefinery sector: Northern Europe (Borregaard, Norway) and Northern America (Envergent Technologies, Canada/United States).

7.4.1 BALI™ Process (Borregaard, Norway)

7.4.1.1 The Borregaard Company

Borregaard is a Norwegian company, established in 1889, in Sarpsborg in the Østfold County. Originally, the Borregaard industrial activities started with pulp and paper processing, and the main products were cellulose and paper. In the late 1930s, the company started the production of bioethanol by the fermentation of the sugar content extracted from spruce wood [175].

Currently, the company is positioning itself as one of the world's leading industries in the lignocellulosic biorefining sector. It is organized into five production divisions, each one targeting specific products and markets [176].

- ChemCell for the production of specialty cellulose and bioethanol,
- LignoTech for the production of lignin-based binding and dispersing agent, as well as trading activities,
- Borregaard Synthesis for the production of fine chemicals for the pharmaceutical sector,
- Borregaard Ingredients division for vanillin production for the food sector,
- Exilva for the production of microfibrillated cellulose (MFC). The company has constructed the world's first commercial-scale MFC production facility.

In their effort to assess the sustainably of their various biorefining processes, the company commissioned several life cycle assessment (LCA) analyses to be conducted for many of its products including cellulose, ethanol, lignin, and vanillin from the Borregaard biorefinery in Sarpsborg [177]. In a recent study, the estimation of several production-related factors impacting the environment was realized including global warming potential, resource consumption in the form of cumulative energy demand, generated wastes, as well as acidification, eutrophication, ozone depletion, and photochemical ozone creation potentials [178].

As for the energy supply, the Norwegian company started in 2000 a strategy to reduce its consumption of fossil energy by lowering its overall energy requirement and gradually replacing the used fossil fuels by renewable energy sources, including bio based ones. With the combination of its own hydropower plant providing electricity, the production of steam through the combustion of municipal waste, bark, and organic side streams, in addition to the biogas from the spray driers for lignins, the Borregaard biorefinery is securing 80% of its total energy consumption from renewable sources [179].

7.4.1.2 The BALI™ Process

The BALI™ (Borregaard Advanced Lignin) process is a technology patented by Booregaard industries in 2009 with the objective to covet low value lignocellulosic biomass into added-value bioproducts including biofuels (ethanol, butanol...) and

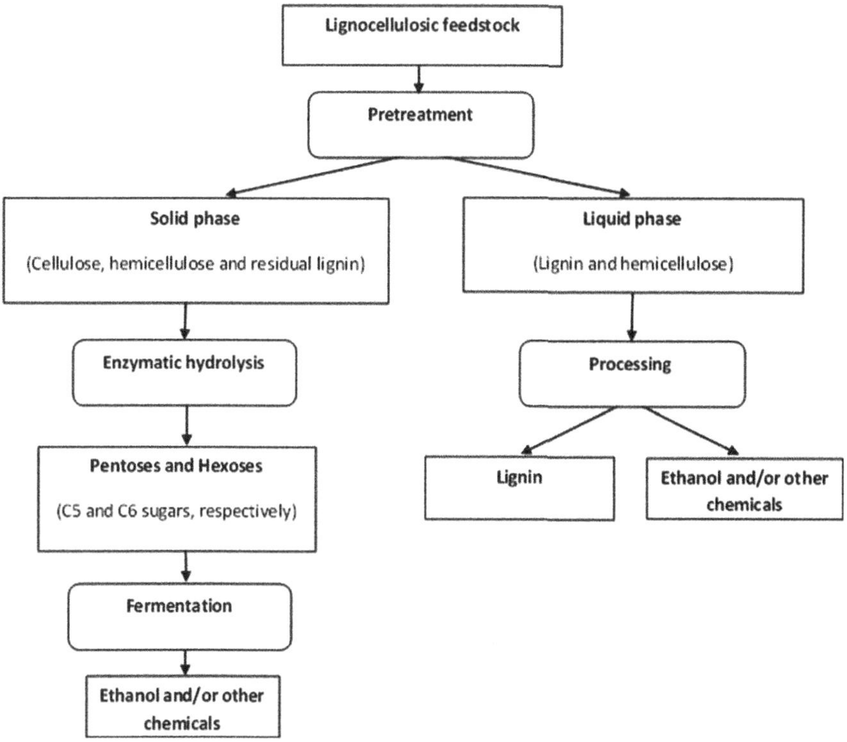

Fig. 7.4 Schematic diagram of the BALI™ process

biomaterials (plastics, single cell proteins...), along with sulfonated lignin. The inventors stated that the BALI™ process enables the valorization of at least 80% of the biomass to marketable products [180].

Figure 7.4 depicts a schematic diagram including the various steps, intermediate compounds, and end products of the BALI™ process.

Overall, this process is based on two key steps:

1. The chemical pretreatment and fractionation of the lignocellulosic biomass aiming at reducing the crystallinity of the cellulose and solubilizing the lignin content. During this pretreatment step, hemicellulosic fraction is either preserved or hydrolyzed, according to the applied operating conditions.
2. After the pretreatment, the water-soluble lignin is processed into various products such as lignosulfonates or oxylignins for dispersing, binders and complexing agents, as well as emulsifiers, corrosion reducers, and soil conditioners. The polysaccharidic content in the solid phase is enzymatically hydrolyzed and the generated simple C5 and C6 sugars are fermented into bioethanol or other added-value biofuels and platform chemicals.

The BALI™ process is exhibiting interesting features for the large-scale biorefining of lignocellulosic bioresources, which could be summarized in the following points [179]:

- High flexibility regarding the feedstock to be processed.
- The remaining lignin fraction in the cellulose, generated after pretreatment, is not inhibiting the hydrolyzing enzymatic activities, which reduces the hydrolysis time and the related investment costs (enzymes and purification procedures).
- High conversion rate of polysaccharides (cellulose and hemicellulose) into monosaccharides (C5 and C6 free sugars).
- The fermentation of monosaccharides is not generating inhibitors, which eliminated the need for the detoxification step.

In 2012, Borregaard launched a demonstration biorefinery plant in Sarpsborg, named Biorefinery Demo, and based on the BALI™ process [181].

7.4.2 RTP™ Technology (Envergent Technologies, Canada/ United States)

7.4.2.1 The Envergent Technologies Company

Envergent Technologies is a joint venture between Honeywell's UOP and Ensyn Corporation, established in 2008 and based in Des Plaines, Illinois. UOP is an international supplier and licensor of process technologies and equipments to the petroleum refining and gas processing industries. Ensyn is the proprietary of the Rapid Thermal Processing (RTP™) technology [182], which enables the rapid conversion of residual biomass from forestry and agro-industrial activities into liquid bio-crude, termed RTP green fuel, via a fast thermal conversion process (i.e., fast pyrolysis).

The combination of the decades-long expertise of both companies qualified Envergent to become the pioneer corporation in designing, building, and commercializing biorefineries based on the RTP technology. So far, seven commercial RTP plants were built in the United States and Canada. These biorefining facilities are converting residual biomass (mainly wood and agricultural wastes) into more than 30 commodities including green fuels, chemicals, food flavorings, adhesive resins for construction, etc. [183].

In Finland, Green Fuel Nordic Oy, a biorefining company established in 2011, has the ambitious goal to build up to 20 biorefineries in the Nordic European country, well known for its thriving forest and wood-processing sectors. The first biorefinery to be construed by the company will be in Iisalmi, Finland, and it will be based on the RTP technology to convert wood residues into second-generation bio-oil [184].

On August 4th, 2016, Envergent Technologies announced the start of the construction of a new RTP-based biorefinery in Port Cartier, Canada. The facility will convert locally available forest residues into liquid green fuel to be used for institutional and industrial heating, as well as the feedstock for further refining steps to produce green fuels for the transportation sector. The biorefinery, to be completed in late 2017, will be able to process 65,000 dry tons of forest residues per year, thus generating an annual yield of RTP green fuel around 40 million liters [185].

7.4.2.2 The RTP Technology

The RTP technology is a fast thermal process where the biomass is fed to a reactor in which the feedstock is mixed with an upward stream of hot sand acting as a heat carrier. Under temperature of 350–600 °C, the biomass is rapidly converted (about 2 s) into a hot vapor stream. The hot stream is quickly quenched (less than 1 s) to a temperature of less than 100 °C, in one or several condensing chambers. This fast cooling condenses the vapor stream into a dark-brown and viscous liquid product, bio-crude or the RTP green fuel [186].

The RTP process also generated several by-products such as the solid biochar and noncondensable gas, which are used to provide heat to the process [187].

Several lignocellulosic biomass were subjected to the RTP process. The production yields of RTP green fuel from selected feedstocks are presented in Table 7.7.

Typically, RTP fuel is an oxygenated water-soluble fuel made of depolymerized components of lignocellulosic biomass. The main properties of this green fuel is a gross calorific value of 16–19 MJ/Kg, a pH fluctuating between 2 and 3, a water content of 15–30%, a solid content of 0.5–2.5%, and flash point of 45–99.

Converting biomass via the RTP technology to produce fuel with such properties is considered to be a sustainable production scheme, based on two main factors: (1) the utilization of biomass with short growth cycles so that, when burnt, the carbon neutrality is ensured and (2) the reduction in the emission of greenhouse gases in the atmosphere. In this context, the use of RTP fuel in burners, instead of fossil fuel oil, was reported to reduce the carbon and sulfur emissions by 90% and more than 95%, respectively [188].

Table 7.7 Estimated production yields of RTP green fuel from selected feedstocks [183]

Biomass	Yield (wt%)	Higher heating value (BTU/LB)
Hardwood	70–75	7400–8200
Softwood	70–80	7300–8000
Hardwood bark	60–65	7180–8680
Softwood bark	55–65	7180–8500
Corn fiber	65–75	7570–8680
Bagasse	70–75	8100–8200

The RTP green fuel can be used for various applications [183, 189], including:

- *Heat*: the RTP fuel could be used as heating oil in industrial burner applications and also in institutional and residential areas. Burning this green fuel generates heat (or steam) with limited greenhouse gas emissions and thus could replace fossil fuel oil or natural gas. RTP green fuel could also be co-fired with coal or other fossil fuels, which helps reducing the emissions of the facility and stabilizing the related energy costs.
- *Power generation*: Modified turbines and stationary diesel engines could run on RTP green fuel to generate electricity in an eco-friendly and cost-effective manner. It revealed that the power production using only RTP green fuel in a stationary diesel engine was possible at around 40% energy conversion efficiency.
- *Transportation fuels*: RTP green fuel can be further refined into several transportation fuels (gasoline, diesel, and jet fuel) using UOP's hydroprocessing technology. Blending those biofuels with the fossil-based ones currently in use in the transportation sector do not require changes to be operated on the vehicle or aircraft engines.

7.5 Outlook

Organic feedstocks, especially biomass residues and algal biomass, are very well positioned to become the raw materials of the future, fulfilling humanity's continuous and increasing need for energy, chemicals, and materials. The development of sustainable conversion processes (i.e., efficient, cost-effective, and eco-friendly) to produce biofuels and platform chemicals from biomass has inherent benefits on various strategic sectors mainly related to energy security (mitigate the dependency on fossil oil generally produced in politically unstable regions) and the preservation of the environment (global warming, toxic wastes and emissions…).

The place to operate such sustainable production schemes from renewable biomass is the biorefinery. In this chapter, various biorefining categories, strategies, and technologies were reported, considering the strategic importance of such production facility in the bioeconomy concept, on the one hand, and the number of issues and challenges that need to be fixed and dealt with, on the other hand. Thus, the biorefinery concept, in both its design and implementation phases, still requires more optimization in order to ensure a wider, more effective, and well-adjusted expansion plan for such facilities around the world.

In this regard, one of the main challenges facing future biorefineries, considering the complexity of such production schemes, is the uncertainties around the available set of data and knowledge. Hence, the necessity for a more efficient designing and operational managing approaches via improved or new simulation tools [114]. The aim is to generate a more flexible biorefining process (various input/output options), reduce the overall production cost, and control the impact on the environment.

The process of biomass fractionation should have a greater role in future biorefineries. Indeed, since the main aim of a biorefinery is to refine and upgrade biomass in order to increase its value, separating the various components in biomass is consistent with the biorefining concept and would lead to a pure stream of each component (cellulose, hemicellulose, lignin, proteins, alginate, etc.). The processing of those steams through a chain of several processes would generate higher yields of pure products (e.g., ethanol or butanol) or chemicals (e.g., organic acids), with more controlled operating conditions and limited inhibitory factors.

To ensure a full energetic autonomy of the biorefinery, which will significantly improve its sustainability, the biorefining facility has to valorize all the potential energy sources from its own production process including production of heat and power via the combustion of the residues. In most cases when lignocellulosic biomass is used as feedstock to produce bioethanol, the residual lignin fraction is burnt to secure part or all of the heat and electricity required by the facility (according to the amount of lignin and the size of the plant) [190].

However, with the R&D innovations in the valorization of lignin for the production of many added-value products (vanillin, polymers resins, nanofibers, etc.), the used-to-be residue became a valuable feedstock. In this case, some researchers stated that the use of external fossil energy (as a replacement for the energy supplied by the combustion of lignin) is possible if it has "*economic benefits and do not unduly burden the life-cycle environmental concern*s" [191]. Such recommendation, even with the existence of many renewable energy sources (solar, wind, geothermal, hydropower, etc.), is not in line with the notion of sustainability and the philosophy behind bioeconomy. This single example shows how profound is the impact of this era of fossil fuels on our way of thinking, even among scientists and researchers. Bottom line, if the economic success of a biorefinery relies on the use of fossil fuels, even to supplement an energetic need, this clearly indicates that we still have work to do.

References

1. Rostrup-Nielsen JR. Making fuels from biomass. Science. 2005;308:1421–2.
2. Ragauskas AJ, Williams CK, Davison BH, Britovsek G, Cairney J, Eckert CA, et al. The path forward for biofuels and biomaterials. Science. 2006;311:484–9.
3. International Energy Agency (IEA). IEA bioenergy Task 42 on biorefineries. Biorefining in a future bioeconomy. Available online at: http://www.ieabioenergy.com/task/biorefining-sustainable-processing-of-biomass-into-a-spectrum-of-marketable-biobased-products-and-bioenergy/
4. Demirbas A. Biorefineries for biomass upgrading facilities. Berlin: Springer; 2009.
5. The European Forum for Industrial Biotechnology and the Bioeconomy (EFIB). What is the Bioeconomy? Available online at: http://www.efibforum.com/about/what-is-the-bioeconomy
6. Detroit Public Library. Samuel Morey's 1826 liquid fuel engine. Available online at: https://digitalcollections.detroitpubliclibrary.org/islandora/object/islandora%3A217484
7. Nature News. Nikolaus August Otto, 1832–1891. Nature. 1932;129:892.

8. Prasad CSN, Reddy KVK, Kumar BSP, et al. Performance and emission characteristics of a diesel engine with castor oil. Ind J Sci Technol. 2009;2:25–31.
9. English A. Ford Model T reaches 100. The Telegraph. Available online at: http://www.telegraph.co.uk/motoring/news/2753506/Ford-Model-T-reaches-100.html
10. Kovarik B. Henry Ford, Charles Kettering and the fuel of the future. Automot Hist Rev. 1998;32:7–27.
11. Koizumi T. Biofuels and food security: biofuel impact on food security in Brazil, Asia and major producing countries. Berlin: Springer; 2014.
12. Koizumi T. Biofuel programs in East Asia: developments, perspectives, and sustainability. In: Bernardes MA, editor. Environmental impact of biofuels. Rijeka: InTech; 2011.
13. Chernov D, Sornette D. Man-made catastrophes and risk information concealment: case studies of major disasters and human fallibility. Berlin: Springer; 2015.
14. US Department of State. The office of the historian. Oil Embargo, 1973–1974. Available online at: https://history.state.gov/milestones/1969-1976/oil-embargo
15. Lane J. BP biofuels: biofuels digest's 2015 5-minute guide. Biofuel digests. Available online at: http://www.biofuelsdigest.com/bdigest/2015/02/16/bp-biofuels-biofuels-digests-2015-5-minute-guide/
16. Chinese Kaidi plans to build a €1 billion biodiesel plant in Kemi. Available online at: http://www.kaidi.fi/uutiset-tiedotteet/2015/12/3/tiedote
17. Abengoa celebrates grand opening of its first commercial-scale next generation biofuels plant. Available online at: http://www.abengoa.com/web/en/novedades/hugoton/noticias/
18. de Jong E, Jungmeier G. Biorefinery concepts in comparison to petrochemical refineries. In: Pandey A, Hofer R, Larroche C, Taherzadeh M, Nampoothiri M, editors. Industrial biorefineries & white biotechnology. Amsterdam: Elsevier; 2015.
19. Manahan SE. Environmental science and technology: a sustainable approach to green science and technology. 2nd ed. Boca Raton, FL: CRC Press; 2006.
20. Zhang CW, Xia SQ, Ma PS. Facile pretreatment of lignocellulosic biomass using deep eutectic solvents. Bioresour Technol. 2016;219:1–5.
21. Mahmood H, Moniruzzaman M, Yusup S, Akil HM. Pretreatment of oil palm biomass with ionic liquids: a new approach for fabrication of green composite board. J Clean Prod. 2016;126:677–85.
22. Zhang X, Yuan Q, Cheng G. Deconstruction of corncob by steam explosion pretreatment: correlations between sugar conversion and recalcitrant structures. Carbohyd Polym. 2017;156:351–6.
23. Yamamoto M, Iakovlev M, van Heiningen A. Kinetics of SO2–ethanol–water (SEW) fractionation of hardwood and softwood biomass. Bioresour Technol. 2014;155:307–13.
24. Li MF, Yang S, Sun RC. Recent advances in alcohol and organic acid fractionation of lignocellulosic biomass. Bioresour Technol. 2016;200:971–80.
25. Piqueras CM, Cabeza Á, Gallina G, Cantero DA, García-Serna J, Cocero MJ. Online integrated fractionation-hydrolysis of lignocellulosic biomass using sub- and supercritical water. Chem. Eng. J. 2017;308:110–25.
26. Kokossis AC, Yang A. On the use of systems technologies: a systematic approach for the synthesis and design of future biorefineries. Comput Chem Eng. 2010;34:1397–405.
27. Moncada J, Tamayo JA, Cardona CA. Integrating first, second, and third generation biorefineries: incorporating microalgae into the sugarcane biorefinery. Chem Eng Sci. 2014;118:126–40.
28. El-Haggar SM. Sustainability and innovation: the next globali revolution. Oxford: Oxford University Press; 2016.
29. Yang ST, El-Ensashy H, Thongchul N. Bioprocessing technologies in biorefinery for sustainable production of fuels, chemicals, and polymers. Hoboken, NJ: Wiley; 2013.
30. Thang VH, Novalin S. Green biorefinery: separation of lactic acid from grass silage juice by chromatography using neutral polymeric resin. Bioresour Technol. 2008;99:4368–79.
31. Mandl MG. Status of green biorefining in Europe. Biofuels Bioprod Biorefin. 2010;4:268–74.

32. Schaffenberger M, Ecker J, Koschuh W, et al. Green biorefinery—production of amino acids from grass silage juice using an ion exchanger device at pilot scale. Chem Eng Trans. 2012;29:505–10.
33. Mikkola JP, Sklavounos E, King AWT, Virtanen P. The biorefinery and green chemistry. In: Bogel-Lukasik R, editor. Ionic liquids in the biorefinery concept: Challenges and perspectives. London: The Royal Society of Chemistry; 2016.
34. Bozell JJ, Moens L, Elliott DC. Production of levulinic acid and use as a platform chemical for derived products. Resour Conserv Recycl. 2000;28:227–39.
35. Kamm B, Gruber PR, Kamm M. Biorefineries—industrial processes and products. Status quo and future directions. Weinheim: Wiley-VCH Verlag GmbH; 2006.
36. Kamm B, Hille C, Schönicke P, Dautzenberg G. Green biorefinery demonstration plant in Havelland (Germany). Biofuels Bioprod Biorefin. 2010;4:253–62.
37. Kromus S, Wachter B, Koschuh W, Mandl M, Krotscheck C, Narodoslawsky M. The green biorefinery Austria—development of an integrated system for green biomass utilization. Chem Biochem Eng Q. 2004;18:7–12.
38. Baier U, Delavy P. UASB treatment of liquid residues from grass bioraffination. Water Sci Technol. 2005;52:405–11.
39. Cong RG, Termansen M. A bio-economic analysis of a sustainable agricultural transition using green biorefinery. Sci Total Environ. 2016;15:153–63.
40. Andersen M, Kiel P. Integrated utilisation of green biomass in the green biorefinery. Ind Crop Prod. 2000;11:129–37.
41. Barcelos CA, Maeda RN, Santa Anna LM, Pereira Jr N. Sweet sorghum as a whole-crop feedstock for ethanol production. Biomass Bioenergy. 2016;94:46–56.
42. Barcelos CA, Maeda RN, Betancur GJV, Pereira N Jr. Ethanol production from sorghum grains [*Sorghum bicolor* (L.) Moench]: evaluation of the enzymatic hydrolysis and the hydrolysate fermentability. Braz J Chem Eng. 2011;28:597–604.
43. International Starch Institute (ISI). TM 18-5www—ISI Technical memorandum on production of corn starch. Available online at: http://www.starch.dk/isi/starch/tm18www-corn.htm
44. Wronkowska M. Wet-milling of cereals. J Food Process Preserv. 2016;40:572–80.
45. Arifeen N, Wang RH, Kookos IK, Webb C, Koutinas AA. Optimisation and cost estimation of novel wheat biorefining for continuous production of fermentation feedstock. Biotechnol Prog. 2007;23:872–80.
46. Dorado MP, Lina SKC, Koutinas A, et al. Cereal-based biorefinery development: utilisation of wheat milling by-products for the production of succinic acid. J Biotechnol. 2009;143:51–9.
47. Kollaras A, Kavanagh JM, Bell GL, et al. Techno-economic implications of improved high gravity corn mash fermentation. Bioresour Technol. 2011;102:7521–5.
48. Szymanowska-Powałowska D, Lewandowicz G, Kubiak P, Błaszczak W. Stability of the process of simultaneous saccharification and fermentation of corn flour. The effect of structural changes of starch by stillage recycling and scaling up of the process. Fuel. 2014;119:328–34.
49. Pietrzak W, Kawa-Rygielska J. Simultaneous saccharification and ethanol fermentation of waste wheat–rye bread at very high solids loading: effect of enzymatic liquefaction conditions. Fuel. 2015;147:236–42.
50. Meher LC, Vidya SD, Naik SN. Technical aspects of biodiesel production by transesterification—a review. Renew Sust Energ Rev. 2006;10:248–68.
51. Rasimoglu N, Temur H. Cold flow properties of biodiesel obtained from corn oil. Energy. 2014;68:57–60.
52. Bosch M, Hazen SP. Lignocellulosic feedstocks: research progress and challenges in optimizing biomass quality and yield. Front Plant Sci. 2013;4:1–3.
53. Gavrilescu M. Biorefinery systems: an overview. In: Gupta VG, Tuohy M, Kubicek CP, Saddler J, Xu F, editors. Bioenergy research: advances and applications. Amesterdam: Elsevier B.V; 2014.

54. Anwar Z, Gulfraz M, Irshad M. Agro-industrial lignocellulosic biomass a key to unlock the future bio-energy: a brief review. J Rad Res Appl Sci. 2014;7:163–73.
55. Wyman CE, Dale BE, Elander RT, et al. Coordinated development of leading biomass pretreatment technologies. Bioresour Technol. 2005;96:1959–66.
56. Mosier NS, Wyman C, Dale B, et al. Features of promising technologies for pretreatment of lignocellulosic biomass. Bioresour Technol. 2005;96:673–86.
57. Alvira P, Tomás-Pejó E, Ballesteros M, Negro MJ. Pretreatment technologies for an efficient bioethanol production process based on enzymatic hydrolysis: a review. Bioresour Technol. 2010;101:4851–61.
58. de Jong E, Higson A, Walsh P, Wellisch M. Product developments in the bio-based chemicals arena. Biofuels Bioprod Biorefin. 2012;6:606–24.
59. Axegård P, Karlsson M, McKeogh P, et al. A bio-solution to climate change. Final report of the biorefinery taskforce to the forest-based sector technology platform (2007). Available online at: http://www.forestplatform.org/files/FTP_biorefinery_report_part1.pdf
60. Azadi P, Inderwildi OR, Farnood R, King DA. Liquid fuels, hydrogen and chemicals from lignin: a critical review. Renew Sust Energ Rev. 2013;21:506–23.
61. Bouxin FP, McVeigh A, Tran F, et al. Catalytic depolymerisation of isolated lignins to fine chemicals using a Pt/Alumina catalyst: part 1—impact of the lignin structure. Green Chem. 2015;17:1235–42.
62. Sahoo S, Misra M, Mohanty AK. Enhanced properties of lignin-based biodegradable polymer composites using injection moulding process. Compos A Appl Sci Manuf. 2011;42:1710–8.
63. Ma X, Kolla P, Zhao Y, Smirnova AL, Fong H. Electrospun lignin-derived carbon nanofiber mats surface-decorated with MnO2 nanowhiskers as binder-free supercapacitor electrodes with high performance. J Power Sources. 2016;325:541–8.
64. Naseem A, Tabasum S, Zia KM, et al. Lignin-derivatives based polymers, blends and composites: a review. Int J Biol Macromol. 2016;93:296–313.
65. Duval A, Lawoko M. A review on lignin-based polymeric, micro- and nano-structured materials. React Funct Polym. 2014;85:78–96.
66. Nitzsche R, Budzinski M, Gröngröft A. Techno-economic assessment of a wood-based biorefinery concept for the production of polymer-grade ethylene, organosolv lignin and fuel. Bioresour Technol. 2016;200:928–39.
67. Bozell J. Oleochemicals as a feedstock for the biorefinery: high value products from fats and oils. National Renewable Energy Laboratory (NREL) report. Available online at: http://citeseerx.ist.psu.edu/viewdoc/download;jsessionid=22B4360C429672AB215DC604A4E6EBE4?doi=10.1.1.522.3326&rep=rep1&type=pdf
68. Misra RD, Murthy MS. Straight vegetable oil usage in a compression ignition engine—a review. Renew Sust Energ Rev. 2010;14:3005–13.
69. Corsini A, Marchegiani A, Rispoli F, Sciulli F, Venturini P. Vegetable oils as fuels in diesel engine. Engine performance and emissions. Energy Procedia. 2015;81:942–9.
70. National Renewable Energy Laboratory (NREL). Straight vegetable oil as a diesel fuel? 2010. Available online at: http://biodiesel.org/docs/default-source/ffs-engine_manufacturers/clean-cities-fact-sheet-straight-vegetable-oil-as-a-diesel-fuel-.pdf?sfvrsn=6
71. Sirajuddin M, Tariq M, Ali S. Organotin (IV) carboxylates as an effective catalyst for the conversion of corn oil into biodiesel. J Org Chem. 2015;779:30–8.
72. Sajjadi B, Abdul Raman AA, Arandiyan H. A comprehensive review on properties of edible and non-edible vegetable oil-based biodiesel: composition, specifications and prediction models. Renew Sust Energ Rev. 2016;63:62–92.
73. Basinger M, Reding T, Williams C, Lackner KS, Modi V. Compression ignition engine modifications for straight plant oil fueling in remote contexts: modification design and short-run testing. Fuel. 2010;89:2925–38.

74. Abollé A, Kouakou L, Planche H. The viscosity of diesel oil and mixtures with straight vegetable oils: palm, cabbage palm, cotton, groundnut, copra and sunflower. Biomass Bioenergy. 2009;33:1116–21.
75. Reddy MS, Sharma N, Agarwal AK. Effect of straight vegetable oil blends and biodiesel blends on wear of mechanical fuel injection equipment of a constant speed diesel engine. Renew Energy. 2016;99:1008–18.
76. Verma P, Sharma MP. Review of process parameters for biodiesel production from different feedstocks. Renew Sust Energ Rev. 2016;62:1063–71.
77. Pourzolfaghar H, Abnisa F, Daud MAW, Aroua MK. A review of the enzymatic hydroesterification process for biodiesel production. Renew Sust Energ Rev. 2016;61:245–57.
78. Sharma Y, Singh B, Upadhyay S. Advancements in development and characterization of biodiesel: a review. Fuel. 2008;87:2355–73.
79. Bouaid A, Vázquez R, Martinez M, Aracil J. Effect of free fatty acids contents on biodiesel quality—pilot plant studies. Fuel. 2016;174:54–62.
80. Nasaruddin RR, Alam MZ, Jami MS. Evaluation of solvent system for the enzymatic synthesis of ethanol-based biodiesel from sludge palm oil (SPO). Bioresour Technol. 2014;154:155–61.
81. Hayyan A, Alam MZ, Mirghani MES, et al. Sludge palm oil as a renewable raw material for biodiesel production by two-step processes. Bioresour Technol. 2010;101:7804–11.
82. Soler L, Illanes A, Wilson L. Immobilization of *Alcaligenes* sp. lipase as catalyst for the transesterification of vegetable oils to produce biodiesel. Catal Today. 2016;259:177–82.
83. Guldhe A, Singh B, Mutanda T, Permaul K, Bux F. Advances in synthesis of biodiesel via enzyme catalysis: novel and sustainable approaches. Renew Sust Energ Rev. 2015;41:1447–64.
84. Shimada Y, Watanabe Y, Sugihara A, Tominaga Y. Enzymatic alcoholysis for biodiesel fuel production and application of the reaction to oil processing. J Mol Catal B Enzym. 2002;17:133–42.
85. Christopher LP, Zambare VP. Enzymatic biodiesel: challenges and opportunities. Appl Energy. 2014;119:497–520.
86. Ito T, Sakurai Y, Kakuta Y, Sugano M, Hirano K. Biodiesel production from waste animal fats using pyrolysis method. Fuel Process Technol. 2012;94:47–52.
87. Chang JS, Cheng JC, Ling TR. Low acid value bio-gasoline and bio-diesel made from waste cooking oils using a fast pyrolysis process. J Taiwan Inst Chem Eng. doi:10.1016/j.jtice.2016.04.014.
88. Budzianowski WM, Postawa K. Total chain integration of sustainable biorefinery systems. Appl Energy. 2016;184:1432–46.
89. John RP, Anisha GS, Nampoothiri KM, Pandey A. Micro and macroalgal biomass: a renewable source for bioethanol. Bioresour Technol. 2011;102:186–93.
90. Wang Z, Adhikari S, Valdez P, Shakya R, Laird C. Upgrading of hydrothermal liquefaction biocrude from algae grown in municipal wastewater. Fuel Process Technol. 2016;142:147–56.
91. Brown LM, Zeiler KG. Aquatic biomass and carbon dioxide trapping. Energy Convers Manag. 1993;34:1005–13.
92. Nautiyal P, Subramanian KA, Dastidar MG. Production and characterization of biodiesel from algae. Fuel Process Technol. 2014;120:79–88.
93. Demirbas A, Demirbas MF. Importance of algae oil as a source of biodiesel. Energy Convers Manag. 2011;52:163–70.
94. Zhou X, Ge HM, Xia L, Zhang D, Hu CX. Evaluation of oil-producing algae as potential biodiesel feedstock. Bioresour Technol. 2013;134:24–9.
95. Gnansounou E, Raman JK. Life cycle assessment of algae biodiesel and its co-products. Appl Energy. 2016;161:300–8.

96. Ho SH, Huang SW, Chen CY, et al. Bioethanol production using carbohydrate-rich microalgae biomass as feedstock. Bioresour Technol. 2013;135:191–8.
97. Borines MG, de Leon RL, Cuello JL. Bioethanol production from the macroalgae *Sargassum* spp. Bioresour Technol. 2013;138:22–9.
98. Uchida M, Miyoshi T, Kaneniwa M, et al. Production of 16.5% v/v ethanol from seagrass seeds. J Biosci Bioeng. 2014;118:646–50.
99. Huesemann MH, Kuo LJ, Urquhart L, Gill GA, Roesijadi G. Acetone–butanol fermentation of marine macroalgae. Bioresour Technol. 2012;108:305–9.
100. Bai B, Zhou J, Yang M, et al. Efficient production of succinic acid from macroalgae hydrolysate by metabolically engineered *Escherichia coli*. Bioresour Technol. 2015;185:56–61.
101. Tabassum MR, Xia A, Murphy JD. Potential of seaweed as a feedstock for renewable gaseous fuel production in Ireland. Renew Sust Energ Rev. 2017;68:136–46.
102. Montingelli ME, Tedesco S, Olabi AG. Biogas production from algal biomass: a review. Renew Sust Energ Rev. 2015;43:961–72.
103. Pienkos PT, Darzins A. The promise and challenges of micro-algal derived biofuels. Biofuels Bioprod Biorefin. 2009;3:431–40.
104. van Hal JW, Huijgen WJJ, Lopez-Contreras AM. Opportunities and challenges for seaweed in the biobased economy. Trends Biotechnol. 2014;32:231–3.
105. Kerton FM, Liu Y, Omari KW, Hawboldt K. Green chemistry and ocean-based biorefinery. Green Chem. 2013;15:860–71.
106. Baghel RS, Trivedi N, Gupta V. Biorefining of marine macroalgal biomass for production of biofuel and commodity chemicals. Green Chem. 2015;17:2436–43.
107. Xia A, Jacob A, Tabassum MR, Herrmann C, Murphy JD. Production of hydrogen, ethanol and volatile fatty acids through co-fermentation of macro- and micro-algae. Bioresour Technol. 2016;205:118–25.
108. Teo SH, Islam A, Taufiq-Yap YH. Algae derived biodiesel using nanocatalytic transesterification process. Chem Eng Res Des. 2016;111:362–70.
109. Casoni AI, Zunino J, Piccolo MC, Volpe MA. Valorization of *Rhizoclonium* sp. algae via pyrolysis and catalytic pyrolysis. Bioresour Technol. 2016;216:302–7.
110. Barreiro DL, Prins W, Ronsse F, Brilman W. Hydrothermal liquefaction (HTL) of microalgae for biofuel production: state of the art review and future prospects. Biomass Bioenergy. 2013;53:113–27.
111. Tian C, Li B, Liu Z, Zhang Y, Lu H. Hydrothermal liquefaction for algal biorefinery: a critical review. Renew Sust Energ Rev. 2014;38:933–50.
112. Goetschalckx M. Supply chain engineering. Berlin: Springer; 2011.
113. Fernando S, Adhikari S, Chandrapal C, Murali N. Biorefineries: current status, challenges, and future direction. Energy Fuel. 2006;20:1727–37.
114. Liu Z, Eden MR. Biorefinery principles, analysis, and design. In: Wang L, editor. Sustainable bioenergy production. Boca Raton, FL: CRC Press; 2014.
115. Yue D, You F, Snyder SW. Biomass-to-bioenergy and biofuel supply chain optimization: overview, key issues and challenges. Comput Chem Eng. 2014;66:36–56.
116. Mula J, Peidro D, Díaz-Madronero M, Vicens E. Mathematica lprogramming models for supply chain production and transport planning. Eur J Oper Res. 2010;204:377–90.
117. Awudu I, Zhang J. Uncertainties and sustainability concepts in biofuel supply chain management: a review. Renew Sust Energ Rev. 2012;16:1359–68.
118. De Meyer A, Cattrysse D, Rasinmäki J, Van Orshoven J. Methods to optimise the design and management of biomass-for-bioenergy supply chains: a review. Renew Sust Energ Rev. 2014;31:657–70.
119. Ghaderi H, Pishvaee MS, Moini A. Biomass supply chain network design: an optimization-oriented review and analysis. Ind Crop Prod. 2016;94:972–1000.

120. Kamm B. Integrated biorefineries—a bottom-up approach to biomass fractionation. In: Mascia PN, Scheffran J, Widholm JM, editors. Plant biotechnology for sustainable production of energy and co-products. Berlin: Springer; 2010.
121. Holweg M, Helo P. Defining value chain architectures: linking strategic value creation to operational supply chain design. Int J Prod Econ. 2014;147:230–8.
122. Ramsay J. The real meaning of value in trading relationships. Int J Oper Prod Manag. 2005;25:549–65.
123. Schenkel M, Caniëls MCJ, Krikke H, van der Laan E. Understanding value creation in closed loop supply chains—past findings and future directions. J Manuf Syst. 2015;37:729–45.
124. Daskin MS. Network and discrete location: models, algorithms, and applications. 2nd ed. - New York: Wiley; 2013.
125. Serrano A, Faulin J, Astiz P, Sánchez M, Belloso J. Locating and designing a biorefinery supply chain under uncertainty in Navarre: a stochastic facility location problem case. Transport Res Proced. 2015;10:704–13.
126. Richard TL. Challenges in scaling up biofuels infrastructure. Science. 2010;329:793–6.
127. Sharma B, Ingalls R, Jones C, Khanchi A. Biomass supply chain design and analysis: basis overview, modeling, challenges, and future. Renew Sust Energ Rev. 2013;24:608–27.
128. Kokossis AC, Yang A. On the use of systems technologies and a systematic approach for the synthesis and the design of future biorefineries. Comput Chem Eng. 2010;34:1397–405.
129. Grossmann IE, Guillén-Gosálbez G. Scope for the application of mathematical programming techniques in the synthesis and planning of sustainable processes. Comput Chem Eng. 2010;34:1365–76.
130. Seider WD, Seader JD, Lewin DR, Widagdo S. Product and process design principles—synthesis, analysis, and evaluation. 3rd ed. New York: Wiley; 2010.
131. Bonomi A, Cavalett O, da Cunha MP, Lima MAP. Virtual biorefinery: an optimization strategy for renewable carbon valorization. Berlin: Springer; 2015.
132. Stuart PR, El-Halwagi MM. Integrated biorefineries: design, analysis, and optimization. Boca Raton, FL: CRC Press; 2012.
133. Hytönen E, Stuart P. Integrating bioethanol production into an integrated Kraft pulp and paper mill: Techno-economic assessment. Pulp Pap Can. 2009;110:58–65.
134. Jeaidi J, Stuart P. Techno-economic analysis of biorefinery process options for mechanical pulp mills. J Sci Technol For Prod Process. 2011;1:62–70.
135. Ng LY, Andiappan V, Chemmangattuvalappil NG, Ng DKS. A systematic methodology for optimal mixture design in an integrated biorefinery. Comput Chem Eng. 2015;81:288–309.
136. Kokossis AC, Tsakalova M, Pyrgakis K. Design of integrated biorefineries. Comput Chem Eng. 2015;81:40–56.
137. Mariano AP, Dias MOS, Junqueira TL, et al. Butanol production in a first-generation Brazilian sugarcane biorefinery: technical aspects and economics of greenfield projects. Bioresour Technol. 2013;135:316–23.
138. Pereira LG, Dias MOS, Junqueira TL, et al. Butanol production in a sugarcane biorefinery using ethanol as feedstock. Part II: Integration to a second generation sugarcane distillery. Chem Eng Res Des. 2014;92:1452–62.
139. Giuliano A, Poletto M, Barletta D. Process optimization of a multi-product biorefinery: the effect of biomass seasonality. Chem Eng Res Design. 2016;107:236–52.
140. Santibanez-Aguilar JE, Morales-Rodriguez R, Gonzalez-Campos JB, Ponce-Ortega JM. Stochastic design of biorefinery supply chains considering economic and environmental objectives. J Clean Prod. 2016;136:224–45.
141. Wang WC. Techno-economic analysis of a bio-refinery process for producing hydro-processed renewable jet fuel from jatropha. Renew Energy. 2016;95:63–73.
142. Jones SB, Zhu Y. Techno-economic analysis for the conversion of lignocellulosic biomass to gasoline via the methanol-to-gasoline (MTG) process. Pacific Northwest National Laboratory. 2009. Available online at: http://www.pnl.gov/main/publications/external/technical_reports/PNNL-18481.pdf

143. Sauvée L, Viaggi D. Biorefineries in the bio-based economy: opportunities and challenges for economic research. Bio-based Appl Econ. 2016;5:1–4.
144. Jungmeier G. The biorefinery complexity index (2014). Working document for the International Energy Agency (IEA). IEA bioenergy Task 42. Available online at: http://www.iea-bioenergy.task42-biorefineries.com/upload_mm/6/2/f/ac61fa53-a1c0-4cbc-96f6-c9d19d668a14_BCI%20working%20document%2020140709.pdf
145. World Bank report. Introduction—Opportunities and challenges in the forest sector. 2008. Available online at: http://siteresources.worldbank.org/EXTFORSOUBOOK/Resources/intro.pdf
146. United Nations' Food and Agriculture Organization (FAO). State of the world's forests (2011). Available online at: http://www.fao.org/docrep/013/i2000e/i2000e00.htm
147. Näyhä A, Pesonen HL. Strategic change in the forest industry towards the biorefining business. Technol Forecast Soc Chang. 2014;81:259–71.
148. Janssen M, Stuart P. Drivers and barriers for implementation of the biorefinery. Pulp Pap Can. 2010;111:13–7.
149. Näyhä A, Pesonen HL. Diffusion of forest biorefineries in Scandinavia and North America. Technol Forecast Soc Chang. 2012;79:1111–20.
150. Söderholm P, Lundmark R. Forest-based biorefineries: implications for market behavior and policy. For Prod J. 2009;59:6–15.
151. Seghetta M, Marchi M, Thomsen M, Bjerre AB, Bastianoni S. Modelling biogenic carbon flow in a macroalgal biorefinery system. Algal Res. 2016;18:144–55.
152. Dong T, Knoshaug EP, Davis R. Combined algal processing: a novel integrated biorefinery process to produce algal biofuels and bioproducts. Algal Res. 2016;19:316–23.
153. Wei N, Quarterman J, Jin YS. Marine macroalgae: an untapped resource for producing, fuels and chemicals. Trends Biotechnol. 2013;31:70–7.
154. Suganya T, Varman M, Masjuki HH, Renganathan S. Macroalgae and microalgae as a potential source for commercial applications along with biofuels production: a biorefinery approach. Renew Sust Energ Rev. 2016;55:909–41.
155. Barlow J, Sims RC, Quinn JC. Techno-economic and life-cycle assessment of an attached growth algal biorefinery. Bioresour Technol. 2016;220:360–8.
156. Quinn JC, Davis R. The potentials and challenges of algae based biofuels: a review of the techno-economic, life cycle, and resource assessment modeling. Bioresour Technol. 2015;184:444–52.
157. Slade R, Bauen A. Micro-algae cultivation for biofuels: cost, energy balance, environmental impacts and future prospects. Biomass Bioenergy. 2013;53:29–38.
158. Dahmani S, Zerrouki D, Ramanna L, Rawat I, Bux F. Cultivation of *Chlorella pyrenoidosa* in outdoor open raceway pond using domestic wastewater as medium in arid desert region. Bioresour Technol. 2016;219:749–52.
159. Badvipour S, Eustance E, Sommerfeld MR. Process evaluation of energy requirements for feed production using dairy wastewater for algal cultivation: theoretical approach. Algal Res. 2016;19:207–14.
160. Hernández D, Riaño B, Coca M, et al. Microalgae cultivation in high rate algal ponds using slaughterhouse wastewater for biofuel applications. Chem Eng J. 2016;285:449–58.
161. Adey WH, Kangas PC, Mulbry W. Algal turf scrubbing: cleaning surface waters with solar energy while producing a biofuel. Biosci. 2011;61:434–41.
162. Gross M, Jarboe D, Wen Z. Biofilm-based algal cultivation systems. Appl Microbiol Biotechnol. 2015;99:5781–9.
163. Christenson LB, Sims RC. Rotating algal biofilm reactor and spool harvester for wastewater treatment with biofuels by-products. Biotechnol Bioeng. 2012;109:1674–84.
164. Sirajunnisa AR, Surendhiran D. Algae—a quintessential and positive resource of bioethanol production: a comprehensive review. Renew Sust Energ Rev. 2016;66:248–67.
165. Georgianna DR, Mayfield SP. Exploiting diversity and synthetic biology for the production of algal biofuels. Nature. 2012;488:329–35.

166. George B, Pancha I, Desai C, et al. Effects of different media composition, light intensity and photoperiod on morphology and physiology of freshwater microalgae *Ankistrodesmus falcatus*—a potential strain for bio-fuel production. Bioresour Technol. 2014;171:367–74.
167. Lundquist TJ, Woertz IC, Quinn NWT, Benemann JR. A realistic technology and engineering assessment of algae biofuel production. Report for the Energy Biosciences Institute, University of California. 2010. Available online at: http://www.energybiosciencesinstitute.org/media/AlgaeReportFINAL.pdf
168. Kim J, Yoo G, Lee H, et al. Methods of downstream processing for the production of biodiesel from microalgae. Biotechnol Adv. 2013;31:862–76.
169. Abdo SM, Abo El-Emin SA, El-Khatib KM, et al. Preliminary economic assessment of biofuel production from microalgae. Renew Sust Energ Rev. 2016;55:1147–53.
170. Thomassen G, Vila UE, Van Dael M, Lemmens B, Van Passel S. A techno-economic assessment of an algal-based biorefinery. Clean Techn Environ Policy. 2016;18:1849–62.
171. Rizwan M, Lee JH, Gani R. Optimal design of microalgae-based biorefinery: economics, opportunities and challenges. Appl Energy. 2015;150:69–79.
172. King D. The future of industrial biorefineries. White paper for the World Economic Forum (2010). Available online at: http://www3.weforum.org/docs/WEF_FutureIndustrialBiorefineries_Report_2010.pdf
173. Kamm B, Kamm M. Principles of biorefineries. Appl Microbiol Biotechnol. 2004;64:137–45.
174. BCC Research. Biorefinery technologies: Global markets. Report EGY054C. July 2016. Available online at: http://www.bccresearch.com/market-research/energy-and-resources/biorefinery-technologies-global-markets-markets-egy054c.html
175. European biofuels technology platform (EBTP). Borregaard—commercial plant in Sarpsborg, Norway (2016). Available online at: http://biofuelstp.eu/factsheets/Factsheet_Borregaard_final.pdf
176. Borregaard's business areas. Available online at: http://www.borregaard.com/Business-Areas
177. Modahl IS, Soldal E. The 2015 LCA of products from the wood-based biorefinery at Borregaard, Sarpsborg. Report commissioned by Borregaard Industries Limited, Norway. Available online at: http://ostfoldforskning.no/media/1678/or-1115-the-2015-lca-of-cellulose-ethanol-lignin-and-vanillin-from-borregaard-sarpsborg-verification.pdf
178. Modahl IS, Brekke A, Valente C. Environmental assessment of chemical products from a Norwegian biorefinery. J Clean Prod. 2015;94:247–59.
179. Rødsrud G, Lersch M, Sjöde A. History and future of world's most advanced biorefinery in operation. Biomass Bioenergy. 2012;46:46–59.
180. Sjöde A, Frölander A, Lersch M, Rødsrud G. Lignocellulosic biomass conversion. Patent application WO 2010/078930 A2. 2010.
181. Borregaard LignoTech. Official inauguration of Borregaard's biorefinery demonstration plant. Available online at: http://www.lignotech.com/News/Official-Inauguration-of-Borregaard-s-Biorefinery-Demonstration-Plant/%28language%29/eng-GB
182. UOP and Ensyn join forces to create Envergent Technologies. Available online at: https://www.envergenttech.com/about-us/
183. The practical, proven path to green energy. RTP™ rapid thermal processing from envergent technologies. Available online at: https://www.envergenttech.com/wp-content/uploads/2015/02/rtp-from-envergent-2010.pdf
184. European biofuels technology platform (EBTP). Bio-oil (via pyrolysis/thermochemical conversion) and tall oil for production of advanced biofuels. Available online at: http://www.biofuelstp.eu/bio-oil.html
185. Honeywell's Envergent RTP Technology to be used in new renewable fuels facility in Quebec. Focus Catal 2016;9:3.
186. Freel B. Rapid thermal conversion of biomass. US 7905990 B2. 2011.
187. Fan J, Kalnes TN, Alward M, et al. Life cycle assessment of electricity generation using fast pyrolysis bio-oil. Renew Energy. 2011;36:632–41.

188. RTP green fuel: an overview for renewable heat and power. Envergent Technologies's RTP white paper. 2013. Available online at: https://www.envergenttech.com/wp-content/uploads/2015/02/2013-envergent-burner-applications-white-paper.pdf
189. RTP™ Technology converts low-value biomass into a high-value liquid asset—RTP green fuel. Available online at: https://www.envergenttech.com/technology/
190. Lasure LL, Zhang M. Bioconversion and biorefineries of the future. Available online at: http://www.pnl.gov/biobased/docs/biorefineries.pdf
191. Cherubini F. The biorefinery concept: using biomass instead of oil for producing energy and chemicals. Energy Convers Manag. 2010;51:1412–21.

Chapter 8
Implementing the Bioeconomy on the Ground: An International Overview

Abstract Bioeconomy is a holistic economic model developed and applied to manage the available raw materials from biomass and valorize the biowastes generated for the various agro-industrial activities. The implementation of this sustainable concept on the ground is based on a variety of green technologies related to the acquisition and conversion of biomass to produce a wide range of biofuels, biochemicals, and biomaterials, with little to no competition with food resources.

The decision to implement bioeconomy in any country is well justified and of strategic importance, especially if enough renewable raw materials could be made available to "fuel" the country's economy and the major industrial activities. In this chapter various visions, strategies, resources, and opportunities, as well many industrial cases related to the implementation of bioeconomy, are presented and discussed in various leading countries in the world including the United States, several European countries, and China. A closer look on the Finnish bioeconomy and its related industrial achievements and prospects is also given.

8.1 Introduction

Bioeconomy is a holistic economic model managing most of its raw materials from biomass, which is processed via an arsenal of green technologies to produce wide arrays of energy, chemical, and material products, with little to no competition with food supplies.

Considering the heavy legacy of the petroleum era on the environment, and its unsustainable production system, the concept of bioeconomy is of special interest to producers and consumers around the world. Nevertheless, the implementation of this concept on the ground is affected by many factors, not only the availability of biomass and the development or acquisition of biorefining technologies but most importantly the strategic vision behind such implementation. The matter is complex involving geopolitical, industrial, societal, and environmental perspectives. This multidimensional aspect of bioeconomy will be thoroughly discussed in the next chapter.

In general, the decision to implement bioeconomy in any country is well justified, especially if alternative raw materials (i.e., renewable biomass) are available to

M. Sillanpää, C. Ncibi, *A Sustainable Bioeconomy*,
DOI 10.1007/978-3-319-55637-6_8

"fuel" the country's economy and major industrial activities. The main difference between countries is how they perceive the bioeconomy concept in the first place: is it "another" economic model to generate more profit and/or is it a sustainable economic model for the well-being of future generations?

The current wave of expansion of bioeconomy coincided with recurrent economic crisis around the world. Although the unsustainability of the current economic model, heavily relying on fossil resources, is one of the main causes, the bioeconomy is still regarded as "another" economic model implemented to generate profit. Sustainability in this case became a by-product of bioeconomy, sometimes a mere marketing tool, not a genuine component of bioeconomy.

In a related matter, a closer look on the international scene clearly shows the existence of two different bioeconomies being implemented around the world:

- The first one, where large amounts of biomass are needed to meet the increasing demand for energy, chemicals, and materials, the goal is to set up efficient and highly profitable production systems while reducing the carbon footprint and the dependency on fossil fuels. Large companies and multinational corporations are the main player in this "strong" bioeconomy.
- The second one is generally implemented in countries, although possessing rich ecosystems, which do not have the infrastructure and technological know-how to fully benefit from the available resources. Thus, the main aim of this "weak" bioeconomy is to generate profit from biomass (as much as possible). Sustainability and environmental issues are not a priority.

The most serious aspect of this matter is that it lays the ground for yet another round of inequity. Indeed, the quest for more biomass to feed the growing bioeconomy will inevitably lead countries to seek raw materials from abroad via importation but also by other highly controversial means such as the acquisition of large surfaces of arable land (and freshwater) in less developed countries to produce energy crops or any other crop of industrial interest, while the indigenous population is needing those lands to grow foodstuff [1].

Considering the importance of this issue, we will discuss it further in the next chapter. Let us now explore the visions, strategies, opportunities, and some industrial cases related to the implementation of bioeconomy in various leading countries in the world including the United States, several European countries, and China.

8.2 Bioeconomy in the United States

In order to have the first impression about the bioeconomy concept in the United States, the official definition has to be analyzed. It states that bioeconomy is *the global industrial transition of sustainably utilizing renewable aquatic and terrestrial biomass resources in energy, intermediate, and final products for economic, environmental, social, and national security benefits* [2]. In short, Americans vision bioeconomy as a sustainable industrial transition, which echoes the importance of the industrial sector in the economy of the country and the consciousness about the legacy of the petroleum era, as well as the need for a transition to a sustainable economic model.

The major industrial activities linked to the bioeconomy concept are agriculture, forestry, biorefining (mainly biofuels), biotechnology (mainly based on enzymes), and textile.

8.2.1 Strategic Vision of the Largest Economy in the World

8.2.1.1 Federal Bioeconomy Strategy

By promoting the sustainability and efficiency of the production system in those various agro-industrial activities, the United States is emphasizing on key bio-based products including biofuels and renewable platform chemicals to foster its bio-economy and reach the strategic objectives of energy security, promoting rural economic development and reducing greenhouse gas emissions from the highly polluting sectors (industries and transportation sector).

Currently, the incorporation of the bioeconomy in the American economic system is still at its early stages. As stressed upon in the definition, the United States is gradually implementing bio-based economic concepts to its industrial activities since the dependency of most of those industries on fossil fuels is still advanced and the transition to sustainable production schemes based on renewable biomass needs to be operated gradually in order to maintain the profitability of the involved industries.

Nonetheless, the United States has several important assets that could speed up this crucial transition phase and ensure a wider and successful implementation of the sustainable bio-based economic model. This includes the involvement of the industrial sector, the development of innovative technologies, and the increasing public awareness about climate change, environmental issues, depletion of fossil fuels, etc. Nevertheless, the main asset, so far, remains the active involvement of the federal government in the bioeconomy sector through its numerous agencies, supportive and protective measures (subsidies and tariffs), and various regulatory policies [3, 4].

In this context, the federal government addressed a memorandum (published in 2010) to the heads of executive departments and agencies to emphasize the federal priorities in science and technology and allocate the necessary budgets accordingly. Hence, within the effort to promote sustainable economic growth and job creation, the federal agencies were directed to "support research to establish the foundations for a twenty-first century bioeconomy." As well it was stated that "advances in biotechnology and improvements in [the] ability to develop biological systems have the potential to address critical national needs in agriculture, energy, health and environment" [5].

In 2012, the National Bioeconomy Blueprint was released by the American administration illustrating its strategic plan for bioeconomy in the country [6, 7]. Five major objectives were highlighted including:

1. Strengthening the scientific research and development field and investing more in innovation.
2. Advancing and facilitating the transition of the developed products, technologies, and innovations from research institutions to markets.
3. Reducing and reforming regulatory barriers affecting the expansion of markets and scientific and technological innovations, as well as the measures discouraging investment.
4. Developing a bioeconomy workforce by updating training programs and promoting the partnership between academic institutions and industries.
5. Fostering multiparty partnerships between public sector, private industry, government agencies, and academic institutions. Such cooperation will bring together several expertise, resources, and feedbacks that would build a momentum around the concept of sustainable bioeconomy and its industrial implementation.

Several specific objectives could also be found in this federal document related to four major sectors in the United States: energy, agriculture, health, and environment.

8.2.1.2 The Billion-Ton Bioeconomy Vision

Biomass is definitely the pillar of bioeconomy and its biorefining industry. Thus, securing a continuous flow of these raw materials at low costs to the various industrial activities is a prerequisite for a successful transition to bioeconomy.

Well aware of the strategic position of biomass in the future sustainable production systems, many governmental agencies in the United States cooperated in order to develop a new vision to secure the proper flow of feedstocks which will promote the implementation, and later expansion, of bioeconomy in the country. In 2005, and after assessing the potentialities of the agricultural and forest sectors to supply low-cost biomass for the new bio-based industries, it was estimated that, each year, one billion tons of biomass needs to be sustainably produced and made available for the industries by 2030 [8]. From this estimation came the "Billion-Ton Bioeconomy Vision."

The work to implement this vision on the ground effectively started in 2005 with the billion-ton study commissioned by the US Department of Energy (DoE) [9]. This study was conducted in order to assess if the American agricultural and forest sectors have the potential to generate, in a sustainable and cost-effective manner, one billion dry tons of biomass each year. This amount of biomass was predicted appropriate for the replacement around 30% of the country's petroleum consumption by 2030 (mainly transportation sector). The major drawback of this first attempt was that the assessment of the biomass potentialities, from both agriculture and forestry, did not include the acquisition costs. Thus, although all of the identified bioresources were potentially available, the incorporation of some of them in biomass supply chain was deemed to be economically challenging due to the costs related to logistic matters (harvesting and/or transportation) or the high content of recalcitrant components in the biomass requiring costly pretreatment steps before the main processing.

Therefore, the billion-ton study was recommissioned by the DoE and an updated report was published in 2011 [10]. In this study the availability of biomass supply was reevaluated based of the acquisition cost. Thus, the potentialities of biomass supply chains to respond to market demands were simulated using economic models, as well as the economic availability of biomass feedstocks under various ranges of prices and production yields (simulated scenarios between 2012 and 2030). The updated study based its assumptions on the objective of an annual production of at least one billion dry tons of biomass by 2030. Other assumptions included an average market price of US$60 per dry ton at the farm gate (i.e., after harvest) or roadside (i.e., ready to be delivered to the processing plant).

In the latest update of this billion-ton study (related report published in July 2016), the involved scientists and experts worked on updating the "farm gate/roadside" analysis using the latest available data, along with adding more biomass feedstocks such as algae and specific herbaceous and woody energy crops to the economic estimation [11]. The new study also simulated a new scenario taking into account the cost to transport biomass to the biorefining facility under various logistical conditions.

A wide range of biomass feedstocks were investigated in the course of the three DoE-commissioned billion-ton studies including:

- Agricultural and farming products and by-products such as grains for bioethanol, oilseeds for biodiesel, and manure for biogas. Specific cases of herbaceous and woody crops were also assessed.
- Forest-based products and by-products including woods, logging residues, and forest thinnings for the production of biofuel, heat, and power.
- Organic urban and industrial wastes such as municipal solid wastes (MSWs), digested sludge, black liquor, mill wastes, etc.

Overall, the American bioeconomy vision is firstly based on the assessment of the country's potentialities in terms of biological resources suitable as feedstocks for the biorefining industries and other bio-based processing facilities. The benefits and impacts of various bioeconomy implementation scenarios are evaluated for the entire supply chain, i.e., from the production or harvesting of the biomass feedstock to its conversion to marketable end product(s). After the implementation phase, the strategic vision of the United States in regard to sustainable bioeconomy is to expand its industrial production capacities in order to increase the share of biofuels, biochemicals, and other bio-based products in their respective markets.

The following Table 8.1 illustrates the strategic objective of the US government in terms of market share and consumption of biofuels and biopower and production of bio-based products between 2010 and 2030.

Table 8.1 Strategic vision for bioenergy and bio-based products in the United States between 2010 and 2030 [12]

		2010	2015	2020	2030
Biofuels	Market share (%)	4	6	10	20
	Consumption (billion gasoline-equivalent gallons)	8	12.9	22.7	51
Biopower	Market share (%)	4	5.5	7	7
	Consumption (quadrillion Btu)	3.1	3.2	3.4	3.8
Bio-based products	Production (billion lbs.)	23.7	26.4	35.6	55.3

To concretize this vision on the ground within the targeted frame time (2030–2040), the United States will have to count mainly on its biological resources, along with the previously mentioned factors (governmental support, industrial involvement, education system, etc.).

The large surface of the country (9,831,510 km^2) and its vast and diverse ecosystems (natural and cultivated) constitute the major asset for the United States to be successful in implementing its bioeconomy vision. The following section illustrates the main bioresources available in the United States for bioeconomy.

8.2.2 *Resources and Opportunities*

In the United States, biological feedstocks suitable for biorefining and other industrial activities include a wide range of biomass resources [13]. Many US-based research institutions worked on the conversion of biomass into added-value products using various kinds of terrestrial and aquatic feedstocks, which could be grouped into:

- *Non-food herbaceous and woody energy crops*: This group comprises perennial crops that could be cultivated on marginal lands [14, 15] and harvested for biomass within 2 to 3 years form the plantation. Different species belong to this group such as switchgrass, miscanthus, bamboo, sweet sorghum, wheatgrass, etc. As for the woody crops, this group includes short-rotation trees, such as hybrid poplar, and willow, which could be grown to provide, within 5 to 8 years, valuable lignocellulosic biomass.
- *Agricultural crops* include a variety of plants cultivated for the industrial potentialities of the plant or its component(s) such as corn (starch and oil), soybean (oil and meal), sugarcane and sugar beet (sugar), and oilseeds (vegetable oils). The firsthand products are starch, sugar, oil, juice, etc. Further industrial processing generally produces end products of higher value such as biofuels and biochemicals [16, 17].

- *Agricultural residues* consist of wastes generated either during the harvest or industrial processing of agricultural crops [18]. Biomass materials from this group include straws, stalks, husks, and leaves.
- *Forest residues* comprise any biomass that was not harvested or removed from logging sites in commercial hardwood and softwood plantations, as well as biomass recovered from forest management operations (thinning and removal of dead trees) [13, 19].
- *Aquatic biomass* include a wide range of microalgae, macroalgae (seaweeds), and other aquatic plants (seagrasses), either harvested from natural aquatic ecosystems or artificially cultivated in ponds or bioreactors, to produce bio-based fuels and chemicals [20, 21].
- *Biomass-processing wastes* are by-products and side streams generated during the industrial processing of various biomass to produce firsthand products. This includes sawdust, bark and branches from wood processing [22], and bagasse from sorghum and gave processing [23, 24]. The first valorization scheme of such residues was their combustion to produce heat and power. Lately, their utilization as feedstocks for the production of fuels and fine chemicals and other added-value products enhanced the economic performance of biorefineries.
- *Municipal wastes* are also reported to be interesting feedstock materials for the American biorefining sector. This biomass group includes any organic matter present in municipal waste collection systems, mainly landfills (residential, commercial, and institutional post-consumption wastes) and wastewater treatment plants (sludge) [25].

For the case of low-cost residual biomass, which is of special interest to industrials in the biorefining sector, the National Renewable Energy Laboratory (NREL) of the US Department of Energy published a series of well-illustrated maps showcasing the various bioresources (residues and wastes) available on American soil by county [26]. The collected data about biomass resources were analyzed via a geographic information system (GIS), and several categories of agro-industrial and commercial wastes were evaluated including crop residues, forest residues, primary and secondary mill residues, urban wood wastes, and methane emissions from animal manure, landfills, and wastewater treatment plants.

Table 8.2 summarizes the amounts of currently and potentially available biomass feedstocks in the United States (the latter is based on two assumption scenarios). The related data are collected from the 2016 billion-ton report [11]. The results of a more developed estimation analysis used various biomass prices (US$30–100), agricultural scenarios (2–4% yield), and forest scenarios (combination of low, medium, and high housing and energy demand) [27].

Table 8.2 Estimated yields of natural resources (from forestry and agriculture) and wastes, currently supplied or potentially available at prices ≤ US$60 [11]

	Million dry tons			
Biomass feedstock	2017	2022	2030	2040
Currently supplied bioresources				
• Forestry resources	154	154	154	154
• Agricultural resources	144	144	144	144
• Waste resources	68	68	68	68
• Total	355	3555	355	355
Potentially available bioresources: Base-case scenario				
• Forestry resources	103	109	97	97
• Agricultural residues	104	123	149	176
• Energy crops	–	78	239	411
• Waste resources	137	139	140	142
• Total	344	449	625	826
Potentially available bioresources: High-yield scenario[a]				
• Forestry resources	95	99	87	76
• Agricultural residues	105	135	174	200
• Energy crops	–	110	380	736
• Waste resources	337	483	782	1154
• Total	537	827	1423	2166

[a]High yield (3% annual growth scenario)

8.2.3 *Industrial Study Cases*

In the early stages, the development and implementation of various biorefining facilities in the United States were mainly oriented toward providing alternative renewable fuels from biomass. Besides, based on the high performances of many agricultural sectors, the main feedstocks used to supply biorefineries were mostly dedicated agricultural and energy crops and derived wastes.

Throughout this implementation phase, new biorefineries were built to process new kinds of feedstocks and to produce a wide range of bioproducts, including biofuels, biochemical, and other bio-based products. In this context, several national and multinational industrial companies invested in biorefining on the American soil, either by designing, building, and operating new facilities or utilizing and upgrading existing production platforms in retrofitting scenarios [28]. This includes retrofit isobutanol conversion (BP/DuPont's Butamax, Gevo), bolt-on cellulosic ethanol (POET-DSM, Abengoa), biodiesel from animal residues (Diamond Green Diesel), thermo-catalytic conversion of wood chips into hydrocarbon fuels (Ensyn/UOP's Envergent), and Fischer-Tropsch synthetic gas-to-liquids technology, the production of diesel and jet fuel from wood wastes and bagasse (ClearFuels/Rentech) [29, 30].

In the next section, selected American industries involved in the production of added-value bioproducts are studied in more details to showcase some key features about the industrial implementation of the bioeconomy concept in the United States.

8.2.3.1 DuPont

E. I. du Pont de Nemours and Co. (DuPont) is an American science and technology-based industrial company founded in July 1802 and incorporated in October 1915. The DuPont conglomerate includes over ten businesses in diverse sector such as agriculture, electronics and communications, industrial biosciences, and nutrition and health, as well as safety and protection. The company is also conducting cutting-edge R&D programs in various scientific fields related to its production schemes, mainly agronomic, biological, and chemical sciences [31].

Conscious about the new and lucrative opportunities to produce added-value products from renewable biomass, the company promoted its biotechnological R&D and invested in more sustainable production processes and facilities. Such orientation toward industrial biotechnology is believed to offer new potential for meeting the world's demand for food, feed, fuel, and materials in a more sustainable manner, thus protecting the environment and decreasing the world's dependence on fossil resources such as petroleum. Nonetheless, for DuPont and any other multi-billion-dollar company, the sustainable dimension of bio-based economy is adopted mainly for its profitability and flexibility (i.e., integrating several input/output scenarios) [32].

In this context, and through R&D and innovation efforts, various biofuels and building blocks were derived from renewable resources and used to produce DuPont's Renewably Sourced Materials, which are reported to have smaller environmental footprint than their petroleum-based counterpart. According to the company, any DuPont's product labeled Renewably Sourced Materials should have least of 20% (by weight) of its components derived from renewable sources [33].

Among the various bio-based products produced by DuPont's industrial biotechnology activities, two key products seem to be given special interest: biofuels and building blocks.

- **DuPont's biofuels**: The company firstly developed an enzyme-based technology to enable the production of first-generation bioethanol from starch. Then, with the creation of new markets for second-generation bioethanol, the company produced this fuel from renewable biomass such as corn stover, switchgrass, and other lignocellulosic wastes residues to produce cellulosic ethanol.

  In 2007, DuPont extended its ethanol market share through its commercially available enzyme-based process, Accellerase®, for the production of cellulosic fuel ethanol. In the field of enzymatic conversion of starch into fermentable sugars for bioethanol production, the company has continued its effort to optimize the conversion process and improve the production yields. One of the related breakthroughs is the development of fermentation additives (FermaSure®) able to speed up the fermentation reaction and reducing the consumption of nutrient and energy while increasing the production yield [34]. It is estimated that more than 50 ethanol-producing plants in the United States are using DuPont's FermaSure [35].

Butamax™ Advanced Biofuels is a joint venture of DuPont and BP. The Butamax technology is designed to convert fermentable sugars from various feedstocks such as corn and sugarcane to added-value biobutanol using existing biofuel production facilities [36]. For instance, in October 2013, Butamax and Highwater Ethanol began retrofitting Highwater's ethanol plant in Lamberton, Minnesota, along with installing a corn oil separation process [37]. The biobutanol-producing facility is now fully operational.

- **Dupont's building blocks**: The company is working on developing innovative processes to convert renewable biomass into building block chemicals by integrating its diverse expertise in biology, chemistry, and engineering and applying specifically engineered microbes. In this effort, DuPont has developed a proprietary fermentation process that produces a wide range of building blocks that could be used as feedstocks to produce bio-based products [38]. For instance, DuPont Tate & Lyle BioProducts, a joint venture between DuPont and Tate & Lyle, is producing 1,3-propanediol (Bio-PDO™) from corn glucose [39], which is a valuable chemical compound used in the production of various polymers such as polyesters, polyethers, and polyurethanes, as well as in the synthesis of heterocyclic compounds [40].

8.2.3.2 Renewable Energy Group (REG)

Renewable Energy Group, Inc. (REG) is an American company founded in 1996 to operate in the renewable energy business. Currently, REG is one of the leading North American companies in the field of biofuels and biochemicals production and manufacturing. It is the largest producer of biodiesel in the United States.

With 12 fully operational biorefineries across the United States, REG is basing its industrial activities on a nationwide production, distribution, and logistics system to convert natural fats, oils, and greases into advanced bio-based fuels and chemicals [41]. In this regard, the products' portfolio of the company includes [42]:

- **Biodiesel**: REG is producing B100 (pure biodiesel) or B99.9 (blended with 0.1% of petrodiesel) biodiesel in its biorefineries. For more than a decade, the company is producing and supplying the US biofuel market with its REG-9000™ bio-based diesel, with an annual production capacity around 225 million gallons [43]. REG is also marketing ultralow sulfur diesel and heating oil in Northeastern and Midwestern United States.
- **Renewable hydrocarbon diesel**: REG-9000/RHD is a 100% renewable hydrocarbon fuel produced from fats and oils by hydro-treatment. It can be used (pure or blended with diesel) in conventional diesel engines without operating any modification.
- **Renewable naphtha**: It is a renewable hydrocarbon gasoline blendstock generated as coproduct during the production process of biodiesel from fats and oils. A blendstock is a gasoline blend with a lower volatility and octane rating than the conventionally produced gasoline.

- **Renewable chemicals**: REG's renewable chemicals division was launched in 2014 and is currently working on developing a sustainable biotechnology-based production platform for the conversion of renewable biomass and residual wastes into chemical products, thus enabling the company to operate and compete in the profitable industrial biotechnology business. So far, the company's chemical division is focusing on biological catalysts which are engineered for the production of selected biochemicals from first-generation feedstocks such as corn (starch) and sugarcane (sugars) and second-generation feedstocks including many lignocellulosic bioresources (fermentable sugars). The valorization of crude glycerin from the biodiesel production process to produce specialty chemicals is also developed.

In order to expand its business activities in the United States and substantially improve its biodiesel market share, REG is working on acquiring biorefining companies specializing in biodiesel production. As a result, in August 2015, Renewable Energy Group and Imperium Renewables signed an asset purchase agreement, where REG acquired almost all the assets of Imperium, including a 100 million gallon nameplate capacity biomass-based diesel refinery [44].

The Imperium's Hoquiam facility was renamed REG Grays Harbor, located in Hoquiam, Washington. This takeover marked the end of the Seattle-based Imperium Renewables, which was founded in 2004 and couldn't survive the impact of severe financial crisis [45]. In March 2016, Renewable Energy Group completed a transaction with Sanimax Energy to acquire its 20 million gallon nameplate capacity biodiesel refinery located in DeForest, Wisconsin [46].

8.3 Bioeconomy in Europe

After the first volume published in 2011 in its document entitled "Bio-based economy For Europe: State of play and future potential—Part 2," the European Commission summarized 35 positions reports on matters related to bioeconomy and developed strategic visions about fostering the implementation of bioeconomy in Europe and the potential benefits and risks [47].

Overall, by fostering bioeconomy, Europe is endeavoring to:

- Rationalize the exploitation of resources and optimize the land-use efficiency, thus creating a more sustainable primary production.
- Contribute to the European and global food security by developing more efficient agricultural practices. The sustainable food chain would avoid the competition between food and non-food use of biomass and improve animal health and welfare.
- Develop the European science and technology sector by integrating new structures between European research organizations and funding bodies (governmental and private). The aim is to promote research and innovation excellence in the

continent and assert Europe's leadership through knowledge and technology transfer.
- Reduce the carbon dioxide emissions, and implement a low-carbon economy.
- Generate new high-skilled jobs.
- Induce an economic and employment stimulus to rural and regional development.
- Build a competitive bio-based industrial sector by creating and supporting new business opportunities with higher potential for value creation through cascading use of biomass and reuse of the generated wastes.

The last strategic objective related to the industry is definitely the most important goal regarding the implementation of biorefinery as it encompasses, directly or indirectly, all the other objectives. Overall, the European global market leadership will be commensurate with the degree of implementation of bioeconomy in the continent's industrial sector and the novelty, efficiency, and cost-effectiveness of its biorefining technologies and processes in the one hand and competitiveness of the generated bio-based products, in terms of both quantity and quality, on the other hand.

In this regard, many European experts are emphasizing the importance of incorporation industrial biotechnologies such as green and white biotechnologies in bioeconomy implementation scenarios and the need to substantially invest in such sustainable technologies [48, 49].

A special attention to white biotechnology is clearly distinguished in European scientific and technological circles and research-based industries mainly in the energy, chemicals, and pharmaceuticals sectors [50–52]. This technology, based on the use of enzymes and microorganisms, has promising potentialities to produce various bio-based products from natural resources and agro-industrial and municipal wastes at competitive costs, including biofuels, biochemicals, food and feed, paper and pulp, and textiles [53].

8.3.1 Strategic Visions in Europe

8.3.1.1 European Biorefinery 2030 Vision

The European Biorefinery 2030 Vision is a strategic vision presented and detailed in a report elaborated by the members of the Star-COLIBRI project, entitled "Strategic Research Targets for 2020—Collaboration Initiative on Biorefineries." This project was a 2-year coordination and support action (from November 1, 2009, until October 31, 2011), funded by the European Commission through its Seventh Framework Programme, and involving five European technology platforms and five European research organizations [54].

Such roadmap document was prepared in order to provide European policy and decision makers with a set guidelines, information, and tools considered to be necessary for the development of a sustainable European bioeconomy, with a special emphasis on the need to implement a wide network of biorefineries throughout the continent.

The achievement of this vision within the specified time frame was deemed to be a vital objective enabling the establishment of European bioeconomy on solid ground and then its gradual evolution toward maturity. During the coming decades, the experts who developed this vision are underlining the necessity to promote the development and commercialization of sustainable biomass-processing technologies. In this context, the assessment of key driving forces, opportunities, and challenges shaping the establishment biorefineries was carried out, along with the contribution of the related industrial activities in the development of bioeconomy until 2030.

Overall, the consortium of the Star-COLIBRI project has elaborated a European biorefinery vision for 2030 by focusing on two main issues:

1. Developing and integrating different industrial activities within the concept of sustainable bioeconomy
2. Promoting and reinforcing the role of biorefineries within bioeconomy by addressing the various challenges related to the implementation (designing and building) and the technological aspects (process operations)

8.3.1.2 National Perspectives

After presenting the European vision about the implementation of bioeconomy on the ground via promoting biorefineries and other sustainable industrial activities, and considering the importance of the local and decentralized dimension of bioeconomy, it is worthwhile to study the visions and perspectives about such implementation from different European countries actively involved in bioeconomy and sustainability.

Finland

The Nordic country has all the prerequisites to become one of the leading countries in bioeconomy thanks its knowledge-based economic system, thriving industrial activities and especially the high availability of renewable natural resources. Indeed, the economy of Finland has long been based on the profitability of added-value commodities produced from biomass, mainly from the abundant forest resources. Currently, while the production of traditional crops is about 6 million tons per year, the annual growth of the biomass derived from forest in Finland is around 56 million tons of dry biomass, which shows that the country could easily and reliably base its bioeconomy strategy on forest resources [55], as well as other bioresources, mainly organic wastes.

Nonetheless, according to Sitra, an independent Finnish fund commissioned for the promotion of stable and balanced development, economic growth, and international competitiveness, the current management and business strategies related to the production and consumption of bio-based products are still constituting an obstacle that needs to be overcome to ensure the full-scale development and expansion of bioeconomy in the country [56].

In 2014, the Finnish government officially adopted the bioeconomy concept and published its "Finnish bioeconomy strategy" [57]. The strategic objectives of this strategy are (1) the establishment of a competitive operating environment and creating new businesses from bioeconomy, (2) the achievement of a strong competence base, and (3) the improvement of the accessibility to renewable resources and wastes and the sustainability of the related biomass supply chains.

Thus, the main objective could be summarized in the development of new bioeconomy business opportunities in Finland based on the sustainable production and utilization of biomass in the one hand and the improvement of related technologies, high added-value products, and services on the other hand.

Germany

With the "National Research Strategy BioEconomy 2030", published in 2011, the German Federal Government is laying the ground for its bioeconomy by formulating its strategic visions to implement this sustainable economic model by 2030 [58]. It includes providing global food supplies and producing added-value bioproducts from renewable resources through a knowledge-based and sustainable bioeconomy and based of the highly competitive German industries and skilled workforce. By achieving its bioeconomy vision, Germany is aiming at becoming a leading research and innovation center in bioeconomy.

In this regard, the Bioeconomy Council of Acatech (the German national academy of science and engineering) highlighted three priorities that need to be addressed by the German government in order to achieve its national strategic vision about bioeconomy [59], including breeding new crop plants and farm animals, developing more efficient cultivation technologies while reducing harvesting losses, and rationalizing the utilization of arable lands for a more sustainable exploitation. From those three priorities, it is clear that Germany is currently focusing on increasing its overall biomass resources, thus establishing a solid and reliable renewable feedstock supply chain for the biorefineries.

Considering the importance of biorefineries in the implementation of bioeconomy, the federal government published in 2012 a document entitled "Biorefineries Roadmap" detailing its action plans for the processing of renewable natural resources and wastes into fuels, materials, and chemicals [60]. The report started by illustrating the current status of biorefineries in Germany and the various categories of biorefining facilitates in the country such as the sugar/starch biorefinery in the federal state of Saxony-Anhalt, the lignocellulosic biomass-based biorefineries in Bavaria and Saxony-Anhalt, and the grass-based green biorefineries in Brandenburg and Hesse. Besides, the scientists and experts involved in this study highlighted the main challenges facing the biorefinery sector in Germany including:

1. The shortage of raw materials for first-generation biorefineries due to the increasing demand for sugar and starch from the food industry.
2. Difficult position for the German industries using sugar and starch as raw material to compete with the global players in the sector, mainly Brazil, the United States, and Southeast Asia.
3. Limited cooperation between sugar- and starch-based industries and the chemical industry.
4. The products' portfolio of German biorefineries is not sufficiently diverse, and further processing and refinement of intermediate products into new added-value bioproducts is highly required.
5. The integrated production of bioproducts and bioenergy needs to be expanded.

Netherlands

In 2007, the Dutch government established a new strategic agenda based on the optimal valorization of biomass. It emphasized the need to prioritize the production of added-value products including biofuels and biomaterials from biomass, especially residual wastes, and the generation of heat and power from all the available low-cost biomass [61]. Despite the stagnation period caused by the economic crisis, the bioeconomy concept is gaining ground in the Netherlands, largely in the bioenergy sector and to a lesser degree in the field of added-value and specialized bioproducts. Such tendency is explained by necessity to meet the increasing energy demand in the country. Indeed, while the share of the fossil fuels is expected to decrease over the next decade (oil and natural gas, respectively, fulfilled 38% and 47% of the country's total primary energy requirement in 2010), the demand for both fuels will increase simultaneously with the increase of the total energy demand [62]. Under the circumstances, the Netherlands is planning to increase the share of renewable energy from 4% (2010) to 7–10% by 2020 [63].

Although the strategy to channel more raw materials toward the bioenergy sector is well justified, especially that the related European target is 14% by 2020, some Dutch experts are concerned about this unbalanced implementation of bioeconomy, which favors bioenergy (biofuels, heat, and electricity) over other high added-value bioproducts [64].

In order to stimulate a new sustainable, innovation-based green economic growth, the Dutch government is currently focusing of speeding up and facilitating the transition to bioeconomy by developing the country's asset and analyzing and addressing the challenges related to the implementation of the new economic model. During this transition phase, and in anticipation of the substantial surge in the demand for renewable natural resources, the Dutch Ministry of Infrastructure and the Environment commissioned the PBL, the Netherlands environmental assessment agency, and CE Delft, an independent research and consultancy organization, to conduct a scan analysis to assess the current situation and future scenarios related to this development in biomass requirement [65]. Overall, this study focused on analyzing two main issues: (1) the balance between feedstock demand and potentially available supply of biomass and (2) the large-scale production chains of bio-based products and their impact on greenhouse gas emissions and potential land-use changes to meet the required supply of raw materials by the involved industries.

The prosperous agricultural and biotechnological sectors in the Netherlands [66], and the adhesion of many science-based companies such as Royal Dsm and AkzoNobel, as well as large corporations such as the Royal Dutch Shell to the concept of sustainability, are a major asset for the Dutch bioeconomy [67]. Furthermore, many Dutch experts believe that the key for the development of bioeconomy in the country is a close collaboration between the industry and agriculture. In this context, AkzoNobel, Dutch leading chemical company, and Royal Cosun group, an agro-industrial corporation, developed a partnership to produce new cellulose-based products from the side streams of sugar beet processing [68].

United Kingdom

Although the United Kingdom did not yet develop its national definition for the bioeconomy, the concept itself is well recognized and promoted by the British government [69]. In general, the term is in line with the definition set by the European Commission. Nonetheless, specific definitions were reported in some governmental documents such as the policy paper UK Bioenergy Strategy (2012) where bioeconomy was termed as "the set of economic activities relating to the invention, development, production and use of biological products and processes" [70].

In 2014, the Science and Technology Select Committee, appointed by the House of Lords, published a report entitled "Waste or resource? Stimulating a bioeconomy" where the concept of bioeconomy was referred to as the "use of biological feedstocks or processes involving biotechnology, to generate economic outputs in the form of energy, materials or chemicals" [71].

In 2015, a new report was published by the British government entitled "Building a high value bioeconomy: Opportunities from waste," in which the UK bioeconomy was assessed to be worth £100 billion a year [72]. The report defined bioeconomy as "the economic activity derived from utilising biological resources or bioprocesses to produce products such as food, energy, and chemicals." The strategic objectives of the United Kingdom, developed in this report, for a successful implementation of bioeconomy, included:

- Decreasing the dependence of British industries in the chemical and energetic sectors on fossil raw materials, thus gradually decoupling United Kingdom's economic growth from the use of depleting resources
- Reducing greenhouse gas emissions

- Building advanced biorefining facilities combining decades-long industrial expertise and innovative R&D breakthroughs
- Supporting the expansion of bioeconomy in the country
- Increasing the number of commercial-scale biorefineries using wastes as feedstock
- Positioning the country as a location of choice for global investment in bioeconomy
- Making the United Kingdom a leading exporter of bio-based process technologies and business models

The strategic report also highlighted the economic potentialities of bio-based industries and pointed out the measures that the British government (or any other government for that matter) would take to assist in the implementation of bioeconomy and its subsequent expansion, including:

1. Working with businesses, industry, and local authorities to speed up and facilitate the transition phase toward bioeconomy
2. Supporting every sustainable and efficient procedures to exploit, use, and reuse renewable resources
3. Promoting innovative R&D efforts related to biomass valorization and encouraging more investments in biorefineries
4. Educating and training a highly skilled workforce at all professional levels

The experts who prepared this policy paper specifically emphasized on the opportunities of converting residual waste into added-high-value products. In this regard, it was estimated that substantial amounts of waste feedstock are potentially available for processing in the country. This includes at least 100 million tons per year of organic waste, as well as 14 million tons of residues from the agricultural and forest sectors every year. The conversion of 25 million tons of those wastes generated annually in the United Kingdom was reported to theoretically yield around 5 million tons of bioethanol, valued at about £2.4 billion.

8.3.2 Governance and Coordination

By adopting and promoting the concept of sustainable bioeconomy, the European Commission is setting ambitious economic, environmental, and societal goals for the continent. In the beginning of this transition process, many European experts criticized the practical measures taken for the implementation of bioeconomy, often judged to be unsatisfactory and based on uncoherent and inconsistent policies [73].

Indeed, despite the presence of many political, environmental, and scientific action plans and initiatives in Europe including the eco-innovation and biomass action plans [74, 75], many technology platforms (ETPs) [76], and several EU strategies for biofuels, life sciences and biotechnology, climate change, waste management and sustainable development [77], the coordination between these policies and action plans, all aiming at promoting sustainability and bioeconomy, remained unsatisfactory, especially that the European bioeconomy is roughly valued at €2 trillion and provides around 22 million jobs to EU citizens. Hence, the European Union has an arsenal of bioeconomy-related policies that needs to be integrated and coordinated by the European institutions and the authorities of member states, thus providing stakeholders with a more coherent policy framework and encouraging private investments [78].

In this context, the European Bioeconomy Alliance (EBA or EUBA), which is an alliance of various European industrial corporations and associations actively involved in the bioeconomy including BIC (Bio-Based Industries Consortium), CEFS (European Association of Sugar Producers), CEPI (Confederation of European Paper Industries), COPA/COGECA (European Farmers/European Agri-Cooperatives), ePURE (European Renewable Ethanol Producers Association), EuropaBio (the European Association for Bioindustries), EUBP (European Bioplastics), FEDIOL (the EU Vegetable Oil and Proteinmeal Industry), and FTP (Forest-Based Sector Technology Platform), was formed [79].

EBA called, during its launching at the European Parliament on February 4, 2015, for more predictable policies and long-term strategies paving the way for a more sustainable and competitive bioeconomy in Europe. In order to become a leader in bioeconomy, the alliance deemed necessary for European authorities to provide an integrated, coherent, and harmonized framework in various policies related to key sectors in bioeconomy such as energy, agriculture, forestry, industry, environment, R&D, and regional development [80].

In this regard, four major fronts were emphasized by the European bioeconomy alliance including:

- The execution of the priority recommendations prepared by the Ad Hoc Advisory Group for bio-based products in the framework of the Lead Market Initiative [81]. Such recommendations are expected to enable economic recovery and growth, as well as creating new markets and job opportunities.
- The implementation, by member states, of measures aiming at increasing the sustainable production of biomass in the agricultural and forest sectors and facilitating its supply chains.
- The removal of obstacles facing private investors in commercial operations, mainly the construction of biorefineries in Europe, and the endorsement of partnership between the public and private sectors involved in bio-based industries.
- The support and promotion of bioeconomy-related debates between policy makers, industrialists, farmers, scientists, and civil society at European, national, and regional levels. The active interacting of all those key players will help developing a more competitive and sustainable bioeconomy in Europe [82].

On a related matter, a study conducted at the Lund University's International Institute for Industrial Environmental Economics (IIIEE) reported three major governance challenges facing the still emerging European bioeconomy including (1) the important role of public-private networks; (2) city regions as drivers for the

KBBE, especially through climate governance; and (3) consumer citizens and NGOs as key players in the development of bioeconomy [83].

8.3.3 Resources and Potentialities

The successful implementation of the bio-based economy primarily relies on biomass availability. Thus, the development of biorefineries, processing non-food biomass, is directly influenced by a range of factors such as the potentialities of agricultural and forestry production systems and waste management and the increasing use of biomass for energy (heat and power generation).

Clearly, food and feed production have to be given top priority when elaborating strategies and action plans for biomass use. Nonetheless, improvements in agricultural productivity, land management, logistics, and storage techniques can increase the efficiency of the food and feed chain and thus make more lands or agricultural by-products available to be used as feedstock for the various categories of biorefineries.

Overall, the ethical conflict of food vs. non-food biomass is unlikely to occur in Europe if the productivity of the agricultural sector can be increased and large areas of marginal lands can be recovered and valorized, mainly via innovative scientific and technological breakthroughs and supporting agricultural policies. Nonetheless, the main issue regarding the biomass supply is expected to come from within bioeconomy through the competition of the various bio-based industries for specific bioresources with richer biochemical composition and easier accessibility.

In this regard, many European experts are focusing on this issue in order to identify the priorities for the available bioresources among all the competing users, including bio-based industries producing biofuels, biochemicals, and biomaterials, as well as facilities generating heat and electricity for biomass [73]. Prioritizing the use of biomass in bioeconomy is of special importance when the raw materials are imported; such is the case for various vegetable oils, pellets, wood chips, etc. [84, 85].

Regarding the biomass potential in Europe, several categories on natural resources are potentially available for the various European bio-based industries. This includes:

- *Dedicated energy crops* including many perennial grasses such as miscanthus, switchgrass, and giant reed [86] and short-rotation crops essentially cultivated for the production of sugar (sugarcane, sweet beet, and sweet sorghum [87]), starch (corn, wheat, barley, rye, potato, etc. [88]), and oil (rapeseed, soybean, sunflower, jatropha, etc. [89]), as well as lignocellulosic crops (mainly willow and poplar trees [90]).
- *Forest biomass* including woods [91, 92] and forest residues generated during logging activities [93] and forest thinnings [94].

- *Industrial by-products and residues* mainly produced by forest and food industries, as well as other agro-industrial facilities. These kinds of residual materials include dry lignocellulosic biomass such as sawdust, wood chips, barks, husks, kernels, etc. [95, 96], or wet cellulosic materials such as sugarcane bagasse [97]. Black liquor, a residual by-product of the sulfate pulping process loaded with dissolved lignin and hemicellulose, currently burnt in recovery boilers to generate heat, could be further fractionated to recover platform chemicals and also processed to produce added-value bioproducts [98, 99].
- *Agricultural residues* consist of a wide range of postharvest or postindustrial processing plant materials including cereal straws, vegetable peels, corn stovers, stems and cobs, etc. These low-cost residual bioresources were extensively investigated in European research institutions as potential feedstocks for the production of added-value chemicals [100], biofuels [101], and biomaterials such as cellulose nanofibrils [102] and biochars [103].
- *Livestock wastes* including wet animal manure principally valorized in biogas production, either exclusively [104] or co-digested with other biomass [105]. Dry manure such as poultry litter can be used valorized in European thermal power plants [106], with high energy content and the generation of an ash by-product with good fertilizing properties [107].
- *Aquatic biomass* including the cultivation or harvesting of algae for the production of polysaccharides or oil-derived biofuels and other added-value nutritional, chemical, and pharmaceutical products [108–110]. So far, although a very interesting biomass, the related costs are still high especially as feedstock for energy. Thus, for this kind of biomass, the orientation toward their sustainable conversion into speciality chemicals with high added value is highly recommended [65].

Furthermore, the European biomass potential can be supplemented by importing biomass (natural resources or wastes), either within or from outside the EU, mainly North America, South America, Africa, and Russia [111].

While analyzing the various reports on biomass potential in Europe, an orientation toward the bioenergy sector is very obvious. Indeed, most related studies and assessments are estimating the current and potential biomass availability for bioenergy production, and to be more specific for biofuel production and, to a lesser extent, for heat and electricity generation. Such tendency clearly indicates the strategic position of the energy issue in the European economy and its thriving industrial complexes. Furthermore, it shows the urgency of securing a sustainable energy supply and providing reliable alternatives to the depleting petroleum reserves and increasing the share of bioenergy in the European primary energy supply.

In this context of biomass for energy, the Atlas of EU biomass potentials, published in 2012, estimated that the potentialities of biomass in Europe for bioenergy in the year 2010 was 314 million tons of oil equivalent (Mtoe) [112]. This unit is used to equal the amount of energy released when burning a million tons of crude oil.

Then two scenarios were developed to estimate the potentialities of bioenergy resources. For the first reference scenario, the biomass potentials are expected to increase to 429 Mtoe in 2020 and slightly decrease to 411 Mtoe by 2030. Under the second sustainability scenario, these potentials are predicted to decrease to 375 Mtoe and 353 Mtoe by 2020 and 2930, respectively. The following Table 8.3 gives more details about the estimations in the Atlas of EU biomass potentials with respect to various bioresources and under the two prediction scenarios.

Table 8.3 Potentialities of various bioenergy resources in Mtoe under reference and sustainable scenarios [112]

Biomass category	Availability in 2010	Estimation in 2020 (reference scenario)	Estimation in 2020 (sustainability scenario)	Estimation in 2030 (reference scenario)	Estimation in 2030 (sustainability scenario)
Agricultural residues (straw, manure, etc.)	89	106	106	106	106
Forestry residues (logging, wood-processing residues, construction wastes, etc.)	66	101	79	97	74
Roundwood production	57	56	56	56	56
Wastes (organic municipal wastes, sludge, used fats and oils, etc.)	42	36	36	33	33
Additional harvestable roundwood (i.e., more harvesting of stem wood within sustainable limits)	41	38	35	39	36
Rotational crops	9	17	0	20	0
Landscape care wood (residues from landscaping activities)	9	15	11	12	11
Perennial crops	0	58	52	49	37
Total	314	429	375	411	353

8.3.4 *Industrial Study Cases: The Finnish Experience*

For many decades, the economic success of Finland was mainly related to its forest sector, which continued to prosper with the availability of raw woody materials at low costs, in the one hand, and the increase in the global demand for forest products, on the other hand. The incorporation of new and more efficient technologies for the harvesting and processing of trees and woods, along with the special status of the forest in the Finnish policy making, substantially promoted this sector [113, 114].

Nonetheless, with the increased competition coming mainly from Latin American and Asian countries, which are benefiting for the introduction of new technologies along with significant wood and labor cost advantages, the Finnish forestry was severely challenged. Indeed, with the increasing competition, the prices of forest products started to decrease. Besides, during the same period (roughly the last couple of decades), the forest businesses have been only marginally invested and did not receive stimulus packages [115, 116].

In order to face such serious challenge, several industrial companies in the forest and paper mills sectors started implementing survival strategies. Related actions included many mergers in the Finnish industrial sector such as the merger between UPM and Kajaani Corporation and the one between Kymmene Corporation and Repola Ltd. and its subsidiary United Paper Mills Ltd., which formed in 1996 the UPM-Kymmene Corporation [117]. Other measures included acquisitions and belt tightening which were taken in order to consolidate the assets of the companies and reduce the operating costs [118, 119].

After this challenging episode, the Finnish forest cluster started adopting vanguard and bold strategies through increased investments, wider products' portfolio, and new business models, in which the sustainable production of lignocellulosic biofuels and various wood-derived bioproducts was one of the key orientations of those Finnish industries to diversify their business activities and explore new markets: It was basically a strategic objective for the Nordic country to generate economic growth and pioneer in bioeconomy [120]. One of the more illustrative changes is the integration of the biorefining technologies by the Finnish pulp and paper industry [121].

Several companies in the forest and energy sectors are currently operating in Finland or from Finland to speed up the implementation of bioeconomy in the country and to place it in a leading position globally. In the following sections, selected companies and industries are showcased in order to give a perspective on the Finnish approach to implement bioeconomy while benefiting from the country's abundant forest resources and the combination of decades-long industrial expertise in biomass conversion with well-tailored education and training curricula, as well as a dynamic R&D sector.

8.3.4.1 UPM-Kymmene Oy (UPM)

UPM-Kymmene Oy is a Finnish corporation in the forest industry specialized in wood processing and the production of pulp and paper. Currently, the company contains six business areas: UPM Biorefining, UPM Energy, UPM Raflatac, UPM Specialty Papers, UPM Paper ENA, and UPM Plywood [122].

Overall, the new businesses developed and implemented by UPM to fulfill its Biofore strategy have one major objective: reconciling profitability with sustainability, which is really at the core of the bioeconomy concept.

In this context, the Finnish company planned the success of its strategy, especially increasing the profitability of its biorefining activities, on several competitive advantages including [123]:

- A steady raw material supply including the company's own resources and crude tall oil which is a by-product from the pulp production process. Indirect land-use change impacts of biofuels.
- Avoiding both the food chain value and the indirect land-use change generally linked with biofuels production by other competing companies and in other countries. For instance, croplands cultivated for the production of biofuels were reported to increase greenhouse gases through the emissions from the direct and indirect land-use change in the Europe and the United States [124, 125].
- Efficient production processes based on innovative technologies, sustainable use of feedstocks, integrated synergies, and appropriate industrial infrastructure.
- The production of high-quality biofuel products with reduced carbon footprints. An 80% reduction in greenhouse gas emissions was estimated [126].

The creation of new business structures, especially UPM Biofuels, was decided by the company in order to diversify its portfolio and be the main component in the execution of its new vision about the transformation of the modern forest industry, the so-called Biofore strategy [127]. To implement this strategy, UPM promoted its industrial biorefining activities, focusing on pulp and timber, by starting the production of biodiesel for lignocellulosic wood in the world's first commercial-scale biorefinery located in Lappeenranta, Finland [128].

The construction of this UPM biorefinery started in the summer of 2012 at the company's Kaukas mill site in Lappeenranta, located in the southeastern Finland, about 220 km from the capital Helsinki where UPM has its headquarter. The related investment amounted to €175 million for an annual production of 120 million liters of biodiesel [129]. The biorefinery became operational in the early 2015, and the production of renewable biodiesel was based on UPM's innovation process BioVerno, developed in the company's Biorefinery R&D Centre in Lappeenranta.

The feedstock used to produce BioVerno diesel is crude tall oil, a residue of pulp production, and a large fraction of this raw material comes from UPM's own pulp mills in Finland. The produced biodiesel can be used in conventional diesel engines without modification, as such or blended with regular diesel. Currently, UPM's renewable biodiesel is available at St1 and ABC gas stations in Finland [130].

8.3.4.2 Green Fuel Nordic Oy (GFN)

Green Fuel Nordic Oy is a Finnish biorefining company established in October 2011 and is headquartered in Kuopio. The company is basing its biorefining activities on Envergent' RTP™ process (Rapid Thermal Processing), described in details in Chap. 7. Thus, in October 2011, Envergent Technologies announced a memorandum of understanding, which was signed with GFN, under which the American and Finnish companies would collaborate on projects to convert biomass to renewable fuel for use in district heating systems in Finland [131]. The strategy of GFN to utilize an existing technology (industrial-scale fast pyrolysis) was mainly adopted to benefit from the proven efficacy and applicability of the RTP technology to convert biomass and to accelerate the start-up of biorefining operations and commission the production and distribution of commercial grade bio-oil, a renewable second-generation liquid fuel [132].

The main goal of the company is definitely to make profit from the use of low-cost and locally available feedstocks to produce refined bio-oil at commercial scale. At a national level, it will contribute, alongside with the other industries in the bioenergy sector, in the effort to reach EU's renewable energy sources directive target (RES) for Finland, estimated at 38% renewable energy of final consumption in 2020 [133].

To achieve these objectives, GFN is proposing a six-point strategy, valid or any industry involved in the biorefining sector, which includes [134]:

- Ensuring a controlled and profitable growth
- Developing more effective implementation scenarios for biorefinery project in collaboration with technology suppliers and markets
- Advancing the operational know-how of biorefineries
- Establishing a competitive and sustainable feedstock supply chain
- Providing competitive alternative to industrial energy production and building strategic and long-term partnerships
- Developing and commercializing new applications for liquid biofuels

8.3.4.3 St1 Oy

St1 is a private Finnish company operating in the energy sector though two business groups: St1 Nordic focusing on fuel marketing activities in Finland, Sweden, and Norway and on renewable energy solutions (bioethanol and wind power) and St1 Group targeting refinery operations. The company, headquartered in Helsinki, was founded in 1995 under the name Greenergy Baltic Oy. It has a total turnover in came to 6.6 billion euros (2014 annual report), employs 700 personnel, and owns the chain of St1 service stations in Finland, Sweden, Norway, and Poland [135].

In 2006, St1 Biofuels Oy was established as a subsidiary of the group St1 Nordic Oy. The objective was to create a sustainable bioethanol production concept that could be utilized widely and replace fossil fuels in a profitable and sustainable way.

Thus, the Finnish company built several plants that would give St1 Biofuels a pioneering position in the production of bioethanol fuel from waste-based feedstock, along with its expertise in biochemical processes, technology development, and engineering [136]. The orientation toward the production of bioethanol from wastes and residues, for the transportation sector, offers an interesting opportunity for the company to compete on a national and European level and benefit from its own distribution chain (retail stations in several European countries), which could be expanded with the introduction of new kinds of eco-friendly and renewable biofuels in the company's portfolio.

In 2013, and within the Finnish commitment to the European renewable energy directive, St1 targeted the construction of a network of biorefining plants with a total production yield of 300 million liters of residue-derived biofuels by the 2020s [137]. Currently, the Finnish company, pioneering in waste-based bioethanol production, has five plants in Finland, all plants produce bioethanol from residual biomass, four of which are Etanolix® plants, processing food industry residues rich in starch and sugar, and one Bionolix® facility using municipal and commercial biowastes as feedstock [138].

For instance, these St1 facilities include the Hämeenlinna Bionolix plant using local municipal biowaste and integrated with a biogas plant operated by biowaste from household, retail, and industries as feedstock, thus combining the production of bioethanol and bioelectricity, along with stillage as a by-product. Another location is Lahti Etanolix, which uses brewery, bakery waste, and bread waste as feedstock to produce bioethanol and also a liquid animal feed as a by-product. In the Hamina dehydration plant, the hydrous ethanol produced in the other St1 Biofuels' facilities (both Etanolix and Bionolix plants) is dehydrated, as well as ethanol from third-party producers. The generated biofuel is 99.8% of bioethanol [139].

Lately, a strategic objective of St1 was set to build three to five more facilities for the production of cellulosic biofuels, in addition to the already operational plants using food industry residues (Etanolix®) and biowaste (Bionolix®). Thus, the Finnish company is upgrading its biorefining activities with the Cellunolix® process. Indeed, on behalf of North European Bio Tech Oy (NEB), St1 begun designing, issuing permits, and coordinating with feedstock suppliers (Alholmens Kraft and UPM) to build a sawdust-based ethanol plant with a production capacity of 50 million liters of bioethanol per year in the Alholma industrial area in Pietarsaari, Western Finland. According to the agreement between St1 and NEB, St1 Biofuels Oy will operate the Cellunolix plant, using sawdust and recycled wood as feedstocks, and carry out the production. The produced bioethanol will then be leased to North European Oil Trade Oy (NEOT), which will engage it in its wholesale trade [140]. NEOT itself is a company owned by the Finnish retailing corporation SOK (51%) and the St1 Nordic group (49%) [141].

Currently, St1 is operating the 10 million litre Cellunolix facility, delivered to NEB in Kajaani, Finland. The biorefining plant produces bioethanol from sawdust and is expected to reach full production speed during 2017, with the possibly to build a new Cellunolix bioethanol plant for NEB also in Kajaani, with a production capacity of 50 million liters of bioethanol per year [142].

St1 has also new projects to build Etanolix and Cellunolix facilities outside Finland. The first Etanolix plant delivered to the international market was built in Gothenburg, Sweden, and inaugurated on June 5, 2015 [143]. Currently, St1 is planning to construct a Cellunolix ethanol plant in Hønefoss, Norway, in cooperation with the Norwegian forest corporation Viken Skog and its subsidiary Treklyngen. The planned production capacity of this biorefining facility is 50 million liters of bioethanol for transportation sector, using local residues from the forest industry as feedstock [144].

8.4 Bioeconomy in China

The biomass refining concept in China is dating back to ancient Chinese civilizations including various processing methods to extract proteins and oil from soybeans, to ferment carbohydrates into wine, and to make fertilizers from biowastes in order to improve the production yields of cultivated crops [145]. More recent historical examples of biorefining activities in China relate to the construction of industrial-scale hydrolysis and batch ABE (acetone-butanol-ethanol) fermentation plants, mainly for the production of acetone and ethanol solvents, between the 1960s and 2000s. The feedstocks used in these plants were corn, cassava, potato, and sweet potato. However, the overwhelming competition from the petrochemical industry providing solvents at cheaper cost accelerated the demise of these facilities at the turn of the twentieth century [146].

To meet with the increasing demand for energy of its fast-growing economy, China overtook the United States as the world's largest importer of crude oil since April 2015, with imports approximating 7.4 million barrels per day, which is roughly 200,000 more barrels per day more than the United States [147]. Currently, the most populated country on earth is facing serious challenges mainly related to energy shortage and environmental pollution [148, 149]. In this difficult context, the Chinese government started by adopting measures to rationalize the use of fossil fuels (mainly local coal resources and imported petroleum) such as the G20 directive to end fossil fuels subsidies by 2020 [150]. The dilemma between economic growth and pollution is certainly not restricted to China, but the amplitude of such problem and its impact on the future of the country is of distinctive significance. The same dilemma, under near equal proportions, is also affecting India and its 1.3 billion pollution and ambitious objective of expanding GDP by 8% each year [151].

Considering all those issues, China has an urgent and strategic need to implement the bioeconomy concept in order to start a long process of upgrading its agricultural and industrial sectors by adopting more innovative and sustainable cultivation and production systems while ensuring an improved social development and more protected environment.

8.4.1 Strategic Vision in China

In China, the gradual process toward sustainability is referred to as the Green Revolution, which is basically a Chinese interpretation of bioeconomy mainly focused on matters related to energy and environment, as illustrated in the 12th five-year plan (FYP) [152].

Although China still did not publish a national bioeconomy strategy per say, but the 11th and 12th five-year plans provided enough information to get a clear perception on the strategic vision of China to implement the sustainable bioeconomy concept. Noting that terminology wise, the Chinese government used the term "bioeconomy" in its 11th plan for national strategic emerging industries (2006–2010), and used the term "bio-industry" more frequently, and sometimes instead of bioeconomy, in the 12th plan (2011–2015).

China made some success in the 11th FYP period (2006 to 2010), during which the country made promising progress in meeting its energy-saving and emission-reduction objectives. This constituted the first step in the implementation of policies developed to achieve the Chinese 2020 objective of increasing the share of renewable energy in the primary energy supply from 7% in 2010 to 11.4% in 2015 and 15% by 2020 [153, 154]. Besides, according to the International Renewable Energy Agency (IRENA) and based on projections from the Chinese Renewable Energy Center (CNREC), the share of renewable energy source (excluding traditional uses of biomass) could increase in China up to 16% by 2030 [155].

As the national congress approved the12th FYP on National Emerging Industries of Strategic Importance, the Chinese government started focusing and investing in seven industrial sectors deemed of strategic importance for the future of the country. The list included [156, 157]:

- *Energy-saving and environmental protection* with the promotion of an energy-efficient industry, an advanced environmental protection industry and a resource recycling industry
- *Information technology* with the development of next-generation information network industry, fundamental industry of core electronics, along with high-end software, and new information service industry
- *Biology and biotechnology* involving the biopharmaceutical, biomedical, bio-breeding, and bio-manufacturing industries
- *High-end equipment manufacturing* in the aviation, marine, and rail transportation equipment industries, as well as in the satellite and intelligent equipment-manufacturing industries
- *New energy industries* by developing more mature nuclear power technology, wind power, solar photovoltaics, thermal utilization, biomass electricity generation, and methane gas, thus promoting the industrialization of renewable energy technologies
- *New material industries* for the production of new functional materials, advanced structural materials, and high-performance composite materials industry

– *New energy automobiles industries* in which innovative and advanced technologies will be promoted including high-quality power batteries and electric motors, as well as supporting the industrialization of the new energy automobile industry

Lately, the Chinese economy started decelerating from the double-digit growth that characterized its earlier development during the 10th and 11th FYPs (average of 10.5% between 2000 and 2010 [158]).The main cause is the high spending and investment levels, especially the stimuli during the global financial crisis, which led to substantial local government debts, but still at a manageable level according to experts [159], in addition to an overcapacity in the industrial sector, particularly China's heavy industrial sector which provides 60% of the country's electricity demand [158].

Overall, the 13th FYP has set a, still-belated, growth target of 6.5% per year. The bulk of this growth is intended to come from services, which planners hope will constitute 56% of the economy by 2020. This restructuration of the Chinese economy is expected to enable the Asian country to reach its goal of cutting the energy consumption and carbon emissions per unit of GDP (carbon intensity) by 15% and 18%, respectively, during the period of the 13th FYP [160]. Thus, China has to substantially reduce its fossil resources, mainly coal, and promote clean and renewable energy alternatives. In this regard, the FYP is confirming a 15% share of nonfossil fuels in electricity generation by 2020, which corresponds to a total installed capacity of 250 GW of wind power and 150 GW of solar power, along with and an expansion of hydropower to a total of 350 GW by 2020 and that of nuclear power to 38–49 GW. As for biomass, an increase by 30 GW is also targeted during the periodic of the 13th FYP [161]. What are then the potentialities of biomass in China and its position in the country's strategic vision to ensure a steady economic growth in a sustainable manner?

8.4.2 *Biomass Resources in China*

China has diverse and abundant biomass resources approximating 0.4 billion tons per year, which, in bioenergy production, can meet 10% of the total energy consumption of the country, if valorized in totality [162]. On an annual basis, China is able to make available around 300 million tons of crop straw residues (600 million tons in total) and roughly the same amount of wastes from the forest sector (900 million tons in total), which are easily available for biofuels production [163]. Besides, the municipal solid wastes generated in the most populated country in the world are expected to reach 210 million tons per year by 2020. Thus, it is estimated that China could produce 2 to 10 billion m^3 of methane if 60% of these municipal wastes is valorized in landfill methane utilization throughout the country [164].

Such potentialities of biomass are of strategic importance for China to secure a sustainable supply of energy and future societal and economic development. But before detailing the current biomass resources in China in the next section, a noteworthy remark has to be made. Most of the published literature on the subject of biomass resources, whether from the Chinese government or academia, interestingly relates it almost exclusively to the energy sector. Indeed, the major part of those reports and scientific articles are disusing or presenting data on "biomass energy resources" or "biomass resource for energy." This does not imply that the biorefining activities in China are limited only to biofuels and other bioenergy sources, but obviously directing biomass supply to other valorization scenarios such as added-value fine chemicals and biomaterial is not a priority in China.

Concisely, the world's most energy-consuming country (3101 Mtoe in 2015 [165]) is endeavoring to find a secure and diverse supply of energy to sustain its economic growth while controlling its carbon emissions.

The main biomass resources in China could be categorized into four different groups: (1) agricultural residues, (2) forest residues, (3) animal manure, and (4) industrial and municipal organic wastes (solid and liquid) [166, 167].

8.4.2.1 Agricultural Residues

Agriculture is a vital economic sector in China, particularly in rural areas, despite the fact that it has lost its nationwide leading position to the services and industrial sectors. Indeed, within a decade, the share of workforce in agriculture dropped from 44.8% in 2005 to 28.3% in 2015, while during the same period, the Chinese workforce in service increased from 31.3% to 42.4% [168]. Nonetheless, China remains one of the largest agricultural countries in the world with substantial potential for biomass production.

Overall, China produces a wide array of agricultural food crops including rice, wheat, corn, legumes, tubers, cotton, fiber crops, beetroots, sugarcane, and various oilseed crops such as peanuts, rapeseeds, and sesame. The harvesting and processing of those crops generate various by-products and residual biomass including stalks, straw, husks, and shells [169].

In general, the amount of agricultural residues is estimated based of the crop production, on the one hand, and the respective residue-to-crop ratios, on the other [170]. The following Table 8.4 presents those ratios for several crops cultivated in China.

Various studies presented the potentialities of agricultural crop residues for biofuels and bioenergy (heat and power) production, from various perspectives including theoretical and available amounts, economic value, energy potential, and geographic distribution in China [173–177]. Table 8.5 summarizes the production yields of several crops cultivated in China in 2010, along with the generated residues and estimated energy potential.

Although biomass resources from agricultural residues in China are plentiful, the collection methods are seriously limiting a full valorization of those resources.

Table 8.4 Residue-to-crop ratios for various crops [171, 172]

Cultivated crops	Residue-to-crop ratio
Cotton	3.00–5.51
Rapeseed	1.01–3.00
Corn	1.25–2.00
Fiber crops	1.70–2.00
Sesame	1.01–2.00
Legumes	1.50
Peanuts	1.01–1.50
Wheat	0.73–1.10
Tubers	1.00
Rice	0.68–1.00
Sugarcane	0.10
Beetroots	0.10

Table 8.5 Production yields, amounts of generated residues, and related fuel potential, at coal equivalent, of many agricultural crops in China [178]

Crop	Production yield (mmt[a])	Generated residues (mmt)	Available residues (mmt)	Ratio of coal equivalent (%)	Fuel potential from residues[b] (mmt)
Rice	195.7	215.3	178.7	43	76.7
Maize	177.2	354.5	294.3	53	155.6
Sugar crops	120.1	12.0	10.6	53	5.6
Wheat	115.2	126.7	105.1	50	52.6
Oilseeds crop	32.3	17.9	61.4	53	32.5
Yam	31.1	37.4	29.9	49	14.5
Beans	18.9	37.9	33.4	54	18.1
Cotton	5.9	17.9	16.1	54	8.7

[a]Million metric tons
[b]Fuel potential from residues is calculated by multiplying the available amount of crop residues by the respective ratio of coal equivalent

Indeed, it is estimated that only 23% of the crop residues are used for forage, 4% for industrial production of materials, and 0.5% for biogas production. The major fraction of such residues is either directly combusted by farmers (37%), lost during the harvesting (15%), or discarded or directly burnt in the field (20.5%) [173].

8.4.2.2 Forest Residues

Since the late 1970s, China launched many large-scale projects for afforestation and reforestation in the country, not only for economic reasons and to deal with climate change but also to mitigate the desertification phenomenon which is threatening vast areas of arable lands. These projects, commonly referred to as the "Great Green Wall of China," are targeting the plantation of about 364,217 km^2 of forests by 2050 [179]. Overall, China is characterized by the presence of various kinds of forests, comprising main species in the Northern Hemisphere, along with tropical rain-forests in the south and boreal forests in the north.

In a recent study, regarding the spatial distribution of forests in China, it was revealed that the total forest area is 149 million hectares, which is less than the official estimation from the Sixth National Forest Inventory (169 million hectares). The same study also reported that there are five major types of forest in the country: evergreen broadleaf forests, deciduous broadleaf forests, broadleaf and needleleaf mixed forests, evergreen needleleaf forests, and deciduous needleleaf forests [180].

Forest residues are the biomass generated during forest harvesting and the subsequent wood processing. Thus, it includes wood residues from logging such as stumps, branches, leaves, etc., and wastes from the industrial manufacturing of wood into primary wood products. The wood-processing residues include discarded logs, bark, sawdust, and shavings and are produced, for example, by veneer, plywood, or pulp mills [172]. Overall, among the total forest biomass energy resources in China, three main group of residues are accounting for 81% of the total resources, namely, wood-processing residues (38%), wood felling and bucking residues (37%), and bamboo residues (6%) [181].

Regarding the energy potential of forest biomass (wood fuel and residues), Chinese researchers are estimating that among the 2810 million tons of biomass that could be obtained annually from forests, about 932 million tons could be made available for energy production. Assuming that the totality of firewood and 50% of other forest biomass resource, including residues, could be supplied, China would allocate around 499 million tons of forest biomass to the energy sector, which is equivalent to 285 million tce (ton of coal equivalent) [171] and around 2320 TWh (terawatt hours).

8.4.2.3 Animal Manure

Animal manure in China, including the excrements from livestock and poultry, has a promising potential, especially in rural regions, to be used as feedstock for bioenergy via the anaerobic fermentative production of biogas [182, 183], as well as the conventional valorization scheme of biofertilizers production [184]. In general, the amount of poultry and livestock manures could be calculated by the breeding cycle, daily excretion, and number of livestock, both on hand and sold [185].

The total amount of animal manure in China was estimated to be around 3.99 billion tons, equivalent to 647 million tce. It is mainly composed of pigs manure (0.49 billion tons) and cattle manure (0.14 billion tons). The available amount manure for bioenergy production is estimated at 3.01 billion tons which corresponds to 440 million tons of coal equivalent [171]. During the last years, and based on official data from the ministry of agriculture, China is maintaining an average growth rate of 8% in its animal farming. Thus, with the subsequent numerical increase in poultry and livestock, the amount of animal manure should increase on a yearly basis, along with its share in biogas production in China [186].

In a recent study, a research team from China Agricultural University in Beijing published results about the compositional characteristics and energy potential of Chinese animal manure, by type and as a whole [187]. It was reported that the total methane production potential of the various types of animal manure in China is about 189 billion m^3/year per year, including an annual biomethane yield of 75.5 billion m^3 from pigs manure, 42.84 billion m^3 from beef manure, and 41.36 billion m^3 from broiler manure.

8.4.2.4 Industrial and Municipal Wastes

This group includes municipal solid wastes (MSWs) and industrial and domestic organic wastewaters. MSW are the wastes generated from urban areas (households, markets, small businesses, etc.), periodically collected and dumped in landfills. In China, the daily urban waste emission is around 1.2 kg per capita. On that basis, the amount of MSW is estimated at around 3700 million tons (equivalent to 529 million tce). However, considering the low utilization rate, the municipal solid waste clearance was limited to 152 billion tons, including around 76 million tons of sanitary landfill [171], which would enable the production of 5.87 million m^3 of landfill gas [185].

Table 8.6 presents the energy potential of MSW that could be generated either by direct incineration or combustion of the produced landfill gas, in various regions in China. As shown, a potential electricity supply of 18,862 GWh can be obtained. The highest potential (7064 GWh) was in Eastern China, which corresponds to almost 37.5% of the national total electric power from MSW.

Organic wastewater, on the other hand, can be divided into domestic and industrial wastewater. Domestic sewage is generated from residential areas and urban commercial and service businesses and continuously channeled by the sewage system to municipal wastewater treatment plants to be treated before being discharged into nearby aquatic environments (lakes, rivers, etc.). In China, it was reported that the amount of domestic sewage approximates 31 billion tons. Based on a disposal rate of 42.55%, the available amount of domestic sewage is estimated at 13.2 billion tons, which is equivalent to 2.88 million tce.

Regarding the case of industrial organic wastewaters, they are mainly generated during the industrial processing on various biological feedstocks for the production of commodities such as sugar, paper, meat, etc. In China, the total amount of industrial wastewater was estimated at 24.7 billion tons, with available amount of 22.2 billion tons (equivalent to 3.58 million tce), estimated based on a collection coefficient of 90% [171].

Table 8.6 Amount and energy potential (electricity generation) of MSW in different regions of China [188]

Region	Disposed MSW (million tons)	Sanitary landfill (million tons)	Landfill gases (million m^3)	Electricity from methane (GWh)	Electricity from incineration (GWh)	Total electricity (GWh)
Northern China	21.67	11.99	2400.24	2400.3	255.0	2655.3
Northeastern China	23.02	8.21	1643.61	1643.6	77.5	1721.1
Eastern China	43.34	25.41	5086.55	5086.6	1977.5	7064.1
Central southern China	40.89	18.02	3606.83	3606.8	1108.5	4715.3
Southwestern of China	12.11	8.11	1623.80	1623.8	164.5	1788.3
Northwestern China	11.08	4.56	913.56	913.6	4.0	917.6
Nationwide total	152.14	76.32	15,274.86	15,274.9	3587.8	18,862.6

8.4.3 Industrial Biorefining Companies

Table 8.7 presents the main industrial companies in China involved in biorefining various bioresources and biowastes, for the production of liquid biofuels (bioethanol and biodiesel).

8.5 Outlook

The success of bioeconomy relies, first and foremost, on the genuine willingness of decision makers and industrialists to implement this concept on the ground and their actual commitment to sustainability. Other decisive factors include the availability of bioresources, the necessity of efficient and eco-friendly processing technologies, as well as the other production and marketing tools to transform low-value biomass to added-value bioproducts.

Considering the fact that bioeconomy is a multifaceted concept, each country is a special case for implementing bioeconomy. Indeed, if two countries share the same kind of bioresources, they do not possess the same infrastructure and biorefining facilities. If they do, then the workforce will make the difference in terms of manpower and quality of education and trainings. To this, the geography, the history, geopolitics, and the general public awareness and involvement in the decision-making process will also be of significant impact.

So, should every country develop its own bioeconomy? Definitely not. The singularity of each country to adopt the concept of bioeconomy and to implement

Table 8.7 Chinese biorefining companies producing liquid biofuels: bioethanol and biodiesel

Name	General information	Start-up year	Feedstock	Biofuel production capacity (tons/year)	By-products
(1) Bioethanol production [172, 189–193]					
Jilin Fuel Ethanol Co., Ltd.	Located in Jilin (northeastern China) and owned by the largest oil and gas company in the country, China National Petroleum Corporation	2001	Corn	500,000–600,000	320,000 t/year of distillers dried grains with soluble (DDGS) and 22,500 tons/year of corn oil
Henan Tianguan Fuel Ethanol Co., Ltd.	Located in Nanyang (southern China) and owed by the Henan Tianguan Group, a Chinese company involved in the sector of renewable energy	2003	Corn/wheat	450,000–600,000	120,000 t/year of DDGS and 45,000 t/year of wheat gluten
Anhui Fengyuan Biochemical Co., Ltd.	Located in the Anhui Province (Eastern China) and is a joint venture formed by Anhui BBCA Biochemical Co., Ltd. and China Petrochemical Corporation	2005	Corn	320,000–400,000	170,000 t/year of (DDGS) and 17,000 tons/year of corn oil
Heilongjiang Huarun Alcohol Co., Ltd.	Located in the Heilongjiang (northeastern China) and is a subsidiary of China Resources (Holdings) Co., Ltd.	1999	Corn	180,000–220,000	160,000 t/year of (DDGS) and 15,000 tons/year of corn oil
Guangxi COFCO Bio-energy Co., Ltd.	Located in the autonomous region of Guangxi (south central China) and owned by China Oil and Foodstuffs Corporation (COFCO)	2006	Cassava	200,000	-

(continued)

Table 8.7 (continued)

Name	General information	Start-up year	Feedstock	Biofuel production capacity (tons/year)	By-products
(2) Biodiesel production [192, 194–198]					
Gushan Environmental Energy Ltd.	Located in the Fujian Province (southeastern coast of China). Since 2012, the company was purchased by the Trillion Energy Investments Holdings Ltd.	2006	Vegetable oil, animal fat, and recycled cooking oil	190,000	Asphalt, glycerol, and erucic acid
Wuxi Huahong Biofuel Co., Ltd.	Located in Wuxi, Jiangsu Province	2006	Waste oil and swill-cooked oil	100,000	Glycerol
Longyan Zhuoyue New Energy Development Co., Ltd.	Located in Longyan (southwestern China) and is a subsidiary of China Biodiesel International Holding Co., Ltd.	2003	Waste oil	50,000	5,000 t/year of industrial glycerin and bitumen
Xiamen Zhuoyue Biomass Energy Co., Ltd.	Located in Xiamen (southeastern coast of China) and is also a subsidiary of China Biodiesel International Holding Co., Ltd.	2006	Waste oil	50,000	5,000 t/year of industrial glycerin and bitumen
Henan Xinyang Hongchang Group	Located in the Henan Province (east-central China)	2006	Local wood plant oil and grease trap waste	30,000	Glycerol
Handan Gushan Olea Chemical Ltd.	Located in the Hebei Province (northern China)	2004	Rapeseed oil	25,000–30,000	Glycerol

it on the ground will generate different scenarios. Some countries will brilliantly succeed, others moderately, and some others will struggle. Regardless of the outcome, a collective knowledge will be gained worldwide from both success and failure scenarios. Such knowledge will help developed countries to reconcile profitability with sustainability in their economic models. It will also be a momentous opportunity for developing countries to benefit from this collective knowledge around bioeconomy (biomass supply chain, process design, conversion technologies, etc.), thus enabling a wiser and more profitable exploitation of their natural resources for generations ahead.

Overall, in order to lay a solid ground for bioeconomy, we need to promote its multidisciplinary R&D effort, develop new markets for its new bioproducts, upgrade its infrastructures and manufacturing facilities, effectively train its workforce, wisely adjust its policies and regulations, and meticulously coordinate its financial systems. All these efforts should be carried out in full cooperation between the decision makers, private and public sectors, scientists, and professionals from backgrounds as diverse as biology, economy, chemistry, biotechnology, engineering, genetics, IT, etc.

References

1. van der Hoeven D. One bioeconomy, two worlds. Bio Based press. 2015. http://www.biobasedpress.eu/2015/01/one-bioeconomy-two-worlds/
2. Golden JS, Handfield RB, Daystar J, McConnell TE. An economic impact analysis of the U.S. biobased products: a report to the congress of the United States of America. 2015. https://www.biopreferred.gov/BPResources/files/EconomicReport_6_12_2015.pdf
3. United States Department of Agriculture (USDA). U.S. biobased products market potential and projections through 2025. 2008. http://www.usda.gov/oce/reports/energy/BiobasedReport2008.pdf
4. Wesseler J, Spielman DJ, Demont M. The future of governance in the global bioeconomy: policy, regulation, and investment challenges for the biotechnology and bioenergy sectors. AgBioForum. 2010;13:288–90.
5. Executive Memorandum M-10-30 for the heads of executive departments and agencies. 2010. https://www.whitehouse.gov/sites/default/files/microsites/ostp/fy12-budget-guidance-memo.pdf
6. National Bioeconomy Blueprint. White House, Washington, US. 2012. https://www.whitehouse.gov/sites/default/files/microsites/ostp/national_bioeconomy_blueprint_april_2012.pdf
7. Lane J. The US bioeconomy blueprint: the 10 minute guide. Biofuels Digest. 2012. http://www.biofuelsdigest.com/bdigest/2012/04/27/the-us-bioeconomy-blueprint-the-10-minute-guide/
8. Stokes B. Billion ton study - A historical perspective. 2015. http://energy.gov/sites/prod/files/2015/08/f25/stokes_bioenergy_2015.pdf
9. Perlack RD, Wright LL, Turhollow AF, et al. Biomass as feedstock for a bioenergy and bioproducts industry: the technical feasibility of a billion-ton annual supply. U.S. Department of Energy. 2005. https://www1.eere.energy.gov/bioenergy/pdfs/final_billionton_vision_report2.pdf
10. Perlack RD, Eaton LM, Turhollow AF Jr, et al. U.S. billion-ton update: biomass supply for a bioenergy and bioproducts industry. U.S. Department of Energy. 2011. http://www1.eere.energy.gov/bioenergy/pdfs/billion_ton_update.pdf
11. Langholtz MH, Stokes BJ, Eaton LM, et al. 2016 billion-ton report: advancing domestic resources for a thriving bioeconomy (Vol. 1): Economic availability of feedstocks. U.S. Department of Energy. 2016. http://energy.gov/sites/prod/files/2016/07/f33/2016_billion_ton_report_0.pdf
12. Biomass Research and Development Initiative (BRDI). Vision for bioenergy and biobased products in the United States. 2006. http://biomassboard.gov/pdfs/final_2006_visionkw.pdf
13. Office of Energy Efficiency & Renewable Energy. Biomass resource basics. 2013. http://energy.gov/eere/energybasics/articles/biomass-resource-basics

14. Villamil MB, Alexander M, Silvis AH, Gray ME. Producer perceptions and information needs regarding their adoption of bioenergy crops. Renew Sustain Energy Rev. 2012;16: 3604–12.
15. Skevas T, Hayden NJ, Swinton SM, Lupi F. Landowner willingness to supply marginal land for bioenergy production. Land Use Policy. 2016;50:507–17.
16. Scheffran J, Bendor T. Bioenergy and land use: a spatial-agent dynamic model of energy crop production in Illinois. Int J Environ Pollut. 2009;39:4–27.
17. Kim S, Dale BE. Life cycle assessment of various cropping systems utilized for producing biofuels: bioethanol and biodiesel. Biomass Bioenergy. 2005;29:426–39.
18. Kim S, Dale BE. Global potential bioethanol production from wasted crops and crop residues. Biomass Bioenergy. 2004;26:361–75.
19. Kocoloski M, Griffin WM, Matthews HS. Estimating national costs, benefits, and potential for cellulosic ethanol production from forest thinnings. Biomass Bioenergy. 2011;35: 2133–42.
20. Wu W, Davis RW. One-pot bioconversion of algae biomass into terpenes for advanced biofuels and bioproducts. Algal Res. 2016;17:316–20.
21. Venteris ER, Skaggs RL, Wigmosta MS, Coleman AM. A national-scale comparison of resource and nutrient demands for algae-based biofuel production by lipid extraction and hydrothermal liquefaction. Biomass Bioenergy. 2014;64:276–90.
22. White EM. Woody biomass for bioenergy and biofuels in the United States - A briefing paper (PNW-GTR-825). U.S. Department of Agriculture, Forest Service. 2010. http://www.fsl.orst.edu/lulcd/Publicationsalpha_files/White_pnw_gtr825.pdf
23. Bi Z, Zhang J, Peterson E, et al. Biocrude from pretreated sorghum bagasse through catalytic hydrothermal liquefaction. Fuel. 2017;188:112–20.
24. Perez-Pimienta JA, Flores-Gómez CA, Ruiz HA, et al. Evaluation of agave bagasse recalcitrance using AFEX^{TM}, autohydrolysis, and ionic liquid pretreatments. Bioresour Technol. 2016;211:216–23.
25. Biogas opportunities roadmap. Report commissioned by U.S. Department of Agriculture, U.S. Environmental Protection Agency and U.S. Department of Energy. 2014. http://www.usda.gov/oce/reports/energy/Biogas_Opportunities_Roadmap_8-1-14.pdf
26. National Renewable Energy Laboratory (NREL). Biomass Maps. http://www.nrel.gov/gis/biomass.html. Last update of the content 18 July 2016.
27. US Department of Energy. Bioenergy knowledge discovery framework (BKF). Executive summary/overview. https://bioenergykdf.net/billionton2016/1/1/table
28. Advanced BioFuels USA. USDA announces availability of funding to develop advanced biofuels projects. 2013. http://advancedbiofuelsusa.info/usda-announces-availability-of-funding-to-develop-advanced-biofuels-projects/
29. Haq Z. DOE Perspectives on advanced hydrocarbon-based biofuels. U.S. Department of Energy, Office of Biomass Program. 2012. https://www.eia.gov/biofuels/workshop/presentations/2012/pdf/zia_haq.pdf
30. US Department of Energy. ClearFuels-Rentech pilot-scale biorefinery - Integrated pilot project for fuel production by thermochemical conversion of woodwaste. 2011. http://energy.gov/sites/prod/files/2014/03/f14/ibr_arra_clearfuelstechnology.pdf
31. Reuters. E I du Pont de Nemours and Co (DD) - Overview. http://www.reuters.com/finance/stocks/companyProfile?symbol=DD
32. DuPont^{TM}. Industrial biotechnology - Sustainable solutions for a growing population. http://www.dupont.com/products-and-services/industrial-biotechnology.html
33. DuPont^{TM}. Biotechnology instead of petroleum. http://www2.dupont.com/Renewably_Sourced_Materials/en_US/processing.html
34. Lane J. DuPont industrial biosciences: Biofuels digest's 2014 5-minute guide. Biofuels Digest. 2014. http://www.biofuelsdigest.com/bdigest/2014/03/17/dupont-industrial-biosciences-biofuels-digests-2014-5-minute-guide/

35. Jessen H. Bacterial battle - Antibiotics are the most common tool against bacteria, but alternatives exist. Ethanol Producer Magazine (2012). http://www.ethanolproducer.com/articles/8928/bacterial-battle
36. Butamax™ Technology. http://www.butamax.com/renewable-fuel-technologies.aspx
37. Lane J. Butamax: Biofuels digest's 2015 5-minute guide. 2015. http://www.biofuelsdigest.com/bdigest/2015/02/01/butamax-biofuels-digests-2015-5-minute-guide/
38. DuPont™. Turning plants into building blocks. http://www2.dupont.com/Renewably_Sourced_Materials/en_US/proc-buildingblocks.html
39. DuPont Tate & Lyle Bio Products Company, LLC. DuPont Tate & Lyle Bio Products begin Bio-PDO™ production in Tennessee. http://www.duponttateandlyle.com/news_112706
40. Lee CS, Aroua MK, Daud WMAW, et al. A review: Conversion of bioglycerol into 1,3-propanediol via biological and chemical method. Renew Sustain Energy Rev. 2015;42: 963–72.
41. Renewable Energy Group, Inc. https://www.linkedin.com/company/renewable-energy-group
42. Renewable Energy Group. Products & Services. http://www.regi.com/products-services
43. Environmental XPRT. Renewable Energy Group, Inc. https://www.environmental-expert.com/companies/renewable-energy-group-inc-reg-36065
44. Biofuels International. Renewable Energy Group sign $15m deal to acquire Imperium Renewables. http://biofuels-news.com/display_news/9459/renewable_energy_group_sign_15m_deal_to_acquire_imperium_renewables/. Published 3 Aug 2015.
45. Gonzalez A. Imperium Renewables biodiesel plant sold to Iowa company. The Seattle Times. http://www.seattletimes.com/business/local-business/imperium-renewables-biodiesel-plant-sold-to-iowa-company/. Published and updated 4 Aug 2015.
46. Business Wire. Renewable Energy Group closes acquisition of Sanimax biodiesel plant. http://www.businesswire.com/news/home/20160316005770/en/Renewable-Energy-Group-Closes-Acquisition-Sanimax-Biodiesel. Published 16 Mar 2016.
47. European Commission. Directorate-General for Research and Innovation. Bio-based economy in Europe: state of play and future potential - Part 2. 2011. https://ec.europa.eu/research/consultations/bioeconomy/bio-based-economy-for-europe-part2.pdf
48. Bevan MW, Franssen MCR. Investing in green and white biotech. Nat Biotechnol. 2006;24: 765–7.
49. Scarlat N, Dallemand JF, Monforti-Ferrario F, Nita V. The role of biomass and bioenergy in a future bioeconomy: policies and facts. Environ Devel. 2015;15:3–34.
50. Frazzetto G. White biotechnology. EMBO Rep. 2003;4:835–7.
51. Kirk O, Borchert TV, Fuglsang CC. Industrial enzyme applications. Curr Opin Biotechnol. 2002;13:345–51.
52. Jegannathan KR, Nielsen PH. Environmental assessment of enzyme use in industrial production – a literature review. J Clean Prod. 2013;42:228–40.
53. McCormick K, Kautto N. The bioeconomy in Europe: an overview. Sustainability. 2013;5: 2589–608.
54. Forest-based Sector – Technology Platform. Star-COLIBRI Project: biorefinery clustering. http://www.forestplatform.org/en/about-ftp/ftp-research-projects/star-colibri
55. Gustafsson M, Stoor R, Tsvetkova A. Sustainable bio-economy: potential, challenges and opportunities in Finland. Sitra studies 51. 2011. https://www.sitra.fi/julkaisut/Selvityksi%C3%A4-sarja/Selvityksi%C3%A4%2051.pdf
56. Kokkonen E. Sitra: bioeconomy needs medium-sized plants. http://www.sitra.fi/en/news/bioeconomy/sitra-bioeconomy-needs-medium-sized-plants. Published 30 Sept 2011.
57. The Finnish Bioeconomy Strategy - Sustainable growth from bioeconomy. 2014. http://biotalous.fi/wp-content/uploads/2014/08/The_Finnish_Bioeconomy_Strategy_110620141.pdf
58. National Research Strategy BioEconomy 2030 - Our route towards a biobased economy. 2011. https://biobs.jrc.ec.europa.eu/sites/default/files/generated/files/policy/German%20bioeconomy%20Strategy_2030.pdf

59. Pietschmann B. Germany as a pioneer of a sustainable bioeconomy. Bio-based News. http://news.bio-based.eu/germany-as-a-pioneer-of-a-sustainable-bioeconomy/. Published 14 June 2011.
60. The Federal government. Biorefineries Roadmap. 2012. https://www.bmbf.de/pub/Roadmap_Biorefineries_eng.pdf
61. Asveld L, van Est R, Stemerding D. Getting to the core of the bio-economy: a perspective on the sustainable promise of biomass. The Hague: Rathenau Instituut; 2011.
62. International Energy Agency (IEA). Oil & gas security - The Netherlands. 2012. https://www.iea.org/publications/freepublications/publication/OilGasSecurityNL2012.pdf
63. Energy Research Centre of the Netherlands (ECN). Renewable energy: current Dutch policy is falling short of 2020 target despite strong growth. https://www.ecn.nl/nl/nieuws/item/renewable-energy-current-dutch-policy-is-falling-short-of-2020-target-despite-strong-growth/. Published 24 Aug 2012.
64. Bosman R, Rotmans J. Transition governance towards a bioeconomy: a comparison of Finland and the Netherlands. Sustainability. 2016;8:1017.
65. Ros J, Olivier J, Notenboom J, Croezen H, Bergsma G. Sustainability of biomass in a bio-based economy. PBL Netherlands Environmental Assessment Agency. Publication number 500143001, The Hague. 2012. http://www.pbl.nl/sites/default/files/cms/publicaties/PBL-2012-Sustainability-of-biomass-in-a-BBE-500143001_0.pdf
66. Sanders J, van der Hoeven D. Opportunities for a bio-based economy in the Netherlands. Energies. 2008;1:105–19.
67. Bonaccorso M. The bioeconomy: the Netherlands in pole position. Renewable Matter. http://www.renewablematter.eu/art/189/The_Bioeconomy_The_Netherlands_in_Pole_Position. Published 24 Aug 2012.
68. AkzoNobel media release. AkzoNobel and Royal Cosun to develop sustainable cellulose products from sugar beet processing. Published 4 May 2016.
69. European Commission. National bioeconomy profile - United Kingdom. 2014. https://biobs.jrc.ec.europa.eu/sites/default/files/generated/files/country/National%20Bioeconomy%20Profile%202014%20UK.pdf
70. UK Government, Department of energy & climate change. UK Bioenergy Strategy. 2012. https://www.gov.uk/government/uploads/system/uploads/attachment_data/file/48337/5142-bioenergy-strategy-.pdf
71. House of the Lords, Science and Technology Select Committee. Waste or resource? Stimulating a bioeconomy. 3rd report of session 2013–14. 2014. http://www.publications.parliament.uk/pa/ld201314/ldselect/ldsctech/141/141.pdf
72. UK Government, Department for Business, Innovation & Skills. Building a high value bioeconomy: opportunities from waste. 2015. https://www.gov.uk/government/uploads/system/uploads/attachment_data/file/408940/BIS-15-146_Bioeconomy_report_-_opportunities_from_waste.pdf
73. En route to the knowledge-based bio-economy. Cologne conference paper. 2007. https://dechema.de/dechema_media/Cologne_Paper-p-20000945.pdf
74. European Commission. The Eco-innovation action plan. https://ec.europa.eu/environment/ecoap/about-action-plan/objectives-methodology
75. Knotková M, Sedlačík R, Perutka T. Regional networks for the development of a sustainable bioenergy market in Europe - BioRegions project. 2011. https://ec.europa.eu/energy/intelligent/projects/sites/iee-projects/files/projects/documents/bioregions_czech_action_plan_zlin_en.pdf
76. European Commission. Innovation Union - European Technology Platforms. http://ec.europa.eu/research/innovation-union/index_en.cfm?pg=etp
77. Industrial or white biotechnology - A policy agenda for Europe (2006). http://www.biobased.us/Biobased-EU-Economy-Policy-Agenda.pdf
78. European Commission. Communication from the Commission to the European Parliament, the Council, the European Economic and Social Committee and the Committee of the

Regions - Innovating for sustainable growth: a bioeconomy for Europe. 2012. http://www.ascension-publishing.com/BIZ/EU-Bioeconomy-strategy.pdf
79. Life-Sciences-Europe. European Bioeconomy Alliance (EBA). http://www.life-sciences-europe.com/organisation/european-bioeconomy-alliance-eba-2015-union-west-europe-2001-36460.html. Content updated 16 Feb 2015.
80. The European Bioeconomy Alliance (EBA). Newly-formed Bioeconomy Alliance calls for EU action. http://www.europabio.org/sites/default/files/eba_press_release.pdf. Press release published 4 Feb 2016.
81. Lead Market Initiative's Ad-hoc advisory group for bio-based products. List of priority recommendations. 2011. https://biobs.jrc.ec.europa.eu/sites/default/files/generated/files/policy/2011%20Lead%20Market%20Initiative%20LMI%20Biobased%20Products%20Priority%20Recommendations.pdf
82. The European Bioeconomy Alliance. Joint call for action on the bioeconomy. 2016. http://www.bioeconomyalliance.eu/sites/default/files/EUBA%20Joint%20Call%20for%20Action%20on%20the%20Bioeconomy%20-%20FINAL.pdf
83. McCormick K. The emerging bio-economy in Europe: exploring the key governance challenges. World Renewable Energy Congress. 8–13 May 2011. Linköping, Sweden. http://www.ep.liu.se/ecp/057/vol10/005/ecp57vol10_005.pdf
84. Junginger M, Bolkesjø T, Bradley D, et al. Developments in international bioenergy trade. Biomass Bioenergy. 2008;32:717–29.
85. Clini C, Bauen A, Caserta G, et al. Global Bioenergy Partnership - White paper. 2005. http://www.globalbioenergy.org/fileadmin/user_upload/docs/WhitePaper-GBEP.pdf
86. Alexopoulou E, Zanetti F, Scordia D, et al. Long-term yields of switchgrass, giant reed, and Miscanthus in the Mediterranean basin. Bioenergy Res. 2015;8:1492.
87. European Commission. From the sugar platform to biofuels and biochemicals - Final report for the European Commission Directorate-General Energy N° ENER/C2/423-2012/SI2.673791. 2015. https://ec.europa.eu/energy/sites/ener/files/documents/EC%20Sugar%20Platform%20final%20report.pdf
88. European Biofuels Technology Platform (EBTP). Starch crops for production of biofuels. http://www.biofuelstp.eu/starch_crops.html
89. EBTP. Oil crops for production of advanced biofuels. http://www.biofuelstp.eu/oil_crops.html
90. Bacenetti J, Bergante S, Facciotto G, Fiala M. Woody biofuel production from short rotation coppice in Italy: environmental-impact assessment of different species and crop management. Biomass Bioenergy. 2016;94:209–19.
91. Sustainable conversion of *Pinus pinaster* wood into biofuel precursors: a biorefinery approach. Fuel. 2016;164:51–8.
92. Olsson O, Hillring B. The wood fuel market in Denmark – Price development, market efficiency and internationalization. Energy. 2014;78:141–8.
93. Ranta T. Logging residues from regeneration fellings for biofuel production–a GIS-based availability analysis in Finland. Biomass Bioenergy. 2005;28:171–82.
94. Mangoyana RB. Bioenergy from forest thinning: carbon emissions, energy balances and cost analyses. Renew Energy. 2011;36:2368–73.
95. Linden M. Forecasting forest chip energy production in Finland 2008–2014. Biomass Bioenergy. 2011;35:590–9.
96. Holm-Nielsen J, Ehimen EA. Biomass supply chains for bioenergy and biorefining. Sawston, Cambridge, UK: Woodhead Publishing; 2016.
97. Sambusiti C, Licari A, Solhy A, et al. One-Pot dry chemo-mechanical deconstruction for bioethanol production from sugarcane bagasse. Bioresour Technol. 2015;181:200–6.
98. Wiinikka H, Johansson AC, Wennebro J, Carlsson P, Öhrman OGW. Evaluation of black liquor gasification intended for synthetic fuel or power production. Fuel Process Technol. 2015;139:216–25.

99. Kudahettige-Nilsson RL, Helmerius J, Nilsson RT, et al. Biobutanol production by *Clostridium acetobutylicum* using xylose recovered from birch Kraft black liquor. Bioresour Technol. 2015;176:71–9.
100. Pleissner D, Qi Q, Gao C, et al. Valorization of organic residues for the production of added value chemicals: a contribution to the bio-based economy. Biochem Eng J. 2016;116:3–16.
101. García-Torreiro M, López-Abelairas M, Lu-Chau TA, Lema JM. Fungal pretreatment of agricultural residues for bioethanol production. Ind Crop Prod. 2016;89:486–92.
102. Nechyporchuk O, Belgacem MN, Bras J. Production of cellulose nanofibrils: a review of recent advances. Ind Crop Prod. 2016;93:2–25.
103. Colantoni A, Evic N, Lord R, et al. Characterization of biochars produced from pyrolysis of pelletized agricultural residues. Renew Sustain Energy Rev. 2016;64:187–94.
104. Hagenkamp-Korth F, Ohl S, Hartung E. Effects on the biogas and methane production of cattle manure treated with urease inhibitor. Biomass Bioenergy. 2015;75:75–82.
105. Tsapekos P, Kougias PG, Treu L, Campanaro S, Angelidaki I. Process performance and comparative metagenomic analysis during co-digestion of manure and lignocellulosic biomass for biogas production. Appl Energy. 2017;185:126–35.
106. Cotana F, Coccia V, Petrozzi A, et al. Energy valorization of poultry manure in a thermal power plant: experimental campaign. Energy Proc. 2014;45:315–22.
107. De Filippis P, Scarsella M, Verdone N, Zeppieri M. Poultry litter valorization to energy. WIT Trans Ecol Environ. 2008;109:261–7.
108. Plaza M, Santoyo S, Jaime L, et al. Screening for bioactive compounds from algae. J Pharm Biomed Anal. 2010;51:450–5.
109. Lardon L, Hélias A, Sialve B, Steyer JP, Bernard O. Life-cycle assessment of biodiesel production from microalgae. Environ Sci Technol. 2009;43:6475–81.
110. Herrero M, Sánchez-Camargo AP, Cifuentes A, Ibáñez E. Plants, seaweeds, microalgae and food by-products as natural sources of functional ingredients obtained using pressurized liquid extraction and supercritical fluid extraction. Trends Anal Chem. 2015;71:26–38.
111. Hewitt J. Flows of biomass to and from the EU. Report published by the non-governmental organization FERN. 2011. http://www.fern.org/sites/fern.org/files/Biomass%20imports%20to%20the%20EU%20final_0.pdf
112. Elbersen B, Boywer C, Kretschmer B. Atlas of EU biomass potentials: summary for policy makers. Publication from the Biomass Futures project. 2012. http://www.biomassfutures.eu/public_docs/final_deliverables/WP6/D6.4%20Atlas%20of%20EU%20biomass%20potentials.pdf
113. Ståhls M, Mayer AL, Tikka PM, Kauppi PE. Disparate geography of consumption, production, and environmental impacts: forest products in Finland 1991-2007. J Ind Ecol. 2010;14: 576–85.
114. Kotilainen J, Rytteri T. Transformation of forest policy regimes in Finland since the 19th century. J Hist Geogr. 2011;37:429–39.
115. Chambost V, Stuart PR. Selecting the most appropriate products for the forest biorefinery. Ind Biotechnol. 2007;3:112–9.
116. Janssen M, Chambost V, Stuart PR. Successful partnerships for the forest biorefinery. Ind Biotechnol. 2008;4:352–62.
117. Lamberg JA, Näsi J, Ojala J, Sajasalo P. The evolution of competitive strategies in global forestry industries: comparative perspectives. Berlin: Springer Science & Business Media; 2007.
118. Thorp B. Biorefinery offers industry leaders business model for major change. Pulp Pap. 2005;79:35–9.
119. Ragauskas AJ, Nagy M, Kim DH, et al. From wood to fuels: integrating biofuels and pulp production. Ind Biotechnol. 2006;2:55–65.
120. Hämäläinen S, Näyhä A, Pesonen HL. Forest biorefineries - A business opportunity for the Finnish forest cluster. J Clean Prod. 2011;19:1884–91.
121. UPM in brief. http://www.upm.com/About-us/Pages/default.aspx

122. UPM Oy. This is Biofore. http://www.upm.com/About-us/This-is-biofore/Pages/default.aspx
123. Mokkila K. UPM Bioreofinery. World Biorefinery Conference. Jönköping, Sweden. 2012. http://www.spci.se/shared/files/A_view_from_UPM-Kymmene._Kosti_Mokkila.pdf
124. Ahlgren S, Di Lucia L. Indirect land use changes of biofuel production – A review of modelling efforts and policy developments in the European Union. Biotechnol Biofuels. 2014;7:35–44.
125. Searchinger T, Heimlich R, Houghton RA. Use of U.S. croplands for biofuels increases greenhouse gases through emissions from land-use change. Science. 2008;319:1238–40.
126. Mannonen S. UPM Biofuels: biofuels from wood-based raw materials. European technology platform - Biofuels for low carbon transport & energy security. Brussels. 2014. http://www.biofuelstp.eu/spm6/docs/sari-manonen.pdf
127. UPM Oy. UPM Biorefining. http://www.upm.com/Investors/upm-story/business-areas/upm-biorefining/Pages/default.aspx
128. UPM Oy. UPM Lappeenranta biorefinery wins the commercial scale plant of the year award. http://www.upm.com/About-us/Newsroom/Releases/Pages/UPM-Lappeenranta-Biorefinery-wins-the-Commercial-Scale-Plant-of-the-year-award-001-Tue-03-Mar-2015-09-00.aspx
129. Chemicals-technology.com. UPM Lappeenranta Biorefinery, Finland. http://www.chemicals-technology.com/projects/upm-lappeenranta-biorefinery-biodiesel-biofuels-finland/
130. Bioeconomy.fi. Frontrunner in innovation – UPM produces low emission diesel from crude tall oil. http://www.bioeconomy.fi/frontrunner-in-innovation-upm-produces-low-emission-diesel-from-crude-tall-oil/
131. Honeywell UOP. Honeywell's Envergent Technologies and Green Fuel Nordic Oy to collaborate on biomass processing in Finland. https://www.uop.com/?press_release=honeywells-envergent-technologies-and-green-fuel-nordic-oy-to-collaborate-on-biomass-processing-in-finland. Published 12 Oct 2011.
132. Green Fuel Nordic Oy. About us. http://www.greenfuelnordic.fi/about_us
133. Liukko A. Finland's RES-target for 2030. 2015. http://www.tuulivoimayhdistys.fi/filebank/812-AnjaLiukko.pdf
134. Green Fuel Nordic Oy. Premise of our operations. http://www.greenfuelnordic.fi/strategy
135. St1 Oy. St1 is composed of two legal entities. http://www.st1.eu/company-information
136. St1 Biofuels Oy. This is St1 Biofuels Oy. http://www.st1biofuels.com/company
137. Lane J. St1 completes waste-to-ethanol Etanolix project in Sweden. 2015. http://www.biofuelsdigest.com/bdigest/2015/06/06/st1-completes-waste-to-ethanol-etanolix-project-in-sweden/
138. St1 Biofuels Oy. Learn how our solutions add to your business. http://www.st1biofuels.com/solutions
139. Energy, Oil & Gas. Profiles: St1 Biofuels Oy - Challenging the conventional. Issue 130, Mar 2016. http://www.energy-oil-gas.com/2016/03/14/st1-biofuels-oy/
140. St1 Oy. St1's and SOK's joint venture NEB plans 50-million-litre Cellunolix® bioethanol plant in Pietarsaari. http://www.st1.eu/news/st1s-and-soks-joint-venture-neb-plans-50-million-litre-cellunolix-bioethanol-pla. News published 15 Nov 2016.
141. North European Oil Trade Oy (NEOT). Oil market and logistics excellence. http://neot.fi/en/neot-en/north-european-oil-trade-en
142. Biomass Magazine. St1 plans 50 MMly Cellunolix ethanol plant at UPM site in Finland. http://biomassmagazine.com/articles/13921/st1-plans-50-mmly-cellunolix-ethanol-plant-at-upm-site-in-finland. Published 15 Nov 2016.
143. St1 Oy. St1 built a waste-based Etanolix® ethanol production plant in Gothenburg. http://www.st1.eu/news/st1-built-a-waste-based-etanolix-ethanol-production-plant-in-gothenburg. News published 5 June 2016.
144. St1 Oy. St1 signs a letter of intent with Viken Skog and Treklyngen for a Cellunolix® ethanol plant in Norway. http://www.st1.eu/news/st1-signs-a-letter-of-intent-with-viken-skog-and-treklyngen-for-a-cellunolix-eth. News published 19 Aug 2016.

145. Sun Z, Shi Z. The acetone-butanol (ABE) fermentation industries in China. http://dc.engconfintl.org/cgi/viewcontent.cgi?article=1006&context=bioenergy_i
146. Mikkola JP, Sklavounos E, King AWT, Virtanen P. The biorefinery and green chemistry. In: Bogel-Lukasik R, editor. Ionic liquids in the biorefinery concept: challenges and perspectives. London, UK: Royal Society of Chemistry; 2015.
147. Institute for energy research (IER). China overtakes U.S. as world's largest oil importer. http://instituteforenergyresearch.org/analysis/china-overtakes-u-s-as-worlds-largest-oil-importer/#_edn1. Published 13 May 2015.
148. Chow GC. China's energy and environmental problems and policies. CEPS Working Paper No. 152. Princeton University. 2007. https://www.princeton.edu/ceps/workingpapers/152chow.pdf
149. Song L, Wing TW, editors. China's dilemma: economic growth, the environment and climate change. Canberra: ANU Press; 2008.
150. Chen H. Ending fossil fuel subsidies by 2020: a goal for China G20. Natural Resources Defense Council (NRDC). https://www.nrdc.org/experts/han-chen/ending-fossil-fuel-subsidies-2020-goal-china-g20. Published 29 Aug 2016.
151. The Economist. India and the environment: catching up with China. http://www.economist.com/news/asia/21672359-prime-minister-wants-india-grow-fast-over-next-20-years-china-has-over-past-20. Published 10 Oct 2015.
152. China's green revolution: energy, environment and the 12th Five-Year Plan. 2001. https://assets.documentcloud.org/documents/1005743/china-s-green-revolution-ebook-2001en.pdf
153. Angang H, Jiaochen L. China's green era begins. China dialogue. https://www.chinadialogue.net/article/show/single/en/4149-China-s-green-era-begins-. Published 8 Mar 2011.
154. Chu J. RE100 china analysis - China's fast track to a renewable future. THE climate Group. 2015. https://www.theclimategroup.org/sites/default/files/archive/files/RE100-China-analysis.pdf
155. International Renewable Energy Agency (IRENA). Renewable energy prospects: China, REmap 2030 analysis. 2014. http://irena.org/remap/IRENA_REmap_China_report_2014.pdf
156. Gu X. China releases blueprint to promote seven emerging industries. China Briefing. http://www.china-briefing.com/news/2012/06/01/china-releases-blueprint-to-promote-seven-emerging-industries.html. Published 1 June 2012.
157. Lu Y. China releases 12th Five-Year Plan for National Strategic Emerging Industries. China Briefing. http://www.china-briefing.com/news/2012/07/25/china-releases-12th-five-year-plan-for-national-strategic-emerging-industries.html. Published 25 July 2012.
158. Green F, Stern N. China's "new normal": structural change, better growth, and peak emissions. The Grantham Research Institute on Climate Change and the Environment. Policy brief. 2015. http://www.lse.ac.uk/GranthamInstitute/wp-content/uploads/2015/06/Chinas_new_normal_green_stern_June_2015.pdf
159. Zhang YS, Barnett SA. Fiscal vulnerabilities and risks from local government finance in China. IMF Working Paper No. 14/4. 2014. https://papers.ssrn.com/sol3/papers.cfm?abstract_id=2393604
160. The Economist Corporate Network. China's 13th Five-Year Plan - Opportunities for Finnish companies. Sponsored by Tekes. 2016. https://www.tekes.fi/globalassets/global/ohjelmat-ja-palvelut/kasvajakansainvalisty/future-watch/chinas-13th-five-year-plan.pdf
161. Hilton I. China's 13th Five Year Plan: a green light for a green future. Energy and Climate Intelligence Unit. http://eciu.net/blog/2016/chinas-13th-plan. Published 5 Apr 2016.
162. Xian M. Recent development of bioenergy and biorefinery in China. Trans Renew Energy. 2015;1:129–30.
163. National Development and Reform Commission. Medium and long-term development plan for renewable energy in China. 2007. http://www.china.org.cn/e-news/news070904-11.htm
164. Jiang K. Biomass in China. Global Bioenergy. 2007. http://www.globalbioenergy.org/fileadmin/user_upload/gbep/docs/2007_events/Bali_2007/BiomassinChina1.pdf
165. Enerdata. Global energy statistical - Yearbook 2016. https://yearbook.enerdata.net/

166. Li JF, Hu RQ, Song YQ, Shi JL, Bhattacharya SC, Salam PA. Assessment of sustainable energy potential of non-plantation biomass resources in China. Biomass Bioenergy. 2005;29: 167–77.
167. Chen YH, Li ZH, Shen T. The actuality and development measure in using biomass energy in China. J Agric Mech Res. 2006;1:25–8.
168. Statista. Distribution of the workforce across economic sectors in China from 2005 to 2015. 2016. https://www.statista.com/statistics/270327/distribution-of-the-workforce-across-economic-sectors-in-china/
169. Wang W, Liu Y, Zhang L. The spatial distribution of cereal bioenergy potential in China. GCB Bioenergy. 2013;5:525–35.
170. Han LJ, Yan QJ, Liu XY, Hu JY. Straw resources and their utilization in China. Trans Chin Soc Agric Eng. 2002;18:87–91.
171. Yanli Y, Peidong Z, Wenlong Z, Yongsheng T, Yonghong Z, Lisheng W. Quantitative appraisal and potential analysis for primary biomass resources for energy utilization in China. Renew Sustain Energy Rev. 2010;14:3050–8.
172. Tan T, Shang F, Zhang X. Current development of biorefinery in China. Biotechnol Adv. 2010;28:543–55.
173. Liu H, Jiang GM, Zhuang HY, Wang KJ. Distribution, utilization structure and potential of biomass resources in rural China: with special references of crop residues. Renew Sustain Energy Rev. 2008;12:1402–18.
174. Qiu H, Sun L, Xu X, Cai Y, Bai J. Potentials of crop residues for commercial energy production in China: a geographic and economic analysis. Biomass Bioenergy. 2014;64: 110–23.
175. Niu W, Han L, Liu X, et al. Twenty-two compositional characterizations and theoretical energy potentials of extensively diversified China's crop residues. Energy. 2016;100:238–50.
176. Chen X. Economic potential of biomass supply from crop residues in China. Appl Energy. 2016;166:141–9.
177. Li K, Liu R, Sun C. A review of methane production from agricultural residues in China. Renew Sustain Energy Rev. 2016;54:857–65.
178. Yang J, Wang X, Ma H, Bai J, Jiang Y, Yu H. Potential usage, vertical value chain and challenge of biomass resource: evidence from China's crop residues. Appl Energy. 2014;114: 717–23.
179. Jakobsen TG. More forest in the world's largest nations. European Year for Development. https://europa.eu/eyd2015/en/denmark/posts/more-forest-world-s-largest-nations. Published 11 May 2015.
180. Sun Z, Peng S, Li X, Guo Z, Piao S. Changes in forest biomass over China during the 2000s and implications for management. For Ecol Manage. 2015;357:76–83.
181. Zhang C, Zhang L, Xie G. Forest biomass energy resources in China: quantity and distribution. Forests. 2015;6:3970–84.
182. Chen Y, Hu W, Feng Y, Sweeney S. Status and prospects of rural biogas development in China. Renew Sustain Energy Rev. 2014;39:679–85.
183. Qian MY, Li RH, Li J, et al. Industrial scale garage-type dry fermentation of municipal solid waste to biogas. Bioresour Technol. 2016;217:82–9.
184. Chadwick D, Wei J, Yan'an T, Guanghui Y, Qirong S, Qing C. Improving manure nutrient management towards sustainable agricultural intensification in China. Agric Ecosyst Environ. 2015;209:34–46.
185. Wang FH, Ma WQ, Dou ZX, Liu XL, Xu JX, Zhang FS. The estimation of the production amount of animal manure and its environmental effect in China. China Environ Sci. 2006;26: 614–7.
186. Zhang T, Yang Y, Xie D. Insights into the production potential and trends of China's rural biogas. Int J Energy Res. 2015;39:1068–82.
187. Shen X, Huang G, Yang Z, Han L. Compositional characteristics and energy potential of Chinese animal manure by type and as a whole. Appl Energy. 2015;160:108–19.

188. Zhou X, Wang F, Hu H, Yang L, Guo P, Xiao B. Assessment of sustainable biomass resource for energy use in China. Biomass Bioenergy. 2011;35:1–11.
189. China National Petroleum Corporation (CNPC). Renewable energy. http://www.cnpc.com.cn/en/renewable/common_index.shtml
190. Tianguan Group. The news center – Company brief. http://www.tianguan.com.cn/english/
191. Haiyang H. Energy consumption in forestry industry and environmental prospects of bio-fuels applications in Finland and China. Bachelor's Thesis. Saimaa University of Applied Sciences, Imatra, Finland. 2010. http://www.theseus.fi/bitstream/handle/10024/22813/Hu_Haiyang_0701198_P07E.pdf;sequence=1
192. Global Subsidies Initiative (GSI) of the International Institute for Sustainable Development (IISD). Biofuels – At what cost? Government support for ethanol and biodiesel in China. 2008. https://www.iisd.org/gsi/sites/default/files/China_Biofuels_Subsidies.pdf
193. Biofuel digest. Advanced Biofuels in China: TMO, COFCO, CNOOC New Energy join fast-growing ranks. http://www.biofuelsdigest.com/bdigest/2011/05/10/advanced-biofuels-in-china-tmo-cofco-cnooc-new-energy-join-fast-growing-ranks/. Published 10 May 2011.
194. Tsao GT, Ouyang P, Chen J. Biotechnology in China II: chemicals, energy and environment. Berlin: Springer; 2010.
195. Bloomberg. Company overview of Longyan Zhuoyue New Energy Co., Ltd. http://www.bloomberg.com/research/stocks/private/snapshot.asp?privcapid=38594024. Published 28 Nov 2016.
196. China Biodiesel. Company introduction (updated Apr 2012). http://www.chinabiodiesel.cn/en_about.asp
197. Biofuel.org.uk. Gushan Environmental Energy Ltd, China. http://biofuel.org.uk/Gushan-Environmental-Energy.html
198. Lixin Z, Yanli Z, Yujie F. Liquid biofuels for transportation Chinese potential and implications for sustainable agriculture and energy in the 21st century. Assessment study. 2006. https://energypedia.info/images/7/7a/Biofuels_for_Transportation_in_China.pdf

Chapter 9
Bioeconomy: Multidimensional Impacts and Challenges

Abstract For decades, the reliance on fossil resources for the production of fuels, chemicals, and materials has generated unsustainable economic models which have created complicated economic, environmental and geopolitical circumstances around the world. Hence, in order to become sustainable economic model, bioeconomy has to deal with those challenges, as well as the important social factor related to issues such as accentuated disparities, population growth, and mass migration. During the current transition phase towards sustainability, bioeconomy is affected by various factors, notably the availability of biomass and the development or acquisition of biorefining technologies. Once implemented on the ground, bioeconomy starts to impact various aspects related to sustainable development, the environment, and societies. Simultaneously, it starts to face new sets of challenges mainly related to serious agricultural, industrial, environmental, and social issues.

Thus, in the present chapter, the impacts of bioeconomy and the prospects of its worldwide implementation are thoroughly discussed from a multidimensional outlook including industrial, environmental, social, and geopolitical perspectives. This includes the need for a continuous monitoring of the sustainability of bioproducts and biorefineries via various metrics, as well as the assessment of key environmental and social factors such as greenhouse gas emissions, land-use change, biodiversity, employment, food security, and the dangerous, yet somehow underestimated, problem of corruption.

9.1 Sustainability

Food, energy, and materials are of vital importance to modern societies worldwide. For decades, the reliance on fossil resources and derived fuels, chemicals, and materials has generated unsustainable economic models which has created complicated and dangerous situations around the world [1, 2], not only in the economic sector (recurring economic crises), but it also affected the environment (global warming, soil, air and water pollution, deforestation, etc.), created serious geopolitical tensions (Middle East, Latin America, and Africa), and had an impact, sometimes qualified as irreversible, on societies including accentuated disparities, population growth, mass migration from rural regions to cities, and immigration

M. Sillanpää, C. Ncibi, *A Sustainable Bioeconomy*,
DOI 10.1007/978-3-319-55637-6_9

from poor to rich countries due to economic hardships, armed conflicts, environmental disasters, etc. [3–5].

No matter how advanced the county is, decoupling economic growth from the use of fossil resources is a very delicate task, considering the difficulty to balance the ever-increasing demand for raw materials on the one hand, and the depleting nature of most of the current resources upon which the production of numerous energy, chemicals, and material products still heavily depend.

Based on the estimation made by the US Energy Information Administration (EIA), published in its International Energy Outlook 2016 report, fossil fuels will continue to fulfill most of the world's requirement for energy until 2040. By then, petroleum-derived liquid fuels, natural gas, and coal will account for 78% of total world energy consumption [6]. Continuing relying on those fossil fuels, the way humanity did during the last half century will have dramatic consequences, not on future generations but on us and our kids. We can stop buying furniture, we can trade our fuel-consuming car for a bike, but we cannot stop eating, and, surprisingly, the principal raw material in the modern agriculture is none other than fossil fuel [7]. Further dependency on fossil fuels in the agricultural sector will result in a gradual, but significant, increase in the prices of raw ingredients for food and feedstuffs [8, 9].

Thus, focusing on the energy sector when it comes to fossil resources is a shortsighted perception that would make any future solution vulnerable. What is the solution then? Well, we need to think and operate in a holistic sustainable manner, first by managing the still abundant fossil resources efficiently and ecofriendly, especially coal (reserves forecasted until the end of this century [10]). Simultaneously, sustainable alternatives have to be prepared at commercial scale, not only the raw materials and the end products but *from* raw materials *to* the end products, including supply chain, infrastructure, processing technologies, waste valorization, etc.

9.1.1 Sustainable Development and Bioeconomy

Sustainability is basically the outcome of balancing humanity's needs with the available resources to fulfill the requirement of the former without depleting the latter. Several key factors significantly affect this balance, separately or jointly, including economic necessities, environmental situations, social conditions, and geopolitical circumstances [11, 12]. The major issue with those factors, either helping or constraining his sustainable balance, is the fact that they are all changing factors, both in space and time. Thus, sustainability is not a strategic objective to be reached within a targeted time frame, but rather a continuous and dynamic effort from mankind to leave the world a better place as generations come and go.

Overall, the sustainable development is a paradigm established to fulfill the needs of mankind in an economically viable, environmentally friendly, and socially beneficial manner. In this regard, bioeconomy, as an economic model based on the utilization and conversion of biomass to produce commodities, is at the core of sustainable economic strategies around the world. For instance, among the United Nations' 17 sustainable development goals (adopted in September 2015 and targeted to be achieved within 15 years) [13], more than half are directly related to the concept of bioeconomy including (1) no poverty, (2) zero hunger, (7) affordable and clean energy, (9) industry, innovation and infrastructure, (12) responsible consumption and production, (13) climate action, (14) life below water, (15) life on earth, and (17) partnerships for the goals [14].

Considering the vital objectives of sustainable development, the first measure to be taken is to ensure its proper implementation by, quickly and efficiently, addressing the various challenges and obstructing or slowing this implementation phase, especially in the three major dimensions of sustainable development: economic, social, and environmental. Indeed, according to the UN, more than one billion people are still living in extreme poverty, and income disparities within and between countries are still growing. As well, the still unsustainable consumption and production schemes have generated great economic and social costs and, if continued, may jeopardize life on earth [15].

9.1.2 Challenges to Sustainability

The first challenge to sustainability is to develop a global and authoritative definition of this multidimensional concept, upon which legislations can be drafted and decisions can be made. The main issue in this regard is that among the three main dimensions of sustainably, i.e., economic, environmental, and social, the first two are quantitative notions, whereas the third one is generally estimated in qualitative terms [16]. Besides, the sustainability concept is often reduced to only one of its spheres of influence, mainly economic, then environmental, and to a lesser extent social, depending on how the implementing, private or governmental, body interprets the notion of sustainability [17–20].

For instance, the production of genetically modified organisms (GMOs) is still a hot topic for debates around the world between scientists, decision makers, and environmental activists; the introduction of the concept of bioeconomy gave new perspectives to this debate with highly influential and expected outcome. Indeed, the prospects of using genetically modified crops to produce high yields of biomass or specific compounds, in marginal lands, for the production of added-value nonfood bioproducts are viewed as a sustainable strategy from both economic viewpoint (especially biotech industries) and social perspectives, as it will help rechanneling substantial amounts of food crops back to the food supply chain. Environmentalists, on the other hand, see the same strategy as a threat to biodiversity, with potential health risks. Besides, such process would diminish farmers' self-sufficiency and make them dependent on the crops produced by big biotech companies [21, 22].

Hence, if one single notion (GMOs in this case) could be considered both sustainable and unsustainable, the debate about biorefineries will be much more complicated considering the potential involvement of diverse investors; chemical, agricultural, and biotechnological industries; as well as several governmental bodies and NGOs. Further debates within the framework of bioeconomy are then necessary to overcome such conflicting viewpoints, based on actual scientific fact detailing the economic, environmental, and social advantages and ricks of production procedures or facilities using genetically modified species.

Along with the vital food sector, the sustainability debate is also focusing on another important sector: bioenergy. Indeed, the production of biofuels is viewed around the world as a viable strategy to fulfill future energy needs by gradually replacing currently used fossil fuels, thus ensuring future energy security in a sustainable manner [23]. Nonetheless, even if the production process itself (i.e., raw material to end product) could be qualified as sustainable, the same qualification is not automatically applied to related activities and procedures from the biomass acquisition (crop production or bioresource harvesting) until the storage and marketing of the end products and the management of the generated solid, liquid, and gaseous wastes.

In Europe, for instance, biofuel sustainability is stipulated in the Renewable Energy Directive, which has set the target for EU member states to ensure a renewable energy share of at least 10% for the energy consumption of the transportation sector by 2020, and the use of biofuels must result in an overall cut of greenhouse gas emissions by 35% [24, 25].

The directive also stated that first-generation biofuels (i.e., from energy crops cultivated on agricultural lands) is limited to 7% of final energy consumption in transport by 2020 [26]. Although such measure aims at limiting the impact of the food vs. nonfood and land-use change issues as well as promoting the utilization of nonfood biomass and wastes as feedstocks for biofuel production, the impact on the agricultural landscape in Europe and its potentialities for food and feed production and their prices will still be affected.

In this context, it estimated that in order to implement its renewable energy directive, Europe has to allocate an additional 4.5 million ha of land by 2020 within its borders [27] and support the biofuel sector with subsidies all along the value chain [28]. The subsequent land-use change will have a direct impact on the agricultural sector, but will also induce some indirect risks especially on the greenhouse gas savings because grasslands and forests (to be replaced by energy crop plantations) absorb high levels of carbon dioxide. Thus, transforming those natural lands into agricultural lands would increase the atmospheric CO_2 levels [29, 30]. On a related matter, the introduction of new species (energy crops) to new ecosystems could be of invasive character, which could affect the already established ecological balance. Such environmental threat, although duly emphasized by scientists [31–33], has not yet been properly assumed by policy makers [34].

As well, the sustainability of bioeconomy could be seriously compromised by another important challenge that could generate severe environmental, social, and geopolitical issues, that is, the competition over biomass at national and international levels, and the channeling of certain bioresources for the production of economically profitable products, while more strategic and vital products, with less economic value, will be of lesser interest.

Overall, we are slowly moving away from the food vs. nonfood issue to enter a new and more complicated dilemma: using nonfood biomass but to produce what kind of bioproducts. It's this precise decision that will create the future dilemma around biomass within bioeconomy. Indeed, the response to this simple question will have worldwide economic, environmental, and social repercussions. Therefore, sustainability is a basically a decision, either from the rational but slow top-down process via legislations and regulations or from the spirited but less coherent bottom-up approach. The better scenario is to combine the two approaches in order to generate decisions around the conversion prospective of biomass which are both profitable and nonconflictual and to ensure a good public momentum behind it.

9.1.3 Evaluating the Sustainability of Bioproducts and Biorefineries

As we have seen in Chap. 7, the International Energy Agency (IEA Bioenergy Task 42 Biorefineries) has defined biorefining as "the sustainable processing of biomass into a spectrum of bio-based products (food, feed, chemicals, and/or materials) and bioenergy (biofuels, power and/or heat)" [35]. Thus, by definition biorefineries have to be operated in a sustainable matter from the acquisition of the biomass feedstock until the production and commercialization of the end products. The big question in this regard is how can the sustainability of biorefineries be assessed.

In general, there are many aspects related to sustainability that needs to be considered before qualifying an industrial activity or an end product as sustainable. This includes many environmental considerations such as global warming, pollution, acidification, impact on biodiversity, and land-use change. Other factors include economic aspects such as production costs and profitability, as well as social issues related to employment, health, and human rights. Thus, the sustainability of a product or an industrial and commercial activity can be measured quantitatively and/or qualitatively by jointly evaluating the impacts of the related environmental, economic, and social factors [36, 37].

Commonly, biorefineries are perceived as sustainable production facilities mainly based on the renewable character of the processed biomass. Nonetheless, such perception is currently being readjusted because sustainability cannot be founded solely on one of its three main dimensions [38]. Thus, in order to ensure the sustainability of one of the fundamental components of the bioeconomy concept, biorefining, all the dimensions of sustainability have to be taken into consideration including, but not limited to, by-product valorization (economic dimension), biomass exploitation and waste management (environmental dimension), and food security and workers' rights (social dimension) [39, 40].

Thus, assessing the sustainability of a biorefining facility is a main endeavor during its designing, building, and operating phases. Such assessment is based on the application of several indicators that are used to evaluate the impact of any biorefinery project on all dimensions of sustainability. For instance, the production of biodiesel for cooking waste oil [41] or microalgae [42] is assumed to be a sustainable production mode from environmental and social viewpoints. However, from an economic perspective, biodiesel is not yet a profitable alternative to the relatively cheap conventional diesel [43].

Considering the three-dimensional aspect of sustainability (i.e., economic, environmental, and social), the developed indicators could be used to evaluate either one of those dimensions (mono-dimensional indicators) and two dimensions at the same time (bidimensional indicators) assessing socioeconomic, socio-environmental, or economic-environmental impacts. The last group comprises three-dimensional indicators evaluating the joint impact of all three dimensions [44, 45].

Moreover, metrics based on mass and energy were developed in order to quantify the sustainable performance of production facilities in terms of volatility, flexibility, and robustness [46]. These metrics are often used to measure the economic and environmental sustainability impacts of various biomass conversion procedures and process designs in relation with feedstock use and production-related costs [47–49].

In the following paragraphs, selected metrics are presented for each dimension of sustainability.

9.1.3.1 Economic Metrics

An economically viable biorefinery is a self-sustaining facility that, once it becomes operational, generates enough profitability to self-sustain its activities without requiring governmental assistance or potential reinvestment plans [50]. Several metrics could be applied to assess such important factor. In this context, an interesting study monitored the sustainability of several categories of biorefineries from economic, environmental, and social perspectives [51].

Regarding the economic assessment, the return on investment (ROI) indicator was applied as the economic metric within the applied analytical hierarchy process (AHP), which was combined with a sensitivity analysis. The authors adopted this analytical procedure since both capital and operating costs, as well as the sale price of product, impact the ROI. This metric was estimated for each type of biorefinery by updating a previous economic analysis conducted to evaluate the capital and operating costs for various biofuel-producing facilities (ethanol from grains and from cellulosic biomass and biodiesel via the Fischer-Tropsch process) [52].

Other studies evaluated other economic indicators such as maximizing profit and net present value since minimized cost metrics were reported to be less useful in

Table 9.1 Estimated capital costs, operating costs, and ROI for three types of biorefineries [51]

Biorefinery	Size of the facility (10^6 GJ per year)	Capital cost (millions of US dollars)	Operating cost (US dollars per liter of gasoline equivalent	ROI (%)
Ethanol production form grains	12	143	0.61	24.1
Ethanol production from cellulosic feedstock	12	585	0.53	11.1
Production of Fischer-Tropsch biodiesel	12	786	0.54	7.9

cases like the production of biodiesel where the high production cost is mostly related to the cost of the oily feedstock [53]. Moreover, considering the market volatility and the fluctuations in prices of products, experts deem necessary to include other factors, such as broadening of the products' portfolio and the selling generated by-products, in such economic assessment related to sustainability [54].

Table 9.1 illustrates the capital and operating costs and the estimated ROI for the three studied biorefining facilities.

9.1.3.2 Environmental Indicators

Among the multiple environmental indictors related to sustainability, life cycle assessment (LCA) is frequently used by researchers assessing the environmental impact of products or production processes [55–57]. The environmental factors that could be monitored include the quality of air, soil, and water, management of solid wastes and wastewaters, emissions of greenhouse gases, and conservation of biodiversity and wildlife [58–60].

Other important environmental metrics include energy demand, greenhouse gas emissions, SOx and NOx emissions, potential for eutrophication, and efficient water and energy utilization [61–63]. A more detailed analysis of the mutual relation between bioeconomy and the environment will be given in Sect. 9.9.2.

9.1.3.3 Social Metrics

Few studies tried to tackle the issue of sustainability for various bio-based production processes from a social perspective [64]. Indeed, although such essential observation was made by scientists since the early 1970s [65], sustainability-related studies are still being analyzed from a bi-dimensional perspective (i.e., economic and environmental) [66].

The economic dimension is the most studied aspect for obvious reasons. Then, during the last decade, a clear emphasis on the environmental dimension was made by both the scientific and industrial communities [67]. Currently, the incorporation of the social dimension is gaining more interest, thus enabling an improved and more reliable assessment of the sustainability of the various products and production procedures [68].

A socially sustainable product is a commodity which has limited negative impact on the society during its production, utilization, recycling, and final disposal. Sustainability indicators related to the social dimension are generally associated with issues such as food security, energy security, employment, disparities, and social welfare [69, 70]. The social dimension also encompasses other factors such as the respect for land property rights, social acceptability, and working conditions [71, 72].

In general, as far as sustainability indicators are concerned, the quantitative assessment is not always possible. Indeed several aspects related to biorefining activities either cannot be quantified or assessed via semiquantitative tools [73] or better described qualitatively. This is especially valid for factors related to the social dimension. In this context, qualitative indicators can be applied to evaluate the impacts for a production facility or procedure via semi-qualitative analyses such as using the 1–10 scale [74].

To conclude this sustainability section, it has to be noted that bioeconomy and sustainable development have to be implemented simultaneously in order to ensure the establishment of a genuine sustainable bioeconomy and not a new economic model using biomass to make profit, with a better, but not sufficient, attention to the environmental impact and little to no consideration to the social aspect. We will get back to this important issue in the coming sections dealing with the bioeconomy impact on both the environment and society. Currently, more industrial companies, including large multinational corporations, are adhering to the sustainability concept which is gradually being implemented [75], either a survival or new marketing strategy. For instance, BASF, the world's leading chemical company, is promoting its industrial activities using the slogan "we create chemistry." Recently, they are trying to "create chemistry for a sustainable future." This is certainly a good prospective for change, which could combine both the well-being of the company's stakeholders in the one hand and the welfare of society and the environment on the other.

9.2 Environmental Considerations

As a sustainable economic model, bioeconomy has to generate growth in an environmentally friendly manner. But the relation between bioeconomy and the environment is much more complex. Indeed, the use of bioresources as feedstocks to produce a wide range of commodities links the concept of bioeconomy to the environment in a reciprocal way.

Thus, from an industrial perspective, the overexploitation of bioresources, extensive land-use change, and unwise management of water resources would affect the biodiversity and carbon balance and induce severe environmental changes in both natural ecosystems and the cultivated agricultural land [76–80], which are the main sources for biomass. Such changes will have direct repercussions on the amount and quality of biomass produced for the biorefining facilities and will ultimately affect the basis of the bioeconomy concept by gradually impairing its sustainable dimension.

This would be the worst case scenario that decision makers and industrialists should be aware of and if necessary reminded by experts and scientists. Fortunately, the current tendency is for the application of technically reliable and environmentally friendly production and conversion procedure.

In order to continuously monitor the environmental impact of any activity related to bioeconomy from the biomass acquisition until the marketing, utilization, and disposal of the end products, an assessment module has to be set [81], which includes:

- The development of reliable environmental indicators related to biomass production, logistics, and potential utilizations
- Conducting comparative life cycle assessments of bioproducts and their supply chains, from the biomass production to end-of-life processes
- Designing minimum sustainability criteria related to biomass (production, channeling, and industrial processing) in terms of resource efficiency, greenhouse gas emissions, land-use change, forest exploitation, etc.
- Tracing those environmental sustainability criteria across the entire supply chain

The impact of competing utilization schemes of both natural resources and arable lands on the environment is also a major endeavor to be undertaken by researchers from various scientific fields in order to anticipate any potential risk on the environment in general and the renewability of those valuable bioresources in specific [82, 83]. Considering the seriousness of this competition issue and its potential worldwide impact, not only on the environment but also on geopolitical and social aspects, we will discuss it further in the coming sections.

So, as far as the impact of bioeconomy-related activities on the environment are concerned, key issues are debated in scientific and decision-making circles and more so in the media and public sector. This includes greenhouse gas emission, land-use change, biodiversity, as well as other environmental issues.

9.2.1 Greenhouse Gas (GHG) Emissions

One of the major driving forces behind a large-scale implementation of biorefineries is their potentialities to reduce GHG emissions. In this regard, many research studies have been conducted worldwide to assess the life cycle GHG emissions of ethanol produced from various bioresources and then compared with gasoline. For instance, a study reported that GHG emissions produced in an integrated biochemical refinery producing ethanol from second-generation feedstocks are negative, which corresponds to equivalent reduction in GHG emissions of −7 g CO_2 eq./MJ for ethanol produced from wheat straw and −19 g CO_2 eq./MJ for poplar as feedstock [84].

Such negative values are mainly related to the valorization of the generated by-products in the production of heat and power, or further processing into added-value chemicals including various organic acids. Thus, if the GHG, which would be emitted if those co-products were to be produced form fossil resources, are higher than the actual biorefinery emissions, then overall balance will be negative.

In another study conducted in the United States, the emissions of various biofuels were compared in order to assess the environmental impact of products within the bio-based circle. It was revealed that the production of biofuels form mixtures of local perennial grass which can generate substantial reduction in GHG emissions when compared with ethanol production from corn grain or biodiesel greatened from soybean. The produced biofuels from the various combinations of 16 perennial herbaceous grassland species were found to be carbon negative since the net CO_2 sequestration in soil and roots (4.4 Mg ha^{-1} $year^{-1}$) is almost 13 times higher than the CO_2 emitted during the production of biofuels (0.32 Mg ha^{-1} $year^{-1}$) [85]. An interesting study, funded by the Dutch ministries of Economic Affairs and Infrastructure and Environment, further developed the potential effect of competing utilization of bioresources for bioenergy and biochemical production on the long-term global CO_2 mitigation [86].

Along with the industrial sector, agricultural practices, especially intensive cultivation and breeding techniques, have significant impact on the environment and its water, soil, and air components. The major issue for bioeconomy is to produce a constant flow of biomass to the biorefining activities without compromising its main and vital task of securing food and feed to the ever-increasing world population. For instance, the farmlands exploited in Canada for the production of food and feed as well as livestock farming are responsible for 8% of the country's total greenhouse gas emissions [87].

According to Canadian scientists, such diverse activities within the agricultural sector along with the developing biofuel industries are complicating the analysis of GHG emissions [88, 89]. Thus, for a reliable assessment of the overall GHG emission budget for the agriculture, a better understating of the potential interactions between the various agricultural production systems has to be developed, along with a consistent quantification method to estimate the GHG emission of the cultivated crops upon which those production systems, either for food or feed, are based [90].

9.2.2 Land-Use Change

The agricultural sector is one of the main sources to provide biomass to the various bio-based industries. However, the current agricultural practices and exploited arable lands are at their optimum potential. Thus, if more biomass has to be produced to secure a constant flow of specific biomass to the industrial sector, the agriculture will face serious challenges that could threaten its sustainability. Such pressure has already led to shifts in the utilization of lands either with the agriculture sector or with other sector, mainly forestry.

These land-use changes could be operated in direct and indirect manner. Regarding the direct land-use change, the land which was used for the cultivation of food crops, for instance, started being cultivated for added-value energy crops. As for the indirect land-use change, it occurs when cultivated crops are displaced to another location in order to cultivate other crops and feedstocks for the bio-based industries [84]. Most of the land-use changes (direct and indirect) are being carried out to start the cultivation of energy crops in order to fulfill the increasing feedstock requirement of the biofuel sector. In this regard, many scientists from around the world proved that the expansion of biofuel production schemes is having a global implication on the utilization of lands [91–93]. In an interesting comparative analysis conducted in the United States by agricultural economists, it was reported that the IEA projections of biofuels growth under current policies would require additional arable lands of about 124 Mha within the 2006–2035 period [94].

The direct and indirect land-use changes have also significant impacts on GHG savings, biodiversity, water resources, etc. Indeed, land-use changes are frequently related to a significant modification in the land cover, which affects the carbon balance [95] and in some cases would generate more GHG than the previous land use (agriculture or forestry).

In a related study, converting forest lands into a cropland in the United Kingdom to cultivate Miscanthus for bioethanol production was proved to release 20 t CO_2 eq. ha^{-1} $year^{-1}$. As a consequence, the GHG emissions of the related ethanol production scheme increased from -11 to 310 g CO_2 eq. MJ^{-1}, which is 3.6 times more than the emissions from gasoline [84]. This study (and many others) clearly shows the importance of assessing the repercussions of land-use change on the environment prior to its implementation in order to have real sustainable production models [96, 97]. In this context, several worldwide research investigations used the input-output approach to estimate the land footprints of various land-use schemes such as cropland, forests, and pastures [98, 99].

On a related issue, preserving agricultural soils is a key endeavor in the bioeconomy concept, since the availability and long-term conservation of fertile soils is the prerequisite for the production of food, feed, and raw materials for the bio-based industries [100]. In this context, land-use change could have serious impacts on the quality of the soils by frequent changes in utilization, along with degradation phenomena [101]. Thus, in addition to the issues of climate change, intensive agricultural practices, and increasing world population [102], bioeconomy has to deal with intrinsic issues in the agricultural sector such as land-use change and competition over fertile soils. Hence, many experts in the agricultural and environmental fields are proposing soil governance measures to address those challenges in a sustainable manner [103–105].

9.2.3 Biodiversity

During the land-use change, the biological diversity in associated ecosystems can be affected, thankfully not always in a negative way since valorizing marginal lands for the cultivation of food or energy crops helps in improving the biodiversity [106, 107]. Nevertheless, most of the land-use changes are carried out on fertile lands, which would alter the already fragile biodiversity, especially if conducted frequently and on large scales. Thus, converting forests and grasslands into croplands for the cultivation of energy crops is generally conduced in large mono-crop areas, which would affect the biological balance in both local fauna and flora [108, 109], because forests and various agricultural crops have lower negative impacts on biodiversity than energy crops [110].

Hence, the introduction of new crops could affect the biodiversity in the ecosystem because of the invasive character of those corps. Indeed, in order to meet the increasing demand for energy crops from the biofuel sector, it was reported that non-native plants are being introduced primarily because of their bioenergy potential (sugar or oil contents) and good adaption to the soil and climate conditions. However, the invasive potential of some energy crops seems to be a factor of a lesser importance (economically but a major one for an environmental perspective) [32, 111].

Several research studies conducted around the world tried to assess the invasive potential of biofuel crops. This includes a work carried out in Florida to monitor such potential using the so-called Australian weed risk assessment tool [112]. In this study, it was concluded that *Arundo donax*, *Eucalyptus camaldulensis*, *Eucalyptus grandis*, *Jatropha curcas*, *Leucaena leucocephala*, *Pennisetum purpureum*, and *Ricinus communis* were species with high invasive potential. On the other hand, *Miscanthus* × *giganteus*, *Saccharum arundinaceum*, *Saccharum officinarum*, and the sweet variety of *Sorghum bicolor* have a low probability to become invasive [113].

This issue of biofuels and invasive species was also analyzed in the African continent. The problem in many African countries is that the decision-making process is, to say the least, shortsighted and ill-advised. For instance, in a related study, the discussion was not about the invasive potential of biofuel crops to be introduced, but rather dealing with post-invasion situations due to the introduction of non-native crops suitable for biofuel production but already known for their invasion potential. For example, introduced *Prosopis* species, such as *P. glandulosa* and *P. juliflora*, have invaded around 4 million hectares in Africa, thus compromising crop and pasture production, along with significant reduction in underground water reserves. Such changes also affected animal species and led to the displacement of millions of people in Africa, directly depending on those natural resources for their survival [114].

Other potential environmental issues caused by bioeconomy-related activities, either agricultural or industrial (mainly biorefineries), include water-use efficiency [115, 116], ocean acidification [117], eutrophication in water bodies [118], and leaching of nutrients from fertilizers in agricultural soils and nearby water streams [119]. The generation of organic and inorganic pollutants during agricultural practices and some industrial processes could cause human and ecotoxicological issues [120–122].

Ultimately, regarding the impact of bioeconomy of the environment, we tend to forget that bioeconomy is an economic concept, aiming primarily to generate profitability to the involved actors and not to preserve the environment. So, intrinsically, bioeconomy is not a sustainable concept unless we make sure that it is implemented in a sustainable manner mainly through strict regulations, continuous R&D, and constant reminder of both private and public sectors that this could be the last chance to mitigate the current global issues and prevent the deterioration of others, as illustrated in the planetary boundaries [123], where the biodiversity boundary was transgressed more than any of the other thresholds related to climate change, chemical pollution, and freshwater use [124].

9.3 Social Reflections

While the economic and the environmental dimensions of sustainability within bioeconomy were given a fair deal of attention from scientists and experts, the social dimension didn't receive the same, or even a close, degree of consideration, based on the respective amount of published materials. This is mainly due to the fact that bioeconomy is an economic model that primarily needs to be assessed economically in terms of feasibility, profitability, and viability. Then, conscious about the disastrous legacy of the fossil-based economy on the environmental, this factor was incorporated in the assessment studies after or alongside the economic factor.

Thus, the main strategy for the implementation of bioeconomy seems to be based on optimizing the profitability and competiveness of bio-based agricultural and industrial activities and then trying to reduce potential environmental risks. If an ongoing environmental issue from the so-called petroleum era can be mitigated while implementing bioeconomy, that would be a bonus. If, in some cases, it tends to complicate certain issues, such as biodiversity loss, then the economic dimension generally overweighs the environmental one; that's why in the previous section about sustainability, we stated that it is a matter of a decision with great consequences. Measures such as the carbon taxes could help balancing those important decisions from an environmental perspective, if instigated in an equitable manner [125, 126].

Now, is bioeconomy a good platform to introduce "social taxes" penalizing activities with negative social impacts? If in such important debate the rational scientific thinking could overcome political rhetoric and influential lobbies, possible breakthrough could be made in order to balance the multidimensional aspect of bioeconomy and its sustainable development aspiration by improving the social factor.

In this regard, many scientists have emphasized the multidimensional character of sustainable development and the need to include the various dimensions in any assessment study [127, 128]. Others scientists however, despite the acknowledged interconnection between these dimensions, are stipulating that related studies need to be conducted by individually analyzing the dimensions associated with sustainability. Thus, following such approach, the social and the economic dimensions, for instance, have to be analyzed in two distinguishing spheres, which, according to the author will help grasping the dimensions of sustainability [129].

Nonetheless, despite the different viewpoints on how to tackle the multidimensional aspect of sustainable development, most experts in the field do agree that all the dimensions have to be included and that neglecting the social factor would marginalize the whole notion of sustainability. Thus, if sustainable bioeconomy manages to build and operate profitable and eco-friendly businesses while members of the society, closely witnessing this achievement, are still living in difficult situations by working long hours for minimal wages, having limited access to freshwater resources, and lacking nearby health-care centers for them and their families or schools for their kids, then sustainability will mean nothing to those people. Unfortunately, there are many of them around the world, even in developed countries.

Overall, at a conceptual level, bioeconomy is supposed to deal with a wide range of social affairs such as food biosecurity, health issues, employment, human and labor rights, rural development, social disparities, etc. In a recent study conducted in Sweden, researchers used social sustainability principles to analyze activities of the extraction life cycle phase related to the conception of products. Several social factors were incorporated in this study including poverty, wage assessment, child labor, working hours, occupation injuries, hazards and deaths, access to drinking water, corruption, gender equity, as well as other factors related to social progress such as the access to basic knowledge and information and years of tertiary education [130].

9.3.1 Employment

Properly implemented bioeconomy has the potential to generate new employment opportunities in a wide range of fields from biomass production to bioproduct marketing and recycling and from plants operation and maintenance to equipment and process design and scientific research.

In Europe, with a total turnover estimated at 2.1 trillion euros, bioeconomy is, directly or indirectly, employing between 18.3 and 19 million people. The major fraction of this workforce is employed in the agricultural sector (53%). Other sectors include the food industry (21.3%), forestry- and forest-based industry (13%), textile manufacturing (4.4%), paper and paper production (3.4%), chemical industry (1.5%), and biofuel production (0.1%) [131, 132]. It was also estimated that every new biorefining facility could generate around 100 direct jobs and approximately 1000 indirect employment opportunities related to transportation, construction, maintenance, and other auxiliary services. The green jobs thus include both highly skilled and low-skilled staff (the latter generally from the local workforce) for the production of biomass and derived bioproducts including food, feed, biofuels, biochemicals, and biomaterials [133, 134].

In the United States, a study funded by the Renewable Fuels Association, estimates that, in 2015, the ethanol industry provided around 86,000 direct jobs and overall 357,000 jobs, in fields such as construction, agriculture, manufacturing, and services [135]. The National Biodiesel Board reported that the biodiesel industry is supporting 62,200 jobs all over the United States [136].

However, there is a significant difference in the estimates of current and expected employment opportunities in bioeconomy. In this regard, it was reported that related academic studies tend to be restricted to certain states or regions. As for the reports highlighting national employment status, the issued numbers have to be confirmed by other sources since they are based on analyses funded by industrial associations [137].

9.3.2 *Food Security*

By developing and implementing advanced and eco-friendly technologies and resource-efficient processes, bioeconomy aims at ensuring a global food security while supplying renewable feedstocks to the bio-based industrial sector via sustainable supply chains. The task is very challenging, especially from a worldwide perspective, and the food vs. nonfood dilemma in the biofuel sector is the best illustration. Indeed, it was reported that the competition between food and energy crops over arable lands and water resources triggered international food price spikes and volatility [138, 139]. For a fair and realistic assessment, it has to be mentioned that such increase or volatility in food prices is also affected by other factors (more so than biofuel competition), including the fluctuations of oil prices and speculation [140].

Agriculture is the main sector for food production, along with fishery. The rapidly increasing world population (9.7 billion by 2050 [141]) is already a big challenge for the current food production potential. If the impact of climate change; the competition for lands (direct or indirect land-use change), water, and energy resources [142]; as well as the overexploitation of fish species [143] will be accentuated over the course on the next decades, the ability to meet the world requirement for food will be seriously compromised, and the infamous starvation episodes could resurface in poor countries already struggling politically, economically, and socially.

So, considering all these challenges, how can more food be produced in a sustainable manner? This vital subject is a hot topic (and will remain so for a long period) for the scientific community, and many related research articles and governmental and nongovernmental reports were published [144–146]. Among those studies, an interesting piece was published in Nature by several British scientists, in which the previous question was addressed in a simple and coherent manner [147].

The main factors and driving forces influencing the global food security could be contextualized in the following points:

- In the beginning, the increasing demand for food was dealt with by exploiting new agricultural lands and fish stocks.
- Gradually, the opportunities to exploit new lands became limited. Hence, during the last half century, arable lands have just increased by a mere 9% around the world. Nonetheless, improving the agricultural practices and selecting more robust cultivars helped doubling the production yields of grains [148].
- Later, although the possibility of exploiting new arable lands for food production continued to be an option, competition with other sectors related to the expansion of human activities, mainly urbanization and the need to cultivate nonfood crops to provide feedstocks for the industrial sector. Other factor affected the potentialities of exploiting the remaining lands for agriculture such as the preservation of biodiversity in large natural parks around the world [149], as well as the serious problems of drought [150], salinization [151], and desertification [152, 153], occurring in various regions such as the African Sahel, Americas, India, and China.
- Lately, the agricultural and fishery sectors are facing a new and serious challenge: adaptation to the climate change. Several studies were (and are being) conduced in this regard, in which the negative impact of climate change on food security was highlighted. In general, short-term variation in food supplies is expected to occur, and the amplitude of this variability will vary from one region to another [154]. Nonetheless, its direct impact on food security will be of global dimension, not only for the nutritional side but also the possible occurrence of cross-border migration waves due to food insecurity [155]. Other interesting studies further detailed this important issue of climate change impact on agriculture and food security [156, 157].

Thus, as far as food security is concerned, bioeconomy has to produce more foodstuffs from roughly the same agricultural areas (even less if the problems of desertification, salinization, drought, and soil pollution continue to damage arable lands). Most experts agree that food security, or food insecurity in many parts of the world, is a delicate and urgent issue that needs to anticipate in order to avoid serious and costly complications. While consulting the related literature, the single keyword that is frequently referred to in those studies is sustainability. Indeed, numerous research investigations and assessment studies are emphasizing on the need to promote the sustainable intensification of agriculture [158, 159], the sustainable management of fisheries [160], and the introduction of new techniques and regulations to ensure a sustainable aquaculture [161].

9.3.3 Menacing Threat of Corruption

This very important and delicate issue was included in the social dimension considering its devastating repercussions on entire countries. It could equally be considered as a major threat to the other economic and environmental dimensions of bioeconomy.

Corruption is a lethal disease for any economic model, and bioeconomy is not going to be an exception. Considering the importance of this issue, a brief reminder about its causes and disastrous impact on the so-called petroleum-based economy is worthwhile in order to grasp the impending threat of corruption on bioeconomy and take the necessary measures to eradicate such deeply rooted problem, especially in developing countries, which would be the main beneficiaries of this global shift towards bioeconomy and sustainable development; besides, many of those countries are potential suppliers of biomass feedstocks, the core component of bioeconomy.

When the subject of petroleum paradox was raised decades ago, most of the political economists and experts proposed several concepts to explain the abnormality that oil-rich countries have weak economies including the *Dutch disease* [162] and the *petroleum curse* [163]. Even when the real problem of corruption is mentioned, it is done in a biased way. The blame is mostly put on oil-producing countries and their predisposition for corruption and violence [164, 165] and rarely on big corporations such as the "seven sisters" or oil-consuming countries, some of them being notorious for their colonial history [166]. Overall, we should always bear in mind that corruption is a two-way street paved with greed, the corrupted giving what he does not possess to the corruptor who does not deserve.

How to avoid this paradox? Self-restrain in the answer and it is possible. Indeed, Norway showed a great deal of self-restrain when discovering fossil fuels (oil and gas) in the North Sea in the late 1960s. This self-restrain, seriously needed in most oil-producing countries, turned the so-called resource curse into a blessing (similar analogy could be made for biomass-producing countries). In the beginning though, Norway lost several manufacturing industries and skilled labor to the lucrative oil sector, lost significant competitive power, and straggled to accommodate the oil revenues with its demanding social welfare system. But the country was almost immune to the lethal disease of corruption (strong political and social traditions of transparency). Thus, although suffering from the severe repercussions of the oil bonanza, which shook the core of its economic system, the Nordic country was able to face this curse and implement the proper strategy by encouraging the private sector and creating a sovereign "oil fund" to deposit the surplus wealth produced by its petroleum income. But again, even if the idea of such "trust fund" interests other countries, it would be worthless if not protected from corruption.

9.4 Final Remarks and Conclusions

Throughout this, hopefully, inspiring book, we tried to tackle the challenging task of discussing the highly anticipated and multidimensional concept of bioeconomy, so that young students could be introduced to this notion, scientists and researcher could explore and compare each other's accomplishments and put more R&D effort to deal with the main issues facing bioeconomy in the agricultural, industrial, technological, environmental, social, and marketing sectors. Reserving an entire chapter to talk about the industrial biorefining activities with study cases from around the word will definitely be of interest to industrialists and investors to assess the situation and think about new business ideas. With this book, decision makers could gain more knowledge about bioeconomy and therefore develop their own perception of this concept.

After 2 years of work on this exciting book and the consultation of thousands of research papers; hundreds of reports from academia, governments, and the industry; as well as tens of books on bioeconomy-related matters, we would like to conclude by stating several takeaway points that we deem necessary to ensure a smooth transition towards bioeconomy and then its worldwide implementation in a sustainable manner.

- First of all, bioeconomy is expected to generate a sustainable economic growth and provide food and energy and other commodities to a growing world population while improving the quality of life and preserving the environment [167, 168]. To be realistic, this perception is a bit deceptive because we are talking about an economic model based on the exploitation of bioresources and not a miraculous therapy to remediate humanity's problems at once. Reaching some of these goals before a further deterioration of the current situation around the globe including climate change, serious pollution issues, and dangerous geopolitical tensions (generally to control the remaining fossil resources) is an achievement on its own.
- Considering the complicated current circumstances all over the world (economically, socially, environmentally, and politically), as well as the fact that bioeconomy is not yet a fully mature concept, a transition phase is necessary to ease the shift from the current economic model mainly based of the exploitation of fossil resources to the more sustainable bio-based economic model.
- Within this transition phase, the catchphrase is "pragmatic raw material change." Nowadays, as the petroleum supply is progressively depleting, a replacement is already planned but then again using other fossil resources including natural gas, coal, and unconventional fossil resources such as tar sand and oil shale [169, 170]. Thus, while bioeconomy is maturing and progressively expanding, the first transition step should aim at gradually reducing the reliance to fossil fuels. At this stage, it is out of question to stop using fossil fuels as raw material because the current economies are too dependent and too weak for such drastic approach. The second phase is a generalized upgrading campaign aiming at substantially increasing the share of bio-based commodities in the markets. By

the time bioeconomy becomes fully functioning (i.e., merging profitability and sustainability), little or no fossil resources will be left to compete renewable bioresources. In order to reach this goal, compromises are more than necessary to make sure that all the contributors could benefit for this thriving economic model.

- Bioeconomy is aknowledge-based concept which makes it closely interconnected with innovative R&D. As a consequence, the implementation of bioeconomy is highly depending on two main factors: the quality of education and training and the amount of funds allocated to bioeconomy-related research. These two factors will make significant difference between competing countries and industrial corporations.
  - Education and training: A coherent educational strategy has to be developed covering all the sectors of bioeconomy and including all potential contributors. Thus, education institutions and training centers should constantly update their curricula, partly based on feedbacks from stakeholders in bioeconomy and also in anticipation for future needs of skilled workforce in emerging fields. Flexible exchange schemes of highly qualified personnel between academia, industry, policy-making establishments, and governmental regulatory establishments will allow a quicker, smoother, more importantly, a coherent implementation of bioeconomy.
  - R&D funding: Obviously more funds need to be made available, under competitive grounds, for researchers involved in bioeconomy and sustainability. Also, there is consensus among the scientific community about the need to simplify the available funding schemes, which, most of the time, are purposely complex. Thus, the funding bodies should develop simple, yet still highly competitive, funding opportunities for R&D activities in bioeconomy.
- As a global concept, international clusters and networks need to be built around bioeconomy [171], connecting scientists, governments, industrialists, unions, as well as representatives from the public and private sectors. In this context, and for various historical and geopolitical reasons, the north-south relationship, respectively between developed and developing nations, was not a success story and even conflicting in some cases. Thus, in order to ensure a real sustainable bioeconomy, genuine north-south cooperation schemes need to be developed based on mutual benefits and reciprocal respect. The reinforcement of the south-south cooperation, especially in the agricultural and industrial sectors, will be a major step forward in the global implementation of bioeconomy.

Ultimately, bioeconomy is initiating a green industrial revolution around the world, and we should all make our contribution in the various dimensions of this sustainable economic model: researchers in their labs, teachers in their classes, farmers in their fields, fishermen in their ships, engineers and workers in their plants, and decision makers in their offices, all aiming at a better and sustainable future for our kids.

References

1. McLaren JS. Crop biotechnology provides an opportunity to develop a sustainable future. Trends Biotechnol. 2005;23:339–42.
2. Smil V. Energy in the twentieth century: resources, conversions, costs, uses, and consequences. Annu Rev Energy Environ. 2000;25:21–51.
3. Ortega-Argilés R. The transatlantic productivity gap: a survey of the main causes. J Econ Surv. 2012;26:395–419.
4. Black R, Adger WN, Arnell NW, Dercon S, Geddes A, Thomas D. The effect of environmental change on human migration. Glob Environ Chang. 2011;21:S3–11.
5. Toth G, Szigeti C. The historical ecological footprint: from over-population to over-consumption. Ecol Indic. 2016;60:283–91.
6. US Energy Information Administration. The International Energy Outlook 2016 (IEO2016). Chapter 1 – World energy demand and economic outlook. http://www.eia.gov/outlooks/ieo/pdf/0484%282016%29.pdf. Full report published 11 May 2016.
7. Parajuli R, Dalgaard T, Jørgensen U, et al. Biorefining in the prevailing energy and materials crisis: a review of sustainable pathways for biorefinery value chains and sustainability assessment methodologies. Renew Sustain Energy Rev. 2015;43:244–63.
8. Marris E. Sugar cane and ethanol: drink the best and drive the rest. Nature. 2006;444:670–2.
9. Polack R, Wood S, Bradley E. Fossil fuels and food security: analysis and recommendations for community organizers. J Community Prac. 2008;16:359–75.
10. Mohr SH, Evans GM. Forecasting coal production until 2100. Fuel. 2009;88:2059–67.
11. Pirages D. Sustainability as an evolving process. Futures. 1994;26:197–205.
12. Mebratu D. Sustainability and sustainable development: historical and conceptual review. Environ Impact Assess Rev. 1998;18:493–520.
13. United Nations (UN). Sustainable development goals – 17 goals to transform our world. http://www.un.org/sustainabledevelopment/sustainable-development-goals/
14. Anand M. Innovation and sustainable development: a bioeconomic perspective. Brief for global sustainable development report, GSDR. 2016. https://sustainabledevelopment.un.org/content/documents/982044_Anand_Innovation%20and%20Sustainable%20Development_A%20Bioeconomic%20Perspective.pdf
15. United Nations' Department of Economic and Social Affairs. World economic and social survey 2013 – sustainable development challenges. United Nations publication. New York. 2013. https://sustainabledevelopment.un.org/content/documents/2843WESS2013.pdf
16. Parada MP, Osseweijer P, Duque JAP. Sustainable biorefineries, an analysis of practices for incorporating sustainability in biorefinery design. Ind Crops Prod. doi:10.1016/j.indcrop.2016.08.052.
17. Kemp R, Martens P. Sustainable development: how to manage something that is subjective and never can be achieved? Sustain Sci Pract Policy. 2007;3:5–14.
18. de Vries BJM, Petersen AC. Conceptualizing sustainable development: an assessment methodology connecting values, knowledge, worldviews and scenarios. Ecol Econ. 2009;68:1006–19.
19. Van Opstal M, Hugé J. Knowledge for sustainable development: a worldviews perspective. Environ Dev Sustain. 2013;15:687–709.
20. Janeiro L, Patel KM. Choosing sustainable technologies. Implications of the underlying sustainability paradigm in the decision-making process. J Clean Prod. 2015;105:438–46.
21. Lotz LAP, Van De Wiel CCM, Smulders MJM. Genetically modified crops and sustainable agriculture: a proposed way forward in the societal debate. NJAS Wagening J Life Sci. 2014;70:95–8.
22. Hedlund-de WA. Rethinking sustainable development: considering how different worldviews envision "development" and "quality of life". Sustainability. 2014;6:8310–28.
23. Robertson GP, Dale VH, Doering OC, et al. Sustainable biofuels reflux. Science. 2008;322:49–50.

24. Directive 2009/28/EC of the European parliament and of the council of 23 April 2009 on the promotion of the use of energy from renewable sources and amending and subsequently repealing Directives 2001/77/EC and 2003/30/EC. 2009.
25. European biofuels technology platform (EBTP). Biofuels and sustainability issues. http://biofuelstp.eu/sustainability.html. Updated 9 Sep 2016.
26. Lane J. EU reshapes its biofuels policy. 2015. http://www.biofuelsdigest.com/bdigest/2015/04/16/eu-reshapes-its-biofuels-policy/. Published 16 Apr 2015.
27. Harrison P. Special report: Europe finds politics and biofuels don't mix. Reuters. 2010. http://www.reuters.com/article/idUSTRE6641FD20100705. Published 5 July 2010.
28. Banse M, van Meijl H, Tabeau A, Woltjer G. Will EU biofuel policies affect global agricultural markets? Eur J Agric Econ. 2008;35:117–41.
29. Gitz V, Ciais P. Amplifying effects of land-use change on future atmospheric CO_2 levels. Glob Biogeochem Cycles. 2003;17:1024.
30. European Commission – Energy. Land use change. 2016. https://ec.europa.eu/energy/en/topics/renewable-energy/biofuels/land-use-change. Updated 12 Dec 2016.
31. Raghu S, Anderson RC, Daehler CC, et al. Adding biofuels to the invasive species fire? Science. 2006;313:1742.
32. Barney JN, DiTomaso JM. Nonnative species and bioenergy: are we cultivating the next invader. Bioscience. 2008;58:64–70.
33. Davis AS, Cousens RD, Hill J, Mack RN, Simberloff D, Raghu S. Screening bioenergy feedstock crops to mitigate invasion risk. Front Ecol Environ. 2010;8:533–9.
34. Sheppard AW, Gillespie I, Hirsch M, Begley C. Biosecurity and sustainability within the growing global bioeconomy. Curr Opin Environ Sustain. 2011;3:4–10.
35. International Energy Agency (IEA). IEA bioenergy Task 42 on biorefineries. Biorefining in a future bioeconomy. http://www.ieabioenergy.com/task/biorefining-sustainable-processing-of-biomass-into-a-spectrum-of-marketable-biobased-products-and-bioenergy/
36. Roseland M. Sustainable community development: integrating environmental, economic, and social objectives. Prog Plan. 2000;54:73–132.
37. Gomes CP. Computational sustainability: computational methods for a sustainable environment, economy, and society. Bridge. 2009;39:5–13.
38. Pfau S, Hagens J, Dankbaar B, Smits A. Visions of sustainability in bioeconomy research. Sustainability. 2014;6:1222–49.
39. Demirbas A. Biorefineries: current activities and future developments. Energy Convers Manag. 2009;50:2782–801.
40. Cambero C, Sowlati T. Assessment and optimization of forest biomass supply chains from economic, social and environmental perspectives – a review of literature. Renew Sust Energy Rev. 2014;36:62–73.
41. Yaakob Z, Mohammad M, Alherbawi M, Alam Z, Sopian K. Overview of the production of biodiesel from waste cooking oil. Renew Sust Energy Rev. 2013;18:184–93.
42. Chisti Y. Biodiesel from microalgae. Biotechnol Adv. 2007;25:294–306.
43. US Department of Energy, Office of Energy Efficiency & Renewable Energy. Biodiesel. 2016. http://www.fueleconomy.gov/feg/biodiesel.shtml
44. Despotovic D, Cvetanovic S, Nedic V, Despotovic M. Economic, social and environmental dimension of sustainable competitiveness of European countries. J Environ Plan Manag. 2016;59:1656–78.
45. Fermeglia M, Longo G, Toma L. Computer aided design for sustainable industrial processes: specific tools and applications. AIChE J. 2009;55:1065–78.
46. Mansoornejad B, Pistikopoulos EN, Stuart P. Metrics for evaluating the forest biorefinery supply chain performance. Comput Chem Eng. 2013;54:125–39.
47. Sacramento-Rivero JC. A methodology for evaluating the sustainability of biorefineries: framework and indicators. Biofuels Bioprod Biorefin. 2012;6:32–44.

48. Ojeda K, Avila O, Suarez J, Kafarov V. Evaluation of technological alternatives for process integration of sugarcane bagasse for sustainable biofuels production-part 1. Chem Eng Res Des. 2011;89:270–9.
49. Sacramento-Rivero JC, Navarro-Pineda F, Vilchiz-Bravo LE. Evaluating the sustainability of biorefineries at the conceptual design stage. Chem Eng Res Design. 2016;107:167–80.
50. Pérez ATE, Camargo M, Rincón PCN, Marchant MA. Key challenges and requirements for sustainable and industrialized biorefinery supply chain design and management: a bibliographic analysis. Renew Sust Energy Rev. 2017;69:350–9.
51. Schaidle JA, Moline CJ, Savage PE. Biorefinery sustainability assessment. Environ Prog Sustain Energy. 2011;30:743–53.
52. Wright M, Brown R. Comparative economics of biorefineries based on the biochemical and thermochemical platform. Biofuels Bioprod Biorefin. 2007;1:49–56.
53. Rincón LE, Valencia MJ, Hernández V, et al. Optimization of the Colombian biodiesel supply chain from oil palm crop based on techno-economical and environmental criteria. Energy Econ. 2015;47:154–67.
54. You F, Tao L, Graziano DJ, Snyder SW. Optimal design of sustainable cellulosic biofuel supply chains: multiobjective optimization coupled with life cycle assessment and input–output analysis. AIChE J. 2012;58:1157–80.
55. Sander K, Murthy GS. Life cycle analysis of algae biodiesel. Int J Life Cycle Assess. 2010;15:704–14.
56. Kloepffer W. Life cycle sustainability assessment of products. Int J Life Cycle Assess. 2008;13:89–94.
57. Tabone MD, Cregg JJ, Beckman EJ, Landis AE. Sustainability metrics: life cycle assessment and green design in polymers. Environ Sci Technol. 2010;44:8264–9.
58. Tanzil D, Beloff BR. Assessing impacts: overview on sustainability indicators and metrics. Environ Qual Manag. 2006;15:41–56.
59. Ruiz-Mercado GJ, Smith RL, Gonzalez MA. Sustainability indicators for chemical processes: I. Taxonomy. Ind Eng Chem Res. 2012;51:2309–28.
60. Bare JC. Life cycle impact assessment research developments and needs. Clean Technol Environ Policy. 2010;12:341–51.
61. Thiede S, Seow Y, Andersson J, Johansson B. Environmental aspects in manufacturing system modelling and simulation – state of the art and research perspectives. CIRP J Manuf Sci Technol. 2013;6:78–87.
62. Nanda S, Azargohar R, Dalai AK, Kozinski JA. An assessment on the sustainability of lignocellulosic biomass for biorefining. Renew Sust Energy Rev. 2015;50:925–41.
63. Simpson T, Sharpley A, Howarth R, Paerl H, Mankin K. The new gold rush: fueling ethanol production while protecting water quality. J Environ Qual. 2008;37:318–24.
64. Mu J, Zhang G, MacLachlan DL. Social competency and new product development performance. IEEE Trans Eng Manag. 2011;58:363–76.
65. Varble DL. Social and environmental considerations in new product development. J Mark. 1972;36:11–5.
66. Gmelin H, Seuring S. Determinants of a sustainable new product development. J Clean Prod. 2014;69:1–9.
67. Simon M, Poole S, Sweatman A, Evans S, Bhamra T, McAloone T. Environmental priorities in strategic product development. Bus Strateg Environ. 2000;9:367–77.
68. Aguilera RV, Rupp DE, Williams CA. Putting the S back in corporate social responsibility: a multilevel theory of social change in organizations. Acad Manag Rev. 2007;32:836–63.
69. Fleurbaey M. On sustainability and social welfare. J Environ Econ Manag. 2015;71:34–53.
70. Martinet V. A characterization of sustainability with indicators. J Environ Econ Manag. 2011;61:183–97.
71. Dempsey N, Bramley G, Power S, Brown C. The social dimension of sustainable development: defining urban social sustainability. Sustain Dev. 2011;19:289–300.

72. Bautista S, Narvaez P, Camargo M, Chery O, Morel L. Biodiesel-TBL+: a new hierarchical sustainability assessment framework of PC&I for biodiesel production – part I. Ecol Indic. 2016;60:84–107.
73. Ataei ME, Asr T, Khalokakai R, Ghanbari K, Mohammadi MRT. Semi-quantitative environmental impact assessment and sustainability level determination of coal mining using a mathematical model. J Min Environ. 2016;7:185–93.
74. Wu RQ, Yang D, Chen JQ. Social life cycle assessment revisited. Sustainability. 2014;6:4200–26.
75. Bakshi BR, Fiksel J. The quest for sustainability: challenges for process systems engineering. AIChE J. 2003;49:1350–8.
76. Rosegrant MW, Ringler C, Zhu T, Tokgoz S, Bhandary P. Water and food in the bioeconomy: challenges and opportunities for development. Agric Econ. 2013;44:139–50.
77. Lewandowski I. Securing a sustainable biomass supply in a growing bioeconomy. Glob Food Secur. 2015;6:34–42.
78. Scarlat N, Dallemand JF, Monforti-Ferrario F, Nita V. The role of biomass and bioenergy in a future bioeconomy: policies and facts. Environ Dev. 2015;15:3–34.
79. Dong XB, Yu BH, Brown MT, et al. Environmental and economic consequences of the overexploitation of natural capital and ecosystem services in Xilinguole league. China Energy Policy. 2014;67:767–80.
80. Aragao LEOC, Poulter B, Barlow JB, et al. Environmental change and the carbon balance of Amazonian forests. Biol Rev. 2014;89:913–31.
81. Nita V, Benini L, Ciupagea C, Kavalov B, Pelletier N. Bio-economy and sustainability: a potential contribution to the bio-economy observatory. European Commission Joint Research Centre Institute for Environment and Sustainability. Report EUR 25743 EN. 2013.
82. Gerssen-Gondelach SJ, Saygin D, Wicke B, et al. Competing uses of biomass – assessment and comparison of the performance of bio-based heat, power, fuels and materials. Renew Sust Energy Rev. 2014;40:964–98.
83. Brunori G. Biomass, biovalue and sustainability: some thoughts on the definition of the bioeconomy. EuroChoices. 2013;12:48–52.
84. Azapagic A. Sustainability considerations for integrated biorefineries. Trends Biotechnol. 2014;32:1–4.
85. Tilman D, Hill J, Lehman C. Carbon-negative biofuels from low-input high-diversity grassland biomass. Science. 2006;314:1598–600.
86. Daioglou V, Wicke B, Faaij APC, van Vuuren DP. Competing uses of biomass for energy and chemicals: implications for long-term global CO_2 mitigation potential. GCB Bioenergy. 2015;7:1321–34.
87. Environment and climate change Canada. National inventory report 1990–2014. Greenhouse gas sources and sinks in Canada. 2016. https://www.ec.gc.ca/ges-ghg/662F9C56-B4E4-478B-97D4-BAABE1E6E2E7/2016_NIR_Executive_Summary_en.pdf
88. Janzen HH, Angers DA, Boehm M, et al. A proposed approach to estimate and reduce net greenhouse gas emissions from whole farms. Can J Soil Sci. 2006;86:401–18.
89. Klein KK, LeRoy DG The biofuels frenzy: what's in it for Canadian agriculture? Green paper prepared for the Alberta Institute of Agrologists. Annual Conference of Alberta Institute of Agrologists. Banf, Alberta. 2007.
90. Dyer JA, Vergé XPC, Desjardins RL, Worth DE, McConkey BG. The impact of increased biodiesel production on the greenhouse gas emissions from field crops in Canada. Energy Sustain Dev. 2010;14:73–82.
91. Miljkovic D, Ripplinger D, Shaik S. Impact of biofuel policies on the use of land and energy in U.S. agriculture. J Policy Model. 2016;38:1089–98.
92. Panichelli L, Gnansounou E. Impact of agricultural-based biofuel production on greenhouse gas emissions from land-use change: key modelling choices. Renew Sust Energy Rev. 2015;42:344–60.

93. Giovannetti G, Ticci E. Determinants of biofuel-oriented land acquisitions in Sub-Saharan Africa. Renew Sust Energy Rev. 2016;54:678–87.
94. Hertel T, Steinbuks J, Baldos U. Competition for land in the global bioeconomy. Agric Econ. 2013;44:129–38.
95. Lambin EF, Geist HJ, Lepers E. Dynamics of land-use and land-cover change in tropical regions. Annu Rev Environ Resour. 2003;28:205–41.
96. Wassell CS, Dittmer TD. Are subsidies for biodiesel economically efficient? Energy Policy. 2006;34:3993–4001.
97. Searchinger T, Heimlich R, Houghton RA, et al. Use of US croplands for biofuels increases greenhouse gases through emissions from land-use change. Science. 2008;319:1238–40.
98. Weinzettel J, Hertwich EG, Peters GP, Steen-Olsen K, Galli A. Affluence drives the global displacement of land use. Glob Environ Chang. 2013;23:433–8.
99. O'Brien M, Schütz H, Bringezu S. The land footprint of the EU bioeconomy: monitoring tools, gaps and needs. Land Use Policy. 2015;47:235–46.
100. Powlson DS, Gregory PJ, Whalley WR, et al. Soil management in relation to sustainable agriculture and ecosystem services. Food Policy. 2011;36:S72–87.
101. Wall DH, Six J. Give soils their due. Science. 2015;347:695.
102. Koch A, McBratney A, Adams M, et al. Soil security: solving the global soil crisis. Glob Policy. 2013;4:434–41.
103. Montanarella L, Vargas R. Global governance of soil resources as a necessary condition for sustainable development. Curr Opin Environ Sustain. 2012;4:559–64.
104. Howard T, Larson A. Soil governance: assessing cross-disciplinary perspectives. Int J Rural Law Policy. 2015;1:1–8.
105. Weigelt J, Müller A, Janetschek H, Töpfer K. Land and soil governance towards a transformational post-2015 development agenda: an overview. Curr Opin Environ Sustain. 2015;15:57–65.
106. Huston MA. The three phases of land-use change: implications for biodiversity. Ecol Appl. 2005;15:1864–78.
107. Plieninger T, Gaertner M. Harnessing degraded lands for biodiversity conservation. J Nat Conserv. 2011;19:18–23.
108. Eppink FV, van den Bergh JCJM. Ecological theories and indicators in economic models of biodiversity loss and conservation: a critical review. Ecol Econ. 2007;61:284–93.
109. Fletcher RJ, Robertson BA, Evans J, et al. Biodiversity conservation in the era of biofuels: risks and opportunities. Front Ecol Environ. 2011;9:161–8.
110. Jeswani HK, Azapagic A. Life cycle sustainability assessment of second generation biodiesel. In: Luque R, Melero JA, editors. Advances in biodiesel preparation – Second generation processes and technologies. Sawston: Woodhead; 2012.
111. Barney JN, DiTomaso JM. Global climate niche estimates for bioenergy crops and invasive species of agronomic origin: potential problems and opportunities. PLoS One. 2011;6:e17222.
112. Pheloung PC, Williams PA, Halloy SR. A weed risk assessment model for use as a biosecurity tool evaluating plant introductions. J Environ Manag. 1999;57:239–51.
113. Gordon DR, Tancig KJ, Onderdonk DA, Gantz CA. Assessing the invasive potential of biofuel species proposed for Florida and the United States using the Australian weed risk assessment. Biomass Bioenergy. 2011;35:74–9.
114. Witt ABR. Biofuels and invasive species from an African perspective – a review. GCB Bioenergy. 2010;2:321–9.
115. Moraes MM, Ringler C, Cai X. Policies and instruments affecting water use for bioenergy production. Biofuels Bioprod Biorefin. 2011;5:431–44.
116. Gheewala SH, Berndes G, Jewitt G. The bioenergy and water nexus. Biofuels Bioprod Biorefin. 2011;5:353–60.
117. Miller CA. Modeling risk in complex bioeconomies. J Respons Innov. 2015;2:124–7.

118. Golden JS, Handfield R. The emergent industrial bioeconomy. Ind Biotechnol. 2014;10:371–5.
119. Venkata MS, Nikhil GN, Chiranjeevi P, et al. Waste biorefinery models towards sustainable circular bioeconomy: critical review and future perspectives. Bioresour Technol. 2016;215:2–12.
120. Cordella M, Torri C, Adamiano A, et al. Bio-oils from biomass slow pyrolysis: a chemical and toxicological screening. J Hazard Mater. 2012;231:26–35.
121. Pimenta AS, Bayona JM, Garcia MT, Solanas AM. Evaluation of acute toxicity and genotoxicity of liquid products from pyrolysis of *Eucalyptus grandis* wood. Arch Environ Contamin Toxicol. 2000;38:169–75.
122. Bernardo M, Lapa N, Barbosa R, et al. Chemical and ecotoxicological characterization of solid residues produced during the co-pyrolysis of plastics and pine biomass. J Hazard Mater. 2009;166:309–17.
123. Rockström J, Steffen W, Noone K, et al. A safe operating space for humanity. Nature. 2009;461:472–5.
124. Mace GM, Reyers B, Alkemade R, et al. Approaches to defining a planetary boundary for biodiversity. Glob Environ Chang. 2014;28:289–97.
125. Roughgarden T, Schneider SH. Climate change policy: quantifying uncertainties for damages and optimal carbon taxes. Energy Policy. 1999;27:415–29.
126. Warren R. Environmental economics: optimal carbon tax doubled. Nat Clim Chang. 2014;4:534–5.
127. Sachs I. Social sustainability and whole development: exploring the dimensions of sustainable development. In: Egon B, Thomas J, editors. Sustainability and the social sciences: a cross-disciplinary approach to integrating environmental considerations into theoretical reorientation. London: Zed Books; 1999.
128. Spangenberg JH, Omannn I. Assessing social sustainability: social sustainability and its multicriteria assessment in a sustainability scenario for Germany. Int J Innov Sustain Dev. 2006;1:318–48.
129. Lehtonen M. The environmental–social interface of sustainable development: capabilities, social capital, institutions. Ecol Econ. 2004;49:199–214.
130. Gould R, Missimer M, Mesquita PL. Using social sustainability principles to analyse activities of the extraction lifecycle phase: learnings from designing support for concept selection. J Clean Prod. 2017;140:267–76.
131. Carrez D. European bioeconomy 2013: € 2.1 trillion turnover and 18.3 million employees. Press release from the bio-based industries consortium (BIC). 2016. http://biconsortium.eu/sites/biconsortium.eu/files/news-image/BIC_PressRelease_Bioeconomy2013_3March2016.pdf. Published 3 Mar 2016.
132. Reinshagen P. Bioeconomy: much more employment in biobased chemicals than in biofuels. Bio Based Press. 2015. http://www.biobasedpress.eu/2015/06/bioeconomy-much-more-employment-in-biobased-chemicals-than-in-biofuels/. Published 2 June 2015.
133. ePURE – European renewable ethanol. Jobs & Growth. 2016. http://epure.org/about-ethanol/ethanol-benefits/jobs-and-growth/
134. Kromus S, Wachter B, Koschuh M, et al. The green biorefinery Austria-development of an integrated system for green biomass utilization. Chem Biochem Eng Q. 2004;18:8–12.
135. Urbanchuk JM. ABF economics – contribution of the ethanol industry to the economy of the United States in 2015. 2016. http://www.ethanolrfa.org/wp-content/uploads/2016/02/Ethanol-Economic-Impact-for-2015.pdf. Published 5 Feb 2016.
136. National Biodiesel Board. Production statistics. 2016. http://biodiesel.org/production/production-statistics
137. US Department of Energy – Energy efficiency and renewable energy. Green jobs in the U.S. bioeconomy DOE/EE-1222. 2015. https://www.energy.gov/sites/prod/files/2015/05/f22/bioenergy_green_jobs_factsheet_2015.pdf

138. Tenenbaum DJ. Food vs fuel: diversion of crops could cause more hunger. Environ Health Perspect. 2008;116:A254–7.
139. Tadasse G, Algieri B, Kalkuhl M, von Braun J. Drivers and triggers of international food price spikes and volatility. In: Kalkuhl M, von Braun J, Torero M, editors. Food price volatility and its implications for food security and policy. Berlin: Springer; 2016.
140. Ajanovic A. Biofuels versus food production: does biofuels production increase food prices? Energy. 2011;36:2070–6.
141. United Nations Department of Economic and Social Affairs. World population projected to reach 9.7 billion by 2050. 2015. http://www.un.org/en/development/desa/news/population/2015-report.html. Published 29 July 2015.
142. Lal R. Food security in a changing climate. Ecohydrol Hydrobiol. 2013;13:8–21.
143. Pauly D, Christensen V, Guénette S, et al. Towards sustainability in world fisheries. Nature. 2002;418:689–95.
144. Smith P, Gregory PJ. Climate change and sustainable food production. Proc Nutr Soc. 2013;72:21–8.
145. Garnett T, Appleby MC, Balmford A, et al. Sustainable intensification in agriculture: premises and policies. Science. 2013;341:33–4.
146. McKenzie FC, Williams J. Sustainable food production: constraints, challenges and choices by 2050. Food Sec. 2015;7:221–33.
147. Godfray HCJ, Beddington JR, Crute IR, et al. Science. 2010;327:812–8.
148. Pretty J. Agricultural sustainability: concepts, principles and evidence. Philos Trans R Soc Lond Ser B Biol Sci. 2008;363:447–65.
149. Balmford A, Green R, Scharlemann JP. Sparing land for nature: exploring the potential impact of changes in agricultural yield on the area needed for crop production. Glob Chang Biol. 2005;11:1594–605.
150. Udmale PD, Ichikawa Y, Kiem AS, Panda SN. Drought impacts and adaptation strategies for agriculture and rural livelihood in the Maharashtra state of India. Open Agric J. 2014;8:41–7.
151. Thomas DSG, Middleton NJ. Salinization: new perspectives on a major desertification issue. J Arid Environ. 1993;24:95–105.
152. D'Odorico P, Bhattachan A, Davis KF, Ravi S, Runyan CW. Global desertification: drivers and feedbacks. Adv Water Resour. 2013;51:326–44.
153. Danfeng S, Dawson R, Baoguo L. Agricultural causes of desertification risk in Minqin. China J Environ Manag. 2006;79:348–56.
154. Wheeler T, von Braun J. Climate change impacts on global food security. Science. 2013;341:508–13.
155. Crush J. Linking food security, migration and development. Int Migr. 2013;51:61–75.
156. Schmidhuber J, Tubiello FN. Global food security under climate change. Proc Natl Acad Sci. 2007;104:19703–8.
157. Lobell DB, Burke MB, Tebaldi C, et al. Prioritizing climate change adaptation needs for food Security in 2030. Science. 2008;319:607–10.
158. Tilman D, Balzer C, Hill J, Befort BL. Global food demand and the sustainable intensification of agriculture. Proc Natl Acad Sci. 2011;108:20260–4.
159. Pretty J, Toulmin C, Williams S. Sustainable intensification in African agriculture. Int J Agric Sustain. 2011;9:5–24.
160. Walters C, Martell SJ. Stock assessment needs for sustainable fisheries management. Bull Mar Sci. 2002;70:629–38.
161. Subasinghe R, Soto D, Jia J. Global aquaculture and its role in sustainable development. Rev Aquac. 2009;1:2–9.
162. Wijnbergen SV. The Dutch disease: a disease after all? Econ J. 1984;94:41–55.
163. Ross ML. The oil curse: how petroleum wealth shapes the development of nations. USA: Princeton University Press; 2012.
164. Jansen AR. Second generation biofuels and biomasses. Essential guide for investors, scientists and decision makers. New Jersey: Wiley; 2013.

165. Ross ML. Blood barrels: why oil wealth fuels conflict. Foreign Aff. 2008;87:1–7.
166. Vidal J. Energy: a crude awakening. Nature. 2012;482:306.
167. Chisti Y. A bioeconomy vision of sustainability. Biofuels Bioprod Biorefin. 2010;4:359–61.
168. Tawfik M. Asia and bioeconomy: growing synergies. Asian Biotechnol Dev Rev. 2004;6:5–8.
169. Clarke T. Tar sands showdown: Canada and the new politics of oil in an age of climate change. Toronto: Lorimer; 2009.
170. Bjørlykke K. Unconventional hydrocarbons: oil shales, heavy oil, tar sands, shale oil, shale gas and gas hydrates. In: Bjørlykke K, editor. Petroleum geoscience – from sedimentary environments to rock physics. Berlin: Springer; 2015.
171. El-Chichakli B, von Braun J, Lang C, Barben D, Philp J. Policy: five cornerstones of a global bioeconomy. Nature. 2016;535:221–3.

The manufacturer's authorised representative in the EU is Springer Nature Customer Service Centre GmbH, Europaplatz 3, 69115 Heidelberg, Germany. If you have any concerns regarding our products, please contact ProductSafety@springernature.com

Printed and bound by CPI Group (UK) Ltd, Croydon, CR0 4YY
30/09/2026
02225234-0012